SCHAUM'S OUTLINE OF

THEORY AND PROBLEMS

OF

HUMAN ANATOMY
AND
PHYSIOLOGY

•

KENT M. VAN DE GRAAFF, Ph.D.

Associate Professor of Zoology
Brigham Young University

and

R. WARD RHEES, Ph.D.

Professor of Zoology
Brigham Young University

•

SCHAUM'S OUTLINE SERIES

McGRAW-HILL, INC.

New York   St. Louis   San Francisco   Auckland   Bogotá   Caracas
Hamburg   Lisbon   London   Madrid   Mexico   Milan   Montreal
New Delhi   Paris   San Juan   São Paulo   Singapore
Sydney   Tokyo   Toronto

KENT M. VAN DE GRAAFF is currently an Associate Professor of Zoology at Brigham Young University in Provo, Utah. He received his B.S. (1965) in Zoology at Weber State College, his M.S. (1969) at University of Utah, and his Ph.D. (1973) at Northern Arizona University. He completed a postdoctorate course in neuromyology (1974), and taught anatomy at University of Minnesota. Van De Graaff is the author or co-author of several college textbooks, including *Human Anatomy* and *Concepts of Human Anatomy and Physiology*.

R. WARD RHEES is a Professor of Zoology at Brigham Young University. He received his B.S. (1967) in Pharmacy at the University of Utah and his Ph.D. (1971) in Physiology at Colorado State University. He taught at Weber State College and has been a Visiting Professor in the Department of Anatomy and the Brain Research Institute at the UCLA School of Medicine. His research on sexual differentiation of the brain has been published in numerous leading scholarly journals and presented at national and international conferences.

Schaum's Outline of Theory and Problems of
HUMAN ANATOMY AND PHYSIOLOGY

3 4 5 6 7 8 9 10 11 12 13 14 15 SH SH 9 8 7 6 5 4 3 2 1 0

ISBN 0-07-066884-1

Sponsoring Editor, David Beckwith
Production Manager, Nick Monti
Editing Supervisor, Marthe Grice

Library of Congress Cataloging-in-Publication Data

Van De Graaff, Kent M. (Kent Marshall)
  Schaum's outline of theory and problems of human
anatomy and physiology.

  (Schaum's outline series)
  Includes index.
  1. Human physiology--Outlines, syllabi, etc.
2. Human physiology--Examinations, questions, etc.
3. Anatomy, Human--Outlines, syllabi, etc. 4. Anatomy,
Human--Examinations, questions, etc. I. Rhees, R. Ward.
II. Title. III. Series, [DNLM: 1. Anatomy--examination
questions. 2. Anatomy--outlines. 3. Physiology--
examination questions. 4. Physiology--outlines.
QS 18 V225s]
QP41.V36   1985   612'.002'02   86-7424
ISBN 0-07-066884-1

Cover design by Amy E. Becker.

To Karen and Karin

Preface

This supplemental book has been written to enable students at all levels in anatomy and physiology to improve their efficiency of study, to measure their progress, and to prepare for examinations. The topic sequence, with discussions of body organization and basic physiological principles followed by chapters devoted to body systems, is the same as in the majority of texts. In addition, because human anatomy and physiology have direct application in maintaining body health and treating dysfunctions, clinical information has been integrated throughout.

Each chapter of the book is composed of objective-survey-problems modules. An *objective* represents a major topic and a level of competency that are addressed by the survey and problems which follow. The *survey* is a carefully phrased body of information (table) that gives the essence of the topic introduced in the objective. The succeeding *problems* test the student's understanding and, in their answers, give further information for meeting the objective at the desired level. Throughout each chapter, clear and accurately labeled illustrations depict body structures or physiological processes. These figures can be enhanced as a learning aid if key structures or body organs are color-coded by the student. At the close of each chapter is a set of review questions whose answers are presented at the end of the book. Key clinical terms are defined at the end of each body system chapter. The book is completed by a particularly comprehensive index.

Several persons assisted in the preparation of this book, and our sincere thanks is extended to each. Chris Creek rendered the illustrations and assisted in their labeling. Kathlyn J. Loveridge typed the manuscript. Cary D. Wasden, Connie C. Erdmann, and Heather Bennion assisted in various aspects of manuscript preparation. Finally, we are extremely appreciative of David Beckwith and Marthe Grice of McGraw-Hill Book Company for their editorial assistance.

KENT M. VAN DE GRAAFF
R. WARD RHEES

Errata

The original sources of certain figures that appear in this work were omitted from the text. Those sources were as follows: Figures 16-6 and 16-7 herein are copied from figures 22-1 and 22-3, respectively, which appear in Weinreb, E.L., ANATOMY AND PHYSIOLOGY, Addison-Wesley Publishing Company, Inc., 1984; and figures 12-4, 7-6, and 7-7 herein are adapted from figures 11-11, 15-6(b), and 15-6(a), respectively which also appear in Weinreb, E.L., ANATOMY AND PHYSIOLOGY, Addison-Wesley Publishing Company, Inc., 1984.

Contents

Chapter 1

Introduction to
the Human Body

OBJECTIVE A *To define the sciences of Anatomy and Physiology and to explain how they are related.*

Survey Anatomy and Physiology are subdivisions of the science of Biology, which is the study of living organisms, both plant and animal. *Anatomy* has to do with structure and the relationships among structures; *physiology*, with the functions of the body parts. In general, function is determined by structure.

1.1 What are the subspecialities of anatomy?

These include: **microscopic anatomy**, the study of structures observed with the aid of a microscope (*cytology* is the study of cells and their organelles, and *histology* is the study of tissues comprising organs); **gross anatomy**, the study of structures observed with the unaided eye; **human anatomy**, the study of the structure of the human body; **comparative anatomy**, the study of differences and similarities in the structure of various animals; **developmental anatomy**, the study of structural changes from conception to birth (*embryology*) and on to physical maturity; **pathological anatomy**, the study of structural changes caused by disease.

1.2 What are the subspecialities of physiology?

These include: **cellular physiology**, the study of cellular homeostasis (equilibrium among the parts) and the specific functions of the organelles and the cells in general; **human physiology**, the study of human body functions from the molecular to the organismic level; **comparative physiology**, the study of functional differences and similarities in comparable body organs of various animals; **developmental physiology**, the study of the functional changes that occur as an organism develops; **pathological physiology**, the study of the functional changes that occur as organs age or become diseased.

OBJECTIVE B *To define the human organism and to classify man with other animals.*

Survey *Homo sapiens*, as we have named ourselves, is a biological organism that has similarities with all living animals. Because we have characteristics unique to us, we are a species within a classification scheme based on similarity of structural features.

1.3 Explain why humans are classed among the animals.

We breathe, eat and digest food, excrete bodily wastes, locomote, and reproduce our own kind, as do all animals. Being composed of organic materials, we decompose in death as other animals (chiefly microorganisms) consume our flesh. The processes by which our bodies produce, store, and utilize energy are similar to those found in all living organisms. The same genetic code that regulates our development is found throughout nature. The fundamental patterns of development observed in many animals are also seen in the formation of the human embryo.

1.4 What are the basic physical requirements for the survival of an organism?

Water, for a variety of metabolic processes; *food*, to supply energy, raw materials for building new living matter, and chemicals necessary in vital reactions; *oxygen*, to release energy from food materials; *heat*, to promote chemical reactions; *pressure*, to allow breathing.

1

1.5   Give the classification scheme of man.

The descending series is shown in Table 1-1. Man is the only extant hominid.

Table 1-1

Taxon	Grouping	Characteristics
Kingdom	Animalia	Cells having a visible nucleus but lacking walls, plastids, and photosynthetic pigments
Phylum	Chordata	Notochord; dorsal hollow nerve cord; pharyngeal pouches
Subphylum	Vertebrata	Cartilaginous or bony endoskeleton; vertebral column
Class	Mammalia	Hair; mammary glands, three ear ossicles; attached placenta; muscular diaphragm
Order	Primata	Prehensile hands with digits modified for grasping; large brains
Family	Hominidae	Large, well-developed cerebrum; bipedal posture and locomotion; well-developed vocal structures; opposable thumb
Genus	*Homo*	
Species	*sapiens*	

OBJECTIVE C   *To describe the levels of organization of the human body.*

Survey   See Fig. 1-1. The chemical and cellular levels are respectively the basic structural and functional levels. Each level represents an association of units from the preceding level. Though the cells number in the trillions, there are only a few hundred specific kinds.

1.6   How are similar cells bound together?

Similar cells are uniformly spaced and bound together as tissue by nonliving *matrix* which the cells secrete. Matrix varies in composition from one tissue to another and may take the form of a liquid, semisolid, or solid. Blood tissue, for example, has a liquid matrix, while bone cells are bound by a solid matrix. Not all similar cells, however, have a binding matrix; secretory cells, for instance, are solitary amidst a tissue of cells of another kind.

1.7   List the four principal types of tissues and describe the function(s) of each.

Epithelia cover body and organ surfaces, line body cavities and lumina, and form various glands. Epithelial tissues are involved with protection, absorption, excretion, and secretion.

Connective tissues bind, support, and protect body parts.

Muscle tissues contract to produce movement of the body and within the body.

Nervous tissues initiate and transmit (electrical) *nerve impulses* that coordinate body activities.

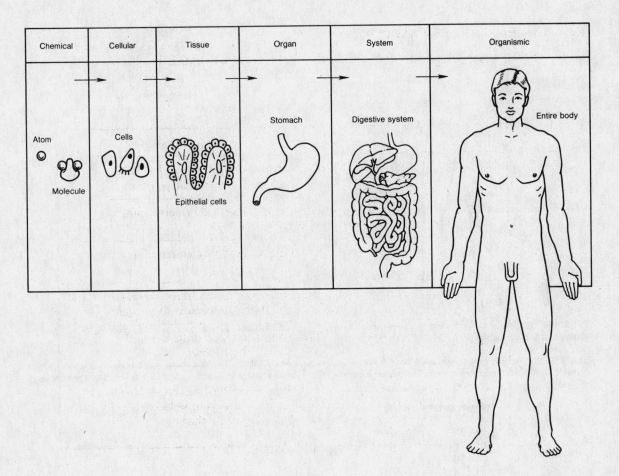

Chemical	Cellular	Tissue	Organ	System	Organismic

Atom

Molecule

Cells

Epithelial cells

Stomach

Digestive system

Entire body

Fig. 1-1

1.8　By means of an example, define an organ and describe its function.

　　　A bone, such as the femur, is an organ because it is composed of several tissue types and cooperates with other organs of the body. The components of the femur include osseous (bone) tissue, nervous tissue, vascular (blood) tissue, and cartilaginous tissue (at a joint). Not only does the femur, as part of the skeletal system, help to maintain body support, but it serves the muscular system as an anchorage for muscles, and also serves the circulatory system by producing blood cells in the bone marrow.

OBJECTIVE D　*To list the* body systems *and to describe the general functions of each.*

Survey　See Figs. 1-2 through 1-11.

1.9　Which body systems function in support and movement?

　　　The muscular and skeletal systems are jointly referred to as the *musculoskeletal system* because of their combined functional role in body support and locomotion. Both systems, along with the movable joints (*diarthroses*), are studied extensively in *kinesiology* (the mechanics of body motion). The integumentary system also provides some support, and its flexibility permits movement.

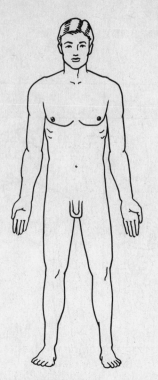

DEFINITION   The integument (skin) and structures derived from it (hair, nails, and oil and sweat glands).

FUNCTIONS   Protects the body, regulates body temperature, eliminates wastes, and receives certain stimuli (tactile, temperature, and pain).

Fig. 1-2   Integumentary System

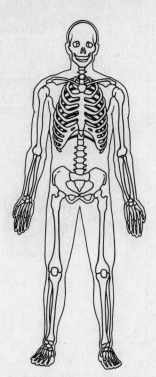

DEFINITION   Bones, cartilage, and ligaments (which guy the bones at the joints).

FUNCTIONS   Provides body support and protection, permits movement and leverage, produces blood cells (*hemopoiesis*), and stores minerals.

Fig. 1-3   Skeletal System

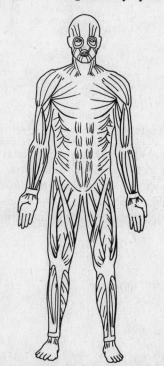

DEFINITION   Skeletal muscles of the body and their tendinous attachments.

FUNCTIONS   Effects body movements, maintains posture, and produces body heat.

Fig. 1-4   Muscular System

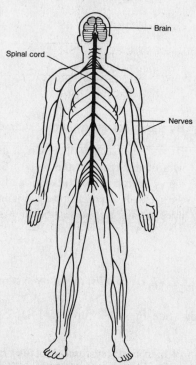

DEFINITION   Brain, spinal cord, nerves, and sensory organs such as the eye and the ear.

FUNCTIONS   Detects and responds to changes in internal and external environments, enables reasoning and memory, and regulates body activities.

Fig. 1-5   Nervous System

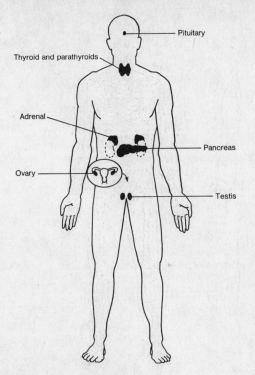

DEFINITION The hormone-producing glands.

FUNCTIONS Controls and integrates body functions via hormones secreted into the bloodstream.

Fig. 1-6 Endocrine System

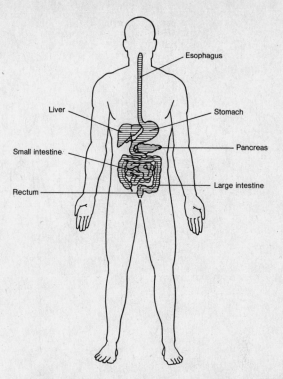

DEFINITION The body organs that render ingested foods absorbable.

FUNCTIONS Mechanically and chemically breaks down foods for cellular use, and eliminates undigested wastes.

Fig. 1-7 Digestive System

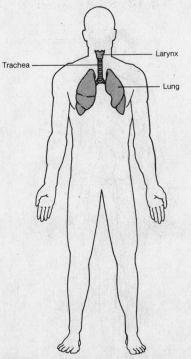

DEFINITION The body organs concerned with movement of respiratory gases (O_2 and CO_2) to and from the pulmonary blood (the blood within the lungs).

FUNCTIONS Supplies oxygen to the blood and eliminates carbon dioxide; also helps to regulate acid-base balance.

Fig. 1-8 Respiratory System

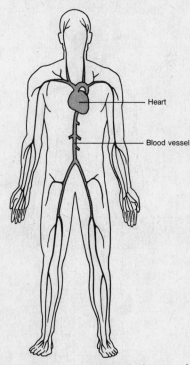

DEFINITION The heart and the vessels that carry blood or blood constituents (lymph) through the body.

FUNCTIONS Transports respiratory gases, nutrients, wastes, and hormones; protects against disease and fluid loss; helps regulate body temperature and acid-base balance.

Fig. 1-9 Circulatory System

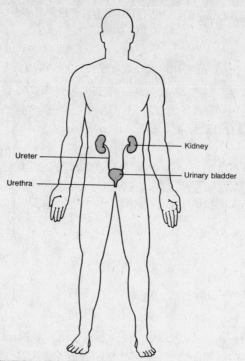

DEFINITION The organs that operate to remove wastes from the blood and to eliminate urine from the body.

FUNCTIONS Removes various wastes from the blood; regulates the chemical composition, volume, and electrolyte balance of the blood; helps to maintain the acid-base balance of the body.

Fig. 1-10   Urinary System

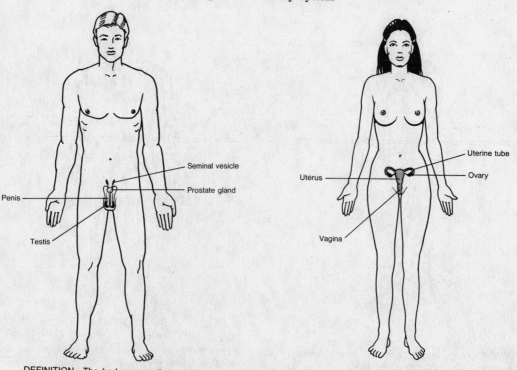

DEFINITION The body organs that produce, store, and transport reproductive cells (*gametes*, or sperm and ova).

FUNCTIONS Reproduce the organism, produce sex hormones.

Fig. 1-11   Reproductive Systems

1.10   Which body systems function in integration and coordination?

The endocrine and nervous systems maintain consistency of body functioning, the one by secretion of chemical substances (hormones) into the bloodstream, and the other by electrochemical signals (nerve impulses) carried via nerve fibers.

1.11   Which body systems are involved with processing and transporting of body substances?

Nutrients, oxygen, and various wastes are processed and transported by the digestive, respiratory, circulatory, *lymphatic*, and urinary systems. (The lymphatic system, which is generally considered part of the circulatory system, is composed of lymphatic vessels, lymph fluid, lymph nodes, the spleen, and the thymus. It transports lymph from tissues to the bloodstream, defends the body against infections, and aids in the absorption of fats.)

OBJECTIVE E   *To define and discuss the term* homeostasis.

Survey   *Homeostasis* is the maintenance of a nearly stable internal environment in the body so that cellular metabolic functions can proceed at maximum efficiency. Homeostasis is maintained through a combination of physiological and biochemical processes.

1.12   What major regulatory process does the body use to maintain homeostasis?

Essentially all the control systems of the body act by *negative feedback*. If a factor becomes excessive or insufficient, then the system which monitors that factor initiates a counterchange (hence "negative") that returns the factor toward normal value. A specific example is indicated in Fig. 1-12.

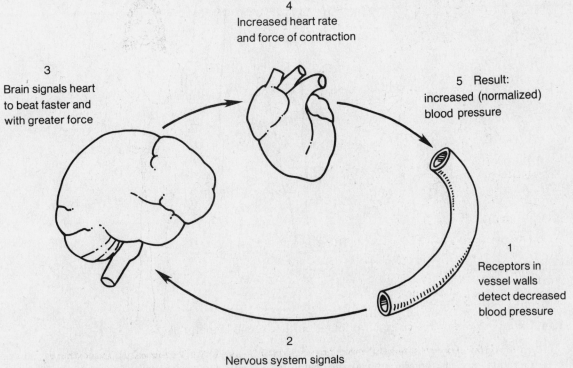

4
Increased heart rate
and force of contraction

3
Brain signals heart
to beat faster and
with greater force

5   Result:
increased (normalized)
blood pressure

1
Receptors in
vessel walls
detect decreased
blood pressure

2
Nervous system signals
brain of low blood pressure

Fig. 1-12

1.13   What is the relationship between homeostasis and *pathophysiology*?

        They are contraries in the sense that health reflects a proper homeostasis, whereas disorder and abnormal function—i.e., pathophysiology—mark a deviation from homeostasis. Pathophysiology is the basis for diagnosis of disease and for instituting treatment intended to restore normal function.

OBJECTIVE F   *To describe the* anatomical position.

Survey   All terms of direction which describe the relationship of one body part to another are made in reference to a standard *anatomical position* (Fig. 1-13). In the anatomical position, the body is erect, feet are parallel and flat on the floor, eyes are directed forward, and arms are at the sides of the body with the palms of the hands turned forward.

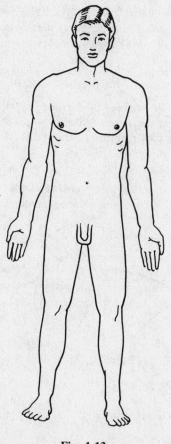

Fig. 1-13

1.14   Why are the palms given an orientation that seems unnatural?

        During early development, the palms are *supine* (facing forward or upward). Later, there is an axial rotation of each forearm so that the palms are *prone* (facing backward or downward). Thus, the anatomical position orients the upper appendages as in early development.

OBJECTIVE G *To list the principal body regions.*

Survey The main body regions are *head*, *neck*, *trunk*, *upper extremity* (two), and *lower extremity* (two). The trunk is frequently divided into *thorax* and *abdomen*.

1.15 Give the regions containing the *brachium*, *popliteal fossa*, and *axilla*.

See Fig. 1-14. Specific structures or clinically important areas within the principal regions have anatomical names. Learning the specific regional terminology provides a foundation for later learning the names of underlying structures.

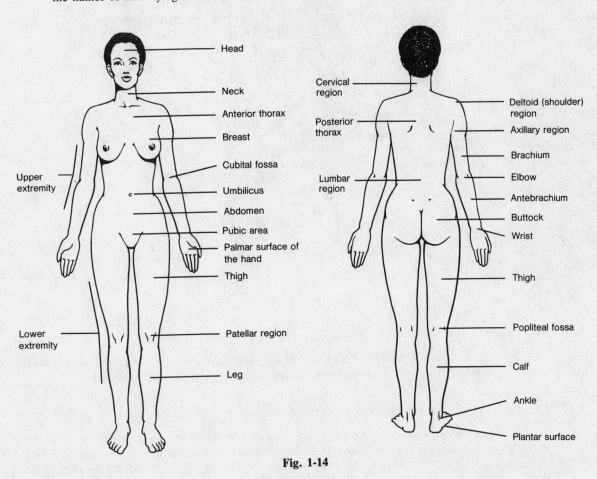

Fig. 1-14

OBJECTIVE H *To name and to locate the principal body cavities and their included organs.*

Survey *Body cavities* are confined spaces that contain organs which are protected, separated, and supported by associated membranes. As shown in Fig. 1-15, the ***dorsal cavity*** includes the *cranial* and *vertebral cavities* (or *vertebral canal*) and contains the brain and spinal cord. The ***ventral cavity*** includes the *thoracic*, *abdominal*, and *pelvic cavities* and contains visceral organs. Body cavities serve to segregate organs and systems by function. The major portion of the nervous system occupies the dorsal cavity; the principal organs of the respiratory and circulatory systems are in the thoracic cavity; the primary organs of digestion are in the abdominal cavity; and the reproductive organs are in the pelvic cavity.

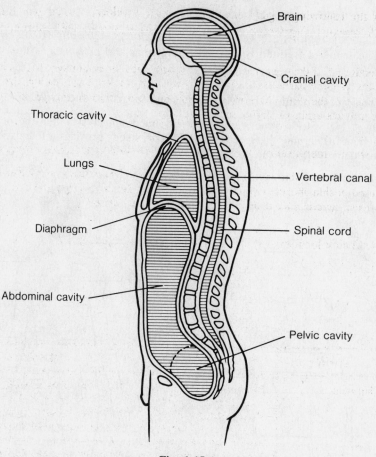

Fig. 1-15

1.16   Where are the *pericardial* and *pleural cavities*?

The thoracic cavity is partitioned into two pleural cavities, one for each lung, and the pericardial cavity, surrounding the heart. The area between the two lungs is known as the *mediastinum*.

1.17   What are the clinical significances of the thoracic viscera's being in separate compartments?

Diseases or infections (*pleurisy*, for example) involving one cavity do not spread readily to another cavity. A *pneumothorax* is the presence of air or other gas in a pleural cavity, causing the lung to collapse. Trauma, such as a knife or bullet wound, could ordinarily produce a pneumothorax in a single cavity, so that the other lung would be spared.

OBJECTIVE 1   *To discuss the types and functions of the various* body membranes.

Survey   *Body membranes* are composed of thin layers of connective and epithelial tissue which serve to cover, lubricate, separate, or support visceral organs, or to line body cavities. The two basic types are *mucous membranes* and *serous membranes*.

1.18   What is the function of mucous membranes?

They secrete a thick, viscid substance, called *mucus*, that lubricates and/or protects the body organs where it is secreted.

1.19 Which of the following organs have their *lumina* (hollow, tubular portions) lined, at least in part, with mucous membranes? (i) trachea, (ii) stomach, (iii) uterus, (iv) oral and nasal cavities.

> The lumina of all the organs listed are lined with mucous membranes. Mucus in the nasal cavity and trachea traps airborne particles; mucus in the oral cavity prevents desiccation (drying); mucus coats and epithelial lining of the stomach to protect against digestive enzymes and hydrochloric acid; and mucus in the uterus protects against entry of pathogens.

1.20 Describe serous membranes, their sites, and their differences from mucous membranes.

> Serous membranes line the thoracic and abdominopelvic cavities, and cover visceral organs. They are composed of thin sheets of epithelial tissue (simple squamous) that lubricate and compartmentalize visceral organs. *Serous fluid* is the watery lubricant they secrete.

1.21 Give the specific locations of the individual serous membranes.

> See Table 1-2.

Table 1-2

Cavity	Serous Membrane	Location
Thoracic	Visceral pleura	Adherent to outer surface of lungs
	Parietal pleura	Adherent to thoracic walls and thoracic surface of diaphragm
	Visceral pericardium (epicardium)	Adherent to outer surface of heart
	Parietal pericardium	Durable covering of heart
Abdominopelvic	Visceral peritoneum	Adherent to abdominal viscera
	Parietal peritoneum	Adherent to abdominal wall
	Mesentery	Double fold of peritoneum connecting parietal to visceral peritoneum

1.22 Define *retroperitoneal*.

> Certain organs, such as the kidneys, adrenal glands, and a portion of the pancreas, are positioned *behing the parietal peritoneum*, but still within the abdominopelvic cavity.

1.23 State the function of the mesenteries.

> The mesenteries support the abdominopelvic viscera in a pendent fashion so that intestinal *peristalsis* (rhythmic waves of muscular contraction) is unimpeded. They also support the vessels and nerves that serve the viscera.

OBJECTIVE J   *To identify the planes of reference used to locate and describe structures within the body.*

Survey   A set of three, mutually perpendicular planes—termed *midsagittal*, *coronal*, and *transverse*— is frequently used to depict structural arrangement.

1.24   Distinguish between the principal body planes.

Refer to Fig. 1-16. The midsagittal plane is the plane of symmetry of the body, dividing it into right and left halves. *Sagittal* (*parasagittal*) planes are parallel to the midsagittal plane; they divide the body into unequal right and left portions. Coronal (*frontal*) planes divide the body into front and back portions. Transverse (*horizontal* or *cross-sectional*) planes divide the body into superior (upper) and inferior (lower) portions.

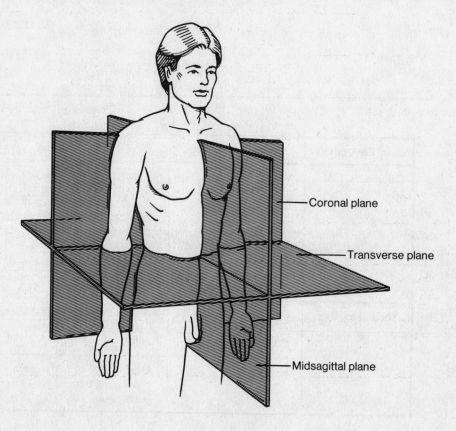

Fig. 1-16

OBJECTIVE K   *To become familiar with the descriptive and directional terms that are applied to anatomical features.*

Survey   These terms are used to fix the position of structures, surfaces, and regions of the body; they are always relative to the anatomical position.

1.25   List, define, and exemplify the chief directional terms.

See Table 1-3.

Table 1-3

Term	Definition	Example
Superior (cranial, cephalic)	Toward the head; toward the top	The thorax is superior to the abdomen
Inferior (caudal)	Away from the head; toward the bottom	The legs are inferior to the trunk
Anterior (ventral)	Toward the front	The navel is on the anterior side of the body
Posterior (dorsal)	Toward the back	The kidneys are posterior to the intestines
Medial	Toward the midline of the body	The heart is medial to the lungs
Lateral	Toward the side of the body	The ears are lateral to the head
Internal (deep)	Away from the surface of the body	The brain is internal to the cranium
External (superficial)	Toward the surface of the body	The skin is external to the muscles
Proximal	Toward the main mass of the body	The knee is proximal to the foot
Distal	Away from the main mass of the body	The hand is distal to the elbow
Visceral	Related to internal organs	The lungs are covered by a thin membrane called the visceral pleura
Parietal	Related to the body walls	The parietal pleura is the inside lining of the thoracic cavity

Review Questions

Multiple Choice

1. Production of secretory materials within cells would be studied as part of the science of: (a) histology, (b) cytology, (c) developmental biology, (d) anatomy.

2. A fingernail is a structure belonging to what body system? (a) skeletal, (b) circulatory, (c) integumentary, (d) lymphatic.

3. Which two body systems are regulatory? (a) endocrine, (b) nervous, (c) lymphatic, (d) skeletal, (e) circulatory.

4. The region of the body between the head and thorax is most appropriately referred to as the: (*a*) lumbar region, (*b*) throat region, (*c*) torso region, (*d*) cervical region.

5. A person in a prone position would be: (*a*) lying face down, (*b*) in anatomical position, (*c*) standing erect, (*d*) in a fetal position.

6. In anatomical position, the thumb is: (*a*) lateral, (*b*) medial, (*c*) proximal, (*d*) superficial.

7. Which is not one of the four principal tissue types? (*a*) nervous, (*b*) bone, (*c*) epithelial, (*d*) muscle, (*e*) connective.

8. Which is not a serous membrane? (*a*) parietal peritoneum, (*b*) mesentery, (*c*) visceral pleura, (*d*) lining of the mouth, (*e*) pericardium.

9. The relationship between structure and function is best described as: (*a*) a negative feedback system, (*b*) the function of an organ is determined by its structure, (*c*) important only during homeostasis of the organ system, (*d*) nonexistent except in certain parts of the body.

10. Which is not a chordate characteristic? (*a*) vertebral column, (*b*) notochord, (*c*) pharyngeal pouches, (*d*) dorsal hollow nerve cord.

11. The abdominal cavity contains the: (*a*) heart, (*b*) lungs, (*c*) spleen, (*d*) trachea.

12. The ventral cavity comprises all of the following body cavities except the: (*a*) spinal, (*b*) pleural, (*c*) thoracic, (*d*) pelvic, (*e*) abdominal.

13. The *antebrachium* is the: (*a*) chest area, (*b*) hand, (*c*) shoulder area, (*d*) arm, (*e*) forearm.

14. Which is positioned retroperitoneally? (*a*) stomach, (*b*) kidney, (*c*) heart, (*d*) appendix, (*e*) liver.

15. The foot is to the thigh as the hand is the: (*a*) brachium, (*b*) shoulder, (*c*) palm, (*d*) digits.

16. Which term best defines the position of the knee relative to the hip? (*a*) lateral, (*b*) medial, (*c*) distal, (*d*) posterior, (*e*) proximal.

17. The thoracic cavity is separated from the abdominopelvic cavity by the: (*a*) mediastinum, (*b*) abdominal wall, (*c*) sternum, (*d*) abdominal septum, (*e*) diaphragm.

18. Long-distance regulation is accomplished via blood-borne chemicals known as: (*a*) blood cells, (*b*) hormones, (*c*) ions, (*d*) motor impulses, (*e*) neurotransmitters.

19. Which serous membrane would be cut first as a physician removes an infected appendix? (*a*) parietal peritoneum, (*b*) dorsal mesentery, (*c*) visceral pleura, (*d*) parietal pleura.

20. If an anatomist wanted to depict the structural relationship of the trachea, esophagus, neck muscles, and a vertebra within the neck, which body plane would be most appropriate? (*a*) sagittal, (*b*) coronal, (*c*) transverse, (*d*) vertical, (*e*) parasagittal.

21. Which pairing of dissectional terms is most appropriate? (*a*) medial and proximal, (*b*) superior and posterior, (*c*) proximal and lateral, (*d*) superficial and deep.

22. A lung is located within the: (*a*) mediastinal, pleural, and thoracic cavities; (*b*) thoracic, pleural, and ventral cavities; (*c*) peritoneal, pleural, and thoracic cavities; (*d*) pleural, pericardial, and thoracic cavities; (*e*) none of the above.

23. Which of the following serous membrane combinations lines the diaphragm? (*a*) visceral pleura–visceral peritoneum, (*b*) visceral pleura–parietal peritoneum, (*c*) parietal pleura–parietal peritoneum, (*d*) parietal pleura–visceral peritoneum.

24. In a negative feedback system: (*a*) input is always maintained constant (*homeostatic*), (*b*) input serves no useful purpose, (*c*) output is partially put back into system, (*d*) output is always maintained constant.

25. Give the proper sequence of body cavities or areas traversed as blood flows from the heart to the uterus through the aorta and the uterine artery: (*a*) thoracic, pericardial, pelvic, abdominal; (*b*) pericardial, mediastinal, abdominal, pelvic; (*c*) pleural, mediastinal, abdominal, pelvic; (*d*) pericardial, pleural, abdominal, pelvic.

True/False

1. Histology is the microscopic examination of tissues.

2. The function of an organ is predictable from its structure.

3. A group of cells cooperating in a particular function is called a tissue.

4. *Genu*, *thigh*, *crus*, *malleolus*, and *popliteal fossa* are all terms that have reference to the lower extremities.

5. A sagittal plane divides the body into right and left *halves*.

6. The thumb is lateral to the other digits of the hand and distal to the forearm.

7. The lungs are kept moist through the secretion of mucus from mucous membranes.

8. Increased body temperature during exercise is an example of a homeostatic feedback mechanism.

9. Mesenteries tightly bind visceral organs to the body wall so that they are protected from excessive movement.

10. A six-inch knife wound lateral to the left nipple of a male would puncture the parietal pleura and cause a pneumothorax.

Matching

1. Toward a central reference point
2. Close to the ground
3. Divides body into right and left halves
4. Toward the back
5. Toward the head
6. Away from the midsagittal plane
7. Upper surface of the body
8. Toward the front
9. Divides body into anterior and posterior portions
10. Toward the feet
11. Away from a central reference point
12. Toward the midsagittal plane
13. Perpendicular to the cranial-caudal axis

(*a*) Dorsal
(*b*) Cranial or superior
(*c*) Transverse plane
(*d*) Distal
(*e*) Ventral
(*f*) Anterior
(*g*) Posterior
(*h*) Caudal or inferior
(*i*) Medial
(*j*) Proximal
(*k*) Coronal plane
(*l*) Midsagittal plane
(*m*) Lateral

Completion

System	Major Organs	Functions
Circulatory		
	Nose, throat, larynx, trachea, lungs	
		Digestion, absorption, and processing of nutrients
	Kidney, bladder, ureter, urethra	
		To support, protect, and move body
Skeletal		
	Brain, spinal cord, nerves, sense organs	
		Control and integration of many activities in body
Reproductive		

Chapter 2

Cellular Chemistry

OBJECTIVE A *To identify by name and symbol the main chemical elements of the body.*

Survey All forms of *matter* (living or nonliving) are made up of building units called the *chemical elements*. Of the 106 different chemical elements, 92 are naturally occurring and, of these, twenty-odd are present in human tissues (Table 2-1).

Table 2-1. Chemical Constitution of the Body

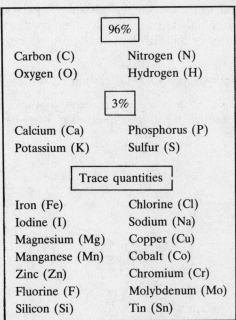

96%	
Carbon (C)	Nitrogen (N)
Oxygen (O)	Hydrogen (H)

3%	
Calcium (Ca)	Phosphorus (P)
Potassium (K)	Sulfur (S)

Trace quantities	
Iron (Fe)	Chlorine (Cl)
Iodine (I)	Sodium (Na)
Magnesium (Mg)	Copper (Cu)
Manganese (Mn)	Cobalt (Co)
Zinc (Zn)	Chromium (Cr)
Fluorine (F)	Molybdenum (Mo)
Silicon (Si)	Tin (Sn)

2.1 Briefly define *atom* and *molecule*.

   Atom—the smallest unit of a chemical element. *Molecule*—a combination of two or more atoms. These may be atoms of the same element (as in the oxygen molecule, O_2) or of different elements (as in the hydrogen sulfide molecule, H_2S). In the latter case, we say that the molecule is the smallest unit of a *chemical compound*.

OBJECTIVE B *To describe the structure of an atom.*

Survey An atom is composed of three kinds of elementary particles: *protons*, *neutrons*, and *electrons*. These are characterized by their *weights* (or *masses*) and *electric charges*, as displayed in Table 2-2; the units of measurement are such that a "normal" carbon atom has weight 12, exactly, and an electron has charge -1, exactly. Protons and neutrons are bound together to make up the *nucleus* of the atom. The number of protons in the nucleus (called the *atomic number*, Z) is the same for all atoms of a given chemical element. Conversely,

17

Table 2-2

Particle (*symbol*)	Weight	Charge
Proton (*p*)	1 (very nearly)	+1
Neutron (*n*)	1 (very nearly)	0
Electron (*e⁻*)	$\frac{1}{1840}$ (very nearly)	−1

different chemical elements have different atomic numbers. Surrounding the nucleus are precisely Z electrons, making the atom as a whole electrically neutral. In the familiar solar-system model, the nucleus is represented as the sun, which is orbited by planetary electrons. However, because electrons have wavelike as well as particlelike properties, it is more useful to speak of the *energy levels* occupied by the electrons. If these energy levels are imagined as organized into successive *shells*, then the chemical properties of the element can be explained in terms of the distribution of the Z electrons among the shells.

2.2   Sketch structures for hydrogen ($Z = 1$), carbon ($Z = 6$), and potassium ($Z = 19$).

See Fig. 2-1, in which the shells are represented by concentric circles. The capacities of the first four shells are 2, 8, 8, 18 electrons. We imagine the atom built up one electron at a time, with a given shell entered only if all interior shells are full.

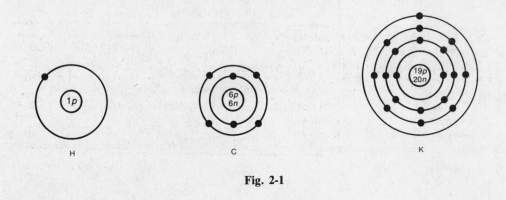

Fig. 2-1

2.3   What are *isotopes* and how are they used in medicine?

Atoms of a given element (all containing the same number, Z, of protons) are said to be *isotopes* of that element if they contain different numbers of neutrons. For example, besides standard 6-neutron carbon, there exist 7-neutron and 8-neutron varieties. The *atomic weight* of an element, as given in chemical tables, is an average of the weights of all isotopes of the element. (The weight of 6-neutron carbon is, by definition, 12.00000; but the atomic weight of carbon is 12.01115.) Since the number of neutrons in the nucleus tends to be close to the number of protons, it follows from Table 2-2 that the atomic weight of an element is *roughly* 2Z.

Because the various isotopes of an element have a common electronic shell structure, they behave identically in ordinary chemical reactions. However, many isotopes are *radioisotopes*, and their radiations can be detected by instruments. A radioisotope may be introduced into the body, and its movement, uptake, distribution, and excretion monitored. Such information is helpful in the diagnosis and treatment of various diseases.

2.4   Compute the molecular weights of water (H_2O), carbon dioxide (CO_2), and glucose ($C_6H_{12}O_6$).

The molecular weight (MW) is the sum of the weights of the atoms composing the molecule.

water (H_2O)	atomic weight of H = 1	$2 \times 1 = 2$
	atomic weight of O = 16	$1 \times 16 = \underline{16}$
		MW = 18
carbon dioxide (CO_2)	atomic weight of C = 12	$1 \times 12 = 12$
	atomic weight of O = 16	$2 \times 16 = \underline{32}$
		MW = 44
glucose ($C_6H_{12}O_6$)	atomic weight of C = 12	$6 \times 12 = 72$
	atomic weight of H = 1	$12 \times 1 = 12$
	atomic weight of O = 16	$6 \times 16 = \underline{96}$
		MW = 180

2.5   What are the major types of *bonds* (attractive forces) that hold atoms together in molecules?

Ionic bonds. If the outer shell of an atom gains or loses electrons, the atom's neutrality is destroyed; such charged atoms are called *ions*. Atoms that gain electrons acquire an overall negative charge and are called *anions*. Atoms that lose electrons acquire an overall positive charge and are called *cations*. An *ionic bond* is the electrical attraction that exists between an atom that has lost an electron (thereby becoming a cation) and the atom that has gained that electron (an anion). Sometimes more than one electron is transferred. The NaCl molecule is held together by ionic bonding (Fig. 2-2).

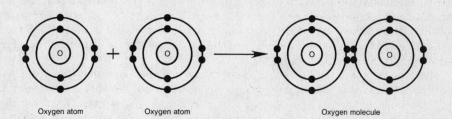

Sodium atom (Na)   Chlorine atom (Cl)   Sodium ion   Chloride ion

Sodium chloride molecule (NaCl)

Fig. 2-2

Covalent bonds result from atoms' *sharing* (rather than transferring) one, two, or three electron *pairs*. A shared pair is indicated by a short line drawn between the chemical symbols. For instance, in the oxygen molecule, O_2, two pairs of electrons are shared (Fig. 2-3), and so the molecule may be indicated as O=O.

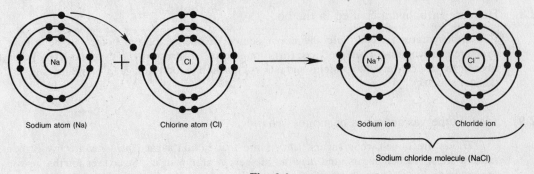

Oxygen atom   Oxygen atom   Oxygen molecule

Fig. 2-3

OBJECTIVE C   *To distinguish between inorganic and organic compounds.*

Survey   *Inorganic compounds* do not contain carbon (exceptions are CO_2 and CO), and are usually small, ionically bonded molecules. *Organic compounds* always contain carbon, are held together by covalent bonds, and are large complex molecules. Both inorganic and organic compounds are essential to life.

2.6   List some important inorganic compounds.

Water; oxygen; carbon dioxide; many salts, acids, and bases; many electrolytes (e.g., Na^+, Cl^-); minerals.

2.7   Name the four major families of organic compounds and give examples of each.

Carbohydrates—glucose, cellulose, glycogen, starch. *Lipids*—phospholipids, steroids, prostaglandins. *Proteins*—enzymes, insulin, albumin. *Nucleic acids*—DNA, RNA.

OBJECTIVE D   *To describe the three types of carbohydrates.*

Survey   All carbohydrates are composed of carbon, hydrogen, and oxygen, with the ratio of hydrogen to oxygen being 2 to 1. They are classified as: *monosaccharides*, or simple sugars; *disaccharides*, or double sugars; or *polysaccharides*, which are complex sugars usually composed of thousands of glucose units.

2.8   How are carbohydrates used in the body?

Carbohydrates (i) constitute the major source of energy for the body; (ii) contribute to cell structure and synthesis of cell products; (iii) form part of the structure of DNA and RNA (deoxyribose, ribose); (iv) are converted to proteins and fats; (v) function in food storage (glycogen storage in liver and skeletal muscles).

2.9   Identify the various types of monosaccharides.

Trioses are three-carbon sugars; *tetroses* are four-carbon sugars; *pentoses* are five-carbon sugars; *hexoses* are six-carbon sugars; and *heptoses* are seven-carbon sugars. Structures for the hexose *glucose* are shown in Fig. 2-4; structures of two important pentoses are shown in Fig. 2-5.

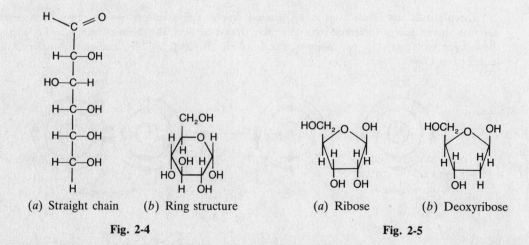

(*a*) Straight chain   (*b*) Ring structure   (*a*) Ribose   (*b*) Deoxyribose

Fig. 2-4   **Fig. 2-5**

2.10　How are disaccharides built up from monosaccharides?

A disaccharide is formed by combining two monosaccharides in the process of *dehydration synthesis* (removal of a water molecule). The synthesis of maltose is indicated in Fig. 2-6.

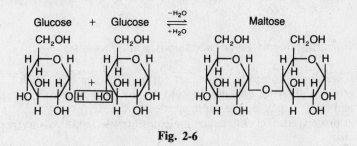

Fig. 2-6

In similar fashion,

$$\text{Glucose} + \text{Galactose} \rightleftharpoons \text{Lactose}$$
$$\text{Glucose} + \text{Fructose} \rightleftharpoons \text{Sucrose} \quad (\textit{table sugar})$$

2.11　In what respect do polysaccharides differ from mono- and disaccharides?

Polysaccharides lack the characteristic sweetness of sugars like sucrose and fructose.

OBJECTIVE E　*To describe the chemical composition of lipids (fats).*

Survey　A lipid is formed when a molecule of glycerol combines with molecules of fatty acids (see Fig. 2-7).

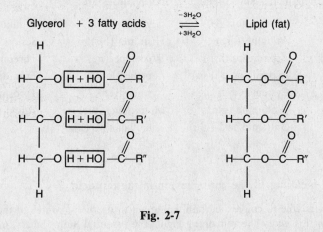

Fig. 2-7

2.12　Distinguish between *saturated* and *unsaturated* fatty acids, and give examples of each.

In a saturated fatty acid, all carbon atoms bond the maximum number of hydrogen atoms (no double bonds between carbons); e.g.,

$$\textit{Butyric acid} \qquad CH_3-(CH_2)_2-COOH$$
$$\textit{Palmetic acid} \qquad CH_3-(CH_2)_{14}-COOH$$
$$\textit{Stearic acid} \qquad CH_3-(CH_2)_{16}-COOH$$

In an unsaturated fatty acid, one or more carbon atoms can still bond one or more hydrogen atoms (at least one double bond between carbons); e.g.,

$$\textit{Oleic acid} \qquad CH_3—(CH_2)_7—CH{=}CH—(CH_2)_7—COOH$$

$$\textit{Linoleic acid} \qquad CH_3—(CH_2)_4—CH{=}CH—CH_2—CH{=}CH—(CH_2)_7—COOH$$

$$\textit{Linolenic acid} \qquad CH_3—CH_2—CH{=}CH—CH_2—CH{=}CH—CH_2—CH{=}CH—(CH_2)_7—COOH$$

OBJECTIVE F   *To describe the chemical composition of proteins.*

Survey   Proteins are large complex structures formed from amino acids. *Peptide bonds* link the amino group (NH_2) of one amino acid to the acid carboxyl group (COOH) of another amino acid (which may be the same species as the first; see Fig. 2-8). The resulting chain is called a protein if its molecular weight exceeds 10,000; otherwise it is known as a *polypeptide*.

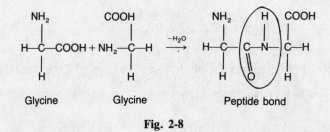

Fig. 2-8

2.13   How many different amino acids make up the proteins in the human body?

Twenty; their names and three-letter abbreviations are:

Alanine (ala)	Glycine (gly)	Proline (pro)
Arginine (arg)	Histidine (his)	Serine (ser)
Asparagine (asn)	Isoleucine (ile)	Threonine (thr)
Aspartic acid (asp)	Leucine (leu)	Tryptophan (trp)
Cysteine (cys)	Lysine (lys)	Tyrosine (tyr)
Glutamine (gln)	Methionine (met)	Valine (val)
Glutamic acid (glu)	Phenylalanine (phe)	

2.14   What is the meaning of the term "essential amino acid"?

The body is able to convert certain amino acids to others; twelve of the twenty amino acids can be synthesized in this way. The remaining eight, the essential amino acids, must be supplied in the diet.

2.15   List some major functions of proteins in the body.

(i) All enzymes in the body are proteins. (ii) Many hormones are proteins. (iii) Proteins are important structural components of cells and tissues. (iv) All antibodies are made of proteins. (v) Most of the clotting factors are proteins. (vi) Proteins play major roles in regulating the concentrations of osmotic solutions in the body.

OBJECTIVE G   *To describe the chemical composition of the* nucleotides, *the primary components of the nucleic acids.*

Survey   As indicated in Fig. 2-9, a nucleotide has three parts: a *phosphate group* (solid circle), a *pentose* (see Problem 2.9), and a *nitrogenous base* (oval). In a nucleic acid molecule, the sugar and the phosphate remain constant from one nucleotide to the next, but the base (in DNA) may be *adenine*, *thymine*, *guanine*, or *cytosine* (abbreviated A, T, G, C). The secondary structures of DNA and RNA are treated in Chapter 3.

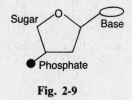

Fig. 2-9

2.16   Of the four nitrogenous bases of DNA, two are called *purine* bases, and two *pyrimidine* bases. Explain.

Figure 2-10 shows two ring structures that contain nitrogen as well as carbon atoms. A comparison with Fig. 2-11 reveals that adenine and guanine are built on the purine ring, while cytosine and thymine are built on the pyrimidine ring.

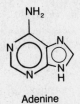

Adenine

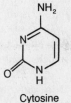

Cytosine

Pyrimidine ring   Purine ring

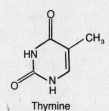

Guanine

Thymine

Fig. 2-10   **Fig. 2-11**

Review Questions

Multiple Choice

1. A neutral atom contains: (*a*) the same number of electrons as it does protons, (*b*) more protons than electrons, (*c*) the same number of electrons as it does neutrons, (*d*) more electrons than protons.

2. The number of protons in an atom is given by the: (*a*) mass number, (*b*) atomic number, (*c*) difference between the atomic number and the mass number, (*d*) atomic weight.

3. A compound is a molecule: (*a*) composed of two or more different atoms, (*b*) composed of only one type of atom, (*c*) linked together only by covalent bonds, (*d*) none of the above.

4. Bonds that result from shared electrons are called: (*a*) ionic bonds, (*b*) covalent bonds, (*c*) peptide bonds, (*d*) covalent or peptide bonds, (*e*) ionic or covalent bonds.

5. Bonds that result from the transfer of electrons are called: (*a*) ionic bonds, (*b*) covalent bonds, (*c*) peptide bonds, (*d*) polar bonds, (*e*) all the above.

6. Molecules composed only of hydrogen and carbon are called: (*a*) carbohydrates, (*b*) inorganic molecules, (*c*) lipids, (*d*) hydrocarbons.

7. It is false that carbohydrates: (*a*) are linked together through dehydration reactions; (*b*) are composed of carbon, hydrogen, and oxygen; (*c*) consist of a carbon chain with an acid carboxyl group at one end; (*d*) are classed as monosaccharides, disaccharides, and polysaccharides.

8. Fats are reaction products of fatty acids and: (*a*) amino acids, (*b*) glycerol, (*c*) monosaccharides, (*d*) nucleic acids.

9. Proteins differ from carbohydrates in that they: (*a*) are not organic compounds, (*b*) are united by covalent bonds, (*c*) contain nitrogen, (*d*) all the above.

10. Which is not a component of a nucleic acid? (*a*) purine base, (*b*) 5-carbon sugar, (*c*) pyrimidine base, (*d*) glycerol, (*e*) phosphate group.

11. The principal solvent in the body is: (*a*) oil (lipids), (*b*) water, (*c*) blood, (*d*) lymph fluid.

12. It is false that acids: (*a*) increase hydrogen ion concentration in solution, (*b*) act as proton donors, (*c*) yield higher hydroxide concentration than hydrogen ion concentration, (*d*) have low pH.

13. *Anabolic* reactions are: (*a*) decomposition reactions, (*b*) synthesis reactions, (*c*) not part of the body's metabolism, (*d*) those that break down molecules for use as energy sources.

14. Deoxyribonucleotides are named according to: (*a*) the base, (*b*) the sugar, (*c*) the phosphate group, (*d*) their position in the macromolecule.

15. The molecular weight is equal to: (*a*) the sum of all the isotopic weights, (*b*) the sum of all the atomic weights, (*c*) the sum of the atomic numbers, (*d*) none of the above.

16. Phospholipids involve a phosphate group and: (*a*) 4 or more fatty acids, (*b*) 3 fatty acids, (*c*) 2 fatty acids, (*d*) 1 fatty acid.

17. Of the following nitrogenous bases, only one is found exclusively in RNA: (*a*) thymine, (*b*) guanine, (*c*) adenine, (*d*) uracil.

18. Which represents the correct sequence in ascending order of size? (*a*) atom, amino acid, polypeptide, protein; (*b*) amino acid, atom, polypeptide, protein; (*c*) atom, amino acid, protein, polypeptide; (*d*) amino acid, atom, protein, polypeptide.

19. Ions have: (*a*) only positive charges, (*b*) only negative charges, (*c*) either positive or negative charges, (*d*) no charge.

20. Atoms of the same atomic number but of different *mass numbers* (different numbers of nuclear particles) are referred to as: (*a*) ions, (*b*) isotopes, (*c*) cations, (*d*) tight atoms.

21. Which of the following substances is not an organic compound? (*a*) starch, (*b*) ribose, (*c*) carbon dioxide, (*d*) lipase.

22. Which of the following is classified as a disaccharide? (*a*) glucose, (*b*) ribose, (*c*) fructose, (*d*) lactose.

23. The eight amino acids that cannot be formed in the body from other amino acids are referred to as the: (*a*) essential enzymes, (*b*) neutral amino acids, (*c*) normal amino acids, (*d*) essential amino acids.

24. Dehydration synthesis: (*a*) requires water, (*b*) results in the splitting of molecules, (*c*) is the means of forming disaccharides, (*d*) occurs when glycogen stores are used by tissue cells.

25. Nucleotides lack: (*a*) a phosphate group, (*b*) an amino group, (*c*) a nitrogenous base, (*d*) a five-carbon sugar.

True/False

1. Protons and electrons each have many times the mass of neutrons.

2. Of the 106 presently known elements, 75% are found in the body.

3. Sodium has atomic number 11 and mass number 23; sodium has 12 neutrons.

4. Positively charged ions are called cations.

5. Unsaturated fatty acids contain only single covalent bonds between carbon atoms.

6. Amino acids are linked by peptide bonds to form polypeptides.

7. The specific nature of a protein is determined mainly by its amino acid sequence and the properties of the respective amino acid R-groups.

8. Substances that increase the hydrogen ion concentration are called *bases*.

9. Covalent bonds are far more important in living organisms than ionic bonds.

10. Hydrogen, carbon, nitrogen, and oxygen account for about half of the body weight.

11. Nucleic acid molecules are small and unspecialized molecules.

12. Purine bases have a single ring of carbon and nitrogen atoms.

Matching

1. Carbohydrates
2. Protons and neutrons
3. Electrons
4. Covalent bonds
5. Nucleic acid
6. Lipids
7. Proteins
8. Hydrogen bonds
9. Peptide bonds
10. Purine bases
11. Pyrimidine bases
12. Cation
13. Anion
14. Acid
15. Base

(a) Proton acceptor
(b) Adenine and guanine
(c) $C_n(H_2O)_n$
(d) Cl^-
(e) Nucleus
(f) Proton donor
(g) Subshells
(h) DNA and RNA
(i) K^+
(j) Cytosine and thymine
(k) Primary structures of proteins
(l) H_2N—CH—COOH
                    |
                    R
(m) Water-insoluble
(n) Shared electrons
(o) Secondary structure of proteins

Chapter 3

Cell Structure and Function

OBJECTIVE A   *To distinguish between* prokaryotic *and* eukaryotic *cells.*

Survey   *Prokaryotic* cells lack a membrane around the nucleus, which is thus indistinct [Fig. 3-1(*a*)]. They contain only one chromosome, and no membrane-limited organelles are present in the cytoplasm. *Eukaryotic* "truly nucleated" cells are larger and more complex; they contain a membrane-limited nucleus and various membrane-limited organelles [Fig. 3-1(*b*)].

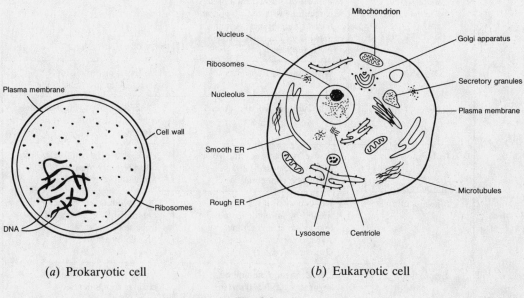

(*a*) Prokaryotic cell   (*b*) Eukaryotic cell

Fig. 3-1

3.1   Give some examples of prokaryotic and eukaryotic life forms.

   Bacteria, blue-green algae, and microplasmas are prokaryotes. Eukaryotic cells are found in protozoa, fungi, green plants, most algae, and all vertebrates.

3.2   What functions do membranes serve within eukaryotic cells?

   Membranes divide the cells into regions where selective metabolic functions can occur under optimal conditions. Specific enzymes become incorporated into the membrane structure and control reactions within the cell.

OBJECTIVE B   *To list the various cell structures and their associated functions.*

Survey   In Table 3-1, dimensions are in nanometers, formerly called "millimicrons" (mμ).

3.3   Which cell types should contain large amounts of smooth ER, and which should contain large amounts of rough ER?

Table 3-1

Organelle/Component	Structure	Function
Plasma membrane	Three-layered: protein layer–lipid bilayer–protein layer	Protects the cell, regulates the flow of molecules in and out
Cytoplasm	Fluid in which the organelles are suspended	Matrix in which chemical reactions occur
Nucleus	Oval organelle, separated from the cytoplasm by a two-fold *nuclear membrane*; contains DNA, genes, chromosomes, and the *nucleolus* (RNA)	Control center that directs all cellular activity
Ribosomes	Bipartite granules in the cytoplasm, roughly 25 nm across, which contain about 65% RNA and 35% protein	Synthesis of proteins to be used in the cell
Endoplasmic reticulum (ER)	Parallel or stacked membranes, continuous with the plasma and nuclear membranes, forming a network of channels through the cytoplasm	
Rough ER	With attached ribosomes	Synthesis of proteins to be secreted for use outside the cell
Smooth ER	Without attached ribosomes	Steroid synthesis, intercellular transport, detoxification processes
Golgi apparatus	Stacked membranes and vessels (*cisternae*), usually in a horseshoe configuration	Packaging of glycoproteins, etc., into secretory vesicles (q.v.); formation of lysosomes (q.v.)
Mitochondria	Rodlike or oval organelles limited by an outer membrane and an inner membrane that folds to form shelflike structures (*cristae*)	ATP production, storage of enzymes of the citric acid cycle and oxidative phosphorylation
Lysosomes	Dense, membrane-limited organelles, 250–500 nm in diameter, containing hydrolytic enzymes	Breakdown of worn-out cellular components or engulfed material
Secretory vesicles	Membrane-bound sacs	Storage of proteins and other synthesized materials destined for secretion
Vesicles/granules	Membrane-bound sacs	Storage of cellular material such as glycogen
Fat droplets	Membrane-bound sacs	Storage of lipids
Microtubules	Long hollow structures, about 25 nm in diameter	Structural members, involved in cell division and movement, and in movement of materials within the cell
Microfilaments	Long solid fibers, 4–6 nm in diameter	Structural members, involved in cell movement
Centrioles	Two short rods or granules, each composed of nine sets of three fused microtubules, located near nucleus	Involved in cell division, movement of chromosomes during mitosis, and contractile activity of *cilia* and *flagella* (Problem 3.8)

Smooth ER appears to be involved in the biosynthesis of steroids such as the androgens, estrogens, glucocorticoids, and mineralocorticoids; it is in fact abundant in such endocrine tissues as the testes, ovaries, and adrenal glands. Rough ER is involved in the synthesis of proteins to be secreted from the cell; it is found in abundance in pancreatic acinar cells, whose secretion contains enzymes (proteins).

3.4 Why are lysosomes sometimes called "little disposals"? Which cell types would have a high number of these organelles?

Lysosomes contain many potent hydrolytic enzymes that can degrade DNA, RNA, proteins, and certain carbohydrates. The enzymes also can digest worn-out or damaged organelles or cells, as well as bacteria and foreign matter, in the process called *phagocytosis* ("cell eating"). (Certain white blood cells which are particularly active in phagocytosis contain an abundance of lysosomes.) For this reason, lysosomes constitute an important defense of the body against invasion by pathogenic organisms.

OBJECTIVE C   *To describe the molecular organization of the plasma membrane.*

Survey   As summarized in Table 3-1, the plasma membrane has two major structural components: (1) a lipid bilayer which is two molecules thick and is composed almost entirely of phospholipids and cholesterol, and (2) large globular protein molecules which both interpenetrate and coat the lipid molecules (see Fig. 3-2).

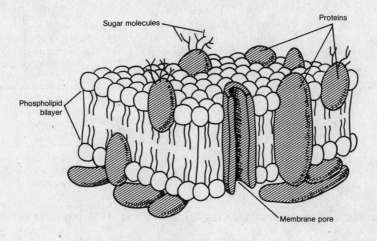

Fig. 3-2

3.5 Explain the selective permeability of the plasma membrane.

The lipid bilayer molecules have a *hydrophilic* ("water-loving"; soluble in water) end and a *hydrophobic* ("water-fearing"; soluble in fat) end. The hydrophobic ends are repelled by water but are attracted to each other, and therefore they tend to line up in the center of the membrane, leaving the hydrophilic ends at the two outer surfaces. Because of the hydrophobic core, the lipid bilayer is almost entirely impermeable to water and to all water-soluble substances. However, fat-soluble substances, such as oxygen, carbon dioxide, alcohols, and steroids, can penetrate the lipid bilayer.

3.6 What are the processes by which materials pass through cell membranes?

Diffusion—substances can migrate across membranes if the particles are small enough to pass through the pores; if they are fat-soluble (Problem 3.5), they can diffuse through the lipid bilayer matrix. *Active transport*—a process in which enzymes and carrier systems "carry" substances across. This process requires energy (from ATP), and, in contrast to diffusion, substances may be transported from a region of lesser to a region of greater concentration. *Endocytosis*—a process by which the membrane engulfs materials. There are two forms: *pinocytosis* ("cell-drinking"), in which the substances ingested are in solution, and *phagocytosis* ("cell-eating"), by which bacteria, dead tissues, or large particulate matter is engulfed. *Exocytosis*—the reverse of pinocytosis: vesicles, vacuoles, or secretory granules fuse with the plasma membrane and then rupture, releasing their contents to the outside.

3.7   What happens to pinocytic or phagocytic vesicles after they enter the cell?

As indicated in Fig. 3-3, lysosomes become attached to the vesicles and release hydrolytic enzymes into them which begin to hydrolyze or digest the contents. The products of digestion diffuse out of the vesicles into the cytoplasm.

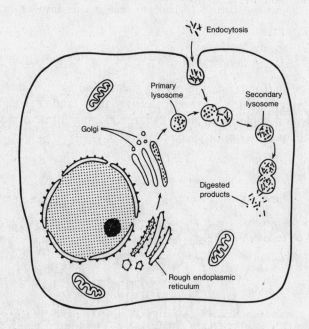

Fig. 3-3

3.8   What are some of the cell surface modifications and what are their functions?

Microvilli are projections of the plasma membrane that increase the surface area of certain cells (e.g., renal tubule or intestinal epithelium); they are mainly involved in absorption. *Cilia* are hairlike motile projections of the plasma membrane of epithelial cells that sweep material past the surface. Each cilium consists of a circumferential double microtubule around two central microtubules. A *flagellum* is the single hairlike motile projection of a spermatozoan which propels the cell forward.

OBJECTIVE D   *To describe the nuclear components and their functions.*

Survey   The nucleus is a large oval or spherical body, usually located near the center of the cell, which exercises control of cellular development and function. It is surrounded by two membranes (the *nuclear envelope*) which are joined at regular intervals to form *nuclear pores*. The nucleus contains the genetic material *deoxyribonucleic acid* (DNA), which, along with histones and other proteins, forms a substance known as *chromatin*. During mitosis, chromatin condenses into a number of elongated rodlike structures called *chromosomes* (human cells have 46). Located within the nucleus is the *nucleolus*, which contains high concentrations of *ribonucleic acid* (RNA).

3.9   What is the function of the nucleolus?

The nucleolus is the region where RNA (especially, ribosomal RNA) is stored prior to migration out through the nuclear pores into the cytoplasm.

3.10   Do all cells contain a full complement of genetic material?

There are a few cell types (mature *erythrocytes*) that do not contain a nucleus and are therefore devoid of genetic material (genes) and unable to grow or divide. Mature sex cells (sperm, ova) contain 23 chromosomes; all other nucleated cells have the full 46 chromosomes.

OBJECTIVE E   *To describe the structure and function of* DNA.

Survey   Recall from Chapter 2 that DNA is made up of tripartite building blocks, called nucleotides (Fig. 2-9). These nucleotides are joined sugar-to-phosphate and phosphate-to-sugar to form two strands, running in opposite directions, that are held together by hydrogen bonding beween pairs of nitrogenous bases. According to the **law of complementary base pairing**, adenine always bonds with thymine, and guanine always bonds with cytosine. This allows four possible purine-pyrimidine pairs: A=T, T=A, G≡C, C≡G. The two strands are entwined into a double helix which makes a full turn every 10 purine-pyrimidine pairs. The entire helix (i.e., the DNA molecule) may be thousands of pairs in length, with the pairs in any order (they *do not* repeat every 10). A typical configuration is suggested in Fig. 3-4.

3.11   What does DNA have to do with genes, protein synthesis, and cellular regulation?

DNA composes genes; genes are responsible for the synthesis of proteins (enzymes); and these enzymes control all cellular functions. (DNA does not manufacture proteins directly, but via RNA—see Problems 3.13–3.16.)

3.12   What is the *genetic code*?

Cellular proteins are composed of some 20 different amino acids. Therefore, "code words" for these amino acids must be present in the structure of genetic DNA. The alphabet available for composing these "words" is obviously: A, G, T, C (the four nucleotide bases). It has been established that *three* consecutive bases along a strand determine a word. The genetic code thus consists of $4 \times 4 \times 4 = 64$ words, which is more than enough to represent the 20 amino acids. (Note that a doublet code, permitting only $4 \times 4 = 16$ words, would not suffice.)

Figure 3-5(*a*) exhibits the genetic code of DNA, and Fig. 3-5(*b*) shows its transcription (Problem 3.13) into (messenger) RNA. Observe the multiple encoding of a given amino acid.

OBJECTIVE F   *To describe the sequence of events involved in copying and applying the genetic code.*

Survey   In the nucleus of the cell, the genetic code of DNA is *transcribed* into ribonucleic acid (RNA). RNA carries the coded message out of the nucleus and into the cytoplasm. There, in association with ribosomes, the RNA code is *translated* into the formation of proteins (enzymes).

3.13   How is the genetic code of DNA transcribed to form RNA?

As shown in Fig. 3-6, the DNA double helix "unzips" under the action of the enzyme *RNA polymerase*. One of its strands becomes the template for the formation of a complementary strand of RNA. The RNA strand is either stored in the nucleolus or passes out into the cytoplasm, and the DNA strands "zip" closed.

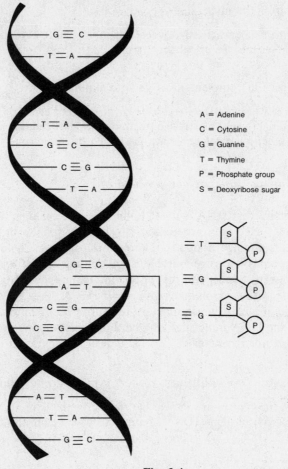

A = Adenine
C = Cytosine
G = Guanine
T = Thymine
P = Phosphate group
S = Deoxyribose sugar

Fig. 3-4

3.14   What types of RNA are formed from DNA?

Ribosomal RNA (rRNA) becomes part of the ribosome. *Transfer RNA* (tRNA) transfers amino acids to ribosomes for use as building blocks in protein synthesis. *Messenger RNA* (mRNA) carries the genetic code that directs the sequencing of the amino acids in protein synthesis. The three-base code words of mRNA [see Fig. 3-5(*b*)] are referred to as *codons*.

3.15   How does RNA differ from DNA?

RNA (i) contains *ribose*, a five-carbon sugar slightly different from deoxyribose; (ii) is usually single-stranded rather than double-stranded; (iii) contains *uracil* (code letter, U) in place of thymine; (iv) is found throughout the cell.

3.16   What are the major events that occur in protein synthesis?

Refer to Fig. 3-7. A strand of mRNA, with its sequence of codons, attaches to a ribosome. A folded strand of tRNA, carrying a specific amino acid at one end and presenting at the opposite end the three-base *anticodon* for that amino acid, moves up to the mRNA, and the anticodon bonds to a complementary codon. In similar fashion, an adjacent codon of the mRNA attracts its complementary amino acid; the two amino acids are joined by a peptide bond (the ribosome furnishes a catalyst for the reaction) and the first tRNA is set free. This process of *translation* (of the mRNA code) continues until a complete protein chain, complementary to the mRNA, has been synthesized.

Second base in DNA triplet

First base in DNA triplet	A	G	T	C	Third base in DNA triplet
A	AAA phenylalanine	AGA serine	ATA tyrosine	ACA cysteine	A
	AAG phenylalanine	AGG serine	ATG tyrosine	ACG cysteine	G
	AAT leucine	AGT serine	ATT "terminator"	ACT "terminator"	T
	AAC leucine	AGC serine	ATC "terminator"	ACC tryptophan	C
G	GAA leucine	GGA proline	GTA histidine	GCA arginine	A
	GAG leucine	GGG proline	GTG histidine	GCG arginine	G
	GAT leucine	GGT proline	GTT glutamine	GCT arginine	T
	GAC leucine	GGC proline	GTC glutamine	GCC arginine	C
T	TAA isoleucine	TGA threonine	TTA asparagine	TCA serine	A
	TAG isoleucine	TGG threonine	TTG asparagine	TCG serine	G
	TAT isoleucine	TGT threonine	TTT lysine	TCT arginine	T
	TAC methionine	TGC threonine	TTC lysine	TCC arginine	C
C	CAA valine	CGA alanine	CTA aspartic acid	CCA glycine	A
	CAG valine	CGG alanine	CTG aspartic acid	CCG glycine	G
	CAT valine	CGT alanine	CTT glutamic acid	CCT glycine	T
	CAC valine	CGC alanine	CTC glutamic acid	CCC glycine	C

(*a*)

Second letter in mRNA triplet

First letter in mRNA triplet	U	C	A	G	Third letter in mRNA triplet
U	UUU phenylalanine	UCU serine	UAU tyrosine	UGU cysteine	U
	UUC phenylalanine	UCC serine	UAC tyrosine	UGC cysteine	C
	UUA leucine	UCA serine	UAA "terminator"	UGA "terminator"	A
	UUG leucine	UCG serine	UAG "terminator"	UGG tryptophan	G
C	CUU leucine	CCU proline	CAU histidine	CGU arginine	U
	CUC leucine	CCC proline	CAC histidine	CGC arginine	C
	CUA leucine	CCA proline	CAA glutamine	CGA arginine	A
	CUG leucine	CCG proline	CAG glutamine	CGG arginine	G
A	AUU isoleucine	ACU threonine	AAU asparagine	AGU serine	U
	AUC isoleucine	ACC threonine	AAC asparagine	AGC serine	C
	AUA isoleucine	ACA threonine	AAA lysine	AGA arginine	A
	AUG methionine	ACG threonine	AAG lysine	AGG arginine	G
G	GUU valine	GCU alanine	GAU aspartic acid	GGU glycine	U
	GUC valine	GCC alanine	GAC aspartic acid	GGC glycine	C
	GUA valine	GCA alanine	GAA glutamic acid	GGA glycine	A
	GUG valine	GCG alanine	GAG glutamic acid	GGG glycine	G

(*b*)

Fig. 3-5

OBJECTIVE G *To give the stages of normal cell development.*

Survey Figure 3-8 shows the time pattern (*cell cycle*) established by human cells grown in tissue cultures with plenty of space and nutrients. A new cell arising from *mitosis* (cell division) undergoes a first growth phase (G_1), then a stage of synthesis (S), then a second growth phase (G_2), and then itself experiences mitosis. The major events in mitosis are pictured in Fig. 3-9.

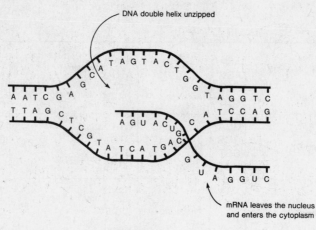

DNA double helix unzipped

mRNA leaves the nucleus
and enters the cytoplasm

Fig. 3-6

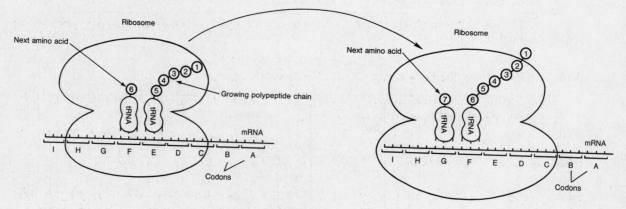

Fig. 3-7

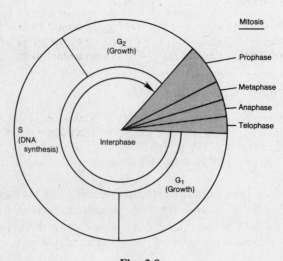

Fig. 3-8

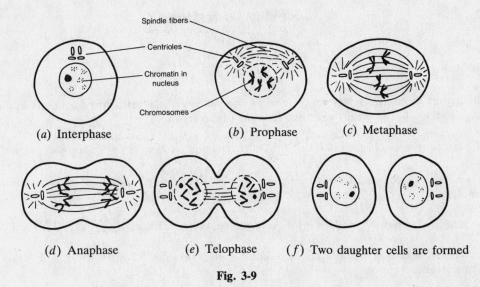

(a) Interphase   (b) Prophase   (c) Metaphase

(d) Anaphase   (e) Telophase   (f) Two daughter cells are formed

Fig. 3-9

3.17 What is the essential importance of mitosis?

Mitosis is the mechanism of body growth and the means of replacing damaged, diseased, or worn-out cells.

3.18 Under mitosis, the body is regenerated "in kind." Explain.

The nuclear material is divided precisely into halves, so that each reproduced cell has the same number and kind of chromosomes as the original parent cell.

OBJECTIVE H *To understand the characteristics of cancer cells.*

Survey *Cancer* is a group of diseases characterized by uncontrolled cell division; that is, cells continue to undergo mitosis at maximum rate when new cells are not needed. *A cancer* is a tumor arising from cancer.

3.19 Distinguish between benign tumors and cancers.

Benign ("harmless") tumors are surrounded by a connective tissue capsule and are not life-threatening (unless they become so large as to interfere with normal function). Cancers are malignant ("deadly") tumors, which grow rapidly and infiltrate surrounding tissues.

3.20 What are some of the causes of cancer?

Examples of *carcinogens*, agents that are capable of causing cancer, are: (i) *chemicals*—cigarette smoke, asbestos, nickel, arsenic, chromates, B-naphthylamine and other dyes, cyclamates, saccharin, and certain fossil-fuel and tar derivatives; (ii) *radiation*—X-rays, solar ultraviolet rays; (iii) some *viruses*.

Review Questions

Multiple Choice

1. The plasma membrane serves to: (*a*) enclose components of the cell, (*b*) regulate absorption, (*c*) give shape to the cell, (*d*) all of the above, (*e*) none of the above.

2. The largest structure in the cell is the: (*a*) Golgi apparatus, (*b*) nucleus, (*c*) ribosome, (*d*) mitochondrion.

3. Which organelle contains hydrolytic enzymes? (*a*) lysosome, (*b*) ribosome, (*c*) mitochondrion, (*d*) Golgi complex.

4. In Question 3, which organelle is involved in protein synthesis?

5. Endoplasmic reticulum with attached ribosomes is called: (*a*) smooth ER, (*b*) Golgi body, (*c*) nodular ER, (*d*) rough ER.

6. Engulfing of solid material by cells is called: (*a*) pinocytosis, (*b*) phagocytosis, (*c*) active transport, (*d*) diffusion.

7. The plasma membrane is a sandwich of: (*a*) lipid–protein–lipid, (*b*) lipid–lipid–protein, (*c*) protein–protein–lipid, (*d*) protein–lipid–protein.

8. The usual function of the Golgi complex is the: (*a*) packaging of materials in membranes for transport out of the cell, (*b*) production of mitotic and meiotic spindles, (*c*) excretion of excess water, (*d*) production of ATP by oxidative phosphorylation.

9. The primary function of mitochondria is the: (*a*) excretion of excess water or waste products from the cell, (*b*) conversion of light energy to chemical energy in the form of ATP, (*c*) synthesis of ATP by oxidative phosphorylation, (*d*) packaging and secretion of the products of the cell.

10. During protein synthesis, amino acids become linked together in a linear chain by: (*a*) hydrogen bonds, (*b*) peptide bonds, (*c*) ionic bonds, (*d*) phosphate bonds, (*e*) amino bonds.

11. Which nucleotide base is absent from DNA? (*a*) adenine, (*b*) cytosine, (*c*) guanine, (*d*) thymine, (*e*) uracil, (*f*) none of the above.

12. Messenger RNA is synthesized in the: (*a*) nucleus, under the direction of DNA; (*b*) cytoplasm, under the direction of the centrioles; (*c*) centrioles, under the direction of DNA; (*d*) Golgi complex, under the direction of DNA.

13. The sequence of nucleotides in a messenger RNA molecule is determined by the: (*a*) sequence of nucleotides in a gene, (*b*) enzyme *RNA polymerase*, (*c*) sequence of amino acids in a protein, (*d*) enzyme *ribonuclease* (RNase), (*e*) sequence of nucleotides in the anticodons of tRNA.

14. The flow of genetic information in most living things may be indicated as: (*a*) protein → DNA → mRNA, (*b*) protein → tRNA → DNA, (*c*) DNA → mRNA → protein, (*d*) DNA → tRNA → protein.

15. The genetic code for a single amino acid consists of: (*a*) one nucleotide, (*b*) two nucleotides, (*c*) three nucleotides, (*d*) four nucleotides.

16. In the DNA molecule, the nitrogenous base adenine always pairs with: (a) uracil, (b) thymine, (c) cytosine, (d) guanine.

17. The "backbone" of DNA consists of repetitive sequences of phosphate and: (a) sugar (glucose), (b) sugar (deoxyribose), (c) nucleic acids, (d) protein (ribose).

18. The molecule to which an amino acid is attached preparatory to protein synthesis is: (a) ribosomal RNA, (b) messenger RNA, (c) transfer RNA, (d) viral RNA, (e) nucleolar RNA.

19. The sequence of amino acids in protein molecules is determined by the sequence of: (a) amino acids in other protein molecules, (b) bases in transfer RNA, (c) bases in messenger RNA, (d) bases in ribosomal RNA, (e) sugars in DNA.

20. A certain gene has 1200 nucleotides (bases) in the coding portion of one strand. The protein coded for by this gene consists of: (a) 400, (b) 600, (c) 1200, (d) 2400, (e) 3600, amino acids.

21. If one strand of a DNA molecule has the base sequence ACGGCAC, the other strand will have the sequence: (a) ACGGCAC, (b) CACGGCA, (c) CATTACA, (d) UGCCGUG, (e) TGCCGTG.

22. A transfer RNA has the anticodon sequence UAC. With which codon in the following messenger RNA will it pair?

    ----GGU----UAC----AUU----CAU----AUG----
        (a)   (b)   (c)   (d)   (e)

23. Chromosome duplication takes place during: (a) telophase, (b) interphase, (c) metaphase, (d) anaphase.

24. In which phase of Question 23 does *cytokinesis* (cleavage) occur?

25. The two daughter cells formed by mitosis have: (a) identical genetic constitutions, (b) exactly half as many genes as the parent cell had, (c) the same amount of cytoplasm as the parent cell had, (d) none of the above.

26. Cancerous growth of cells may be induced by: (a) physical agents, (b) chemical agents, (c) viruses, (d) all the above.

Matching

1. Nucleus
2. Ribosomes
3. Plasma membrane
4. Golgi apparatus
5. Microvilli
6. Lysosomes
7. Nucleolus

(a) Control center of cell
(b) Little disposals
(c) Assembling and packaging
(d) Protein synthesis
(e) Storage area of RNA
(f) Increases surface area
(g) Regulates entrance and exit of substances

Labeling

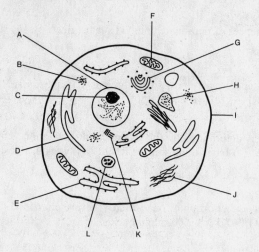

A. _____

B. _____

C. _____

D. _____

E. _____

F. _____

G. _____

H. _____

I. _____

J. _____

K. _____

L. _____

Chapter 4

Tissues

OBJECTIVE A *To define* tissue *and to distinguish between the four major types.*

Survey A *tissue* is an aggregation of similar cells that perform a specific set of functions. *Histology* is the study of tissues within body organs. The body is composed of over 25 kinds of tissues, classified into four principal types: *epithelia*, *connective tissues*, *muscle tissues*, and *nervous tissue*.

4.1 What are the bases for the classification of tissues?

Classification of tissues is based on embryonic development, structural organization, and functional properties. *Epithelia* derive from ectoderm, mesoderm, and endoderm; they cover body and organ surfaces, line body cavities and lumina, and form various glands. They are involved in protection, absorption, excretion, and secretion. *Connective tissues* derive from mesoderm; they bind, support, and protect body parts. *Muscle tissues* derive from mesoderm; they contract to produce movement of the body and within the body. *Nervous tissue* derives from ectoderm; it initiates and conducts nerve impulses that coordinate body activities.

4.2 What part do the tissues play in diagnosis?

In many cases, a particular disease is indicated by the abnormal appearance of tissues removed in biopsy or postmortem examination.

OBJECTIVE B *To describe epithelia on the cellular level and to differentiate between the various kinds.*

Survey An epithelium consists of one or more cellular layers. The outer surface is exposed either to the outside of the body or to a lumen or cavity within the body. The deep inner surface is bound by a *basement membrane* consisting of glycoprotein from the epithelial cells and a meshwork of collagen and reticular fibers from the underlying connective tissue. With few exceptions, epithelial tissue is *avascular* (without blood vessels) and composed of tightly packed cells. Epithelia composed of a single layer of cells are called *simple*; those that are multilayered are *stratified*. According to the shape of the cells on the exposed surface, epithelia are *squamous* (flattened surface cells—"scaly"), *cuboidal*, or *columnar*.

4.3 Catalog the five kinds of simple epithelia as to structure and function, and location in the body.

See Table 4-1; the appearance of the tissues is suggested in Fig. 4-1.

4.4 *True or false*: Endothelium and mesothelium are types of simple epithelia.

True in the sense that the simple squamous epithelium lining lumina of blood and lymphatic vessels is sometimes referred to as *endothelium*, while that covering visceral organs and lining body cavities is called *mesothelium*.

Table 4-1

Type	Structure and Function	Location
Simple squamous epithelium	Single layer of flattened, tightly bound cells; diffusion and filtration	Capillary walls, air sacs of lungs, covering visceral organs, linings of body cavities
Simple cuboidal epithelium	Single layer of cube-shaped cells; excretion, secretion, or absorption	Surface of ovaries; linings of kidney tubules, salivary ducts, and pancreatic ducts
Simple columnar epithelium	Single layer of nonciliated column-shaped cells; protection, secretion, and absorption	Lining of digestive tract
Simple ciliated columnar epithelium	Single layer of ciliated column-shaped cells; transportational role through ciliary motion	Lining the lumina of the uterine tubes
Pseudostratified ciliated columnar epithelium	Single layer of ciliated, irregularly shaped cells; protection, secretion, ciliary movement	Lining the respiratory passageways

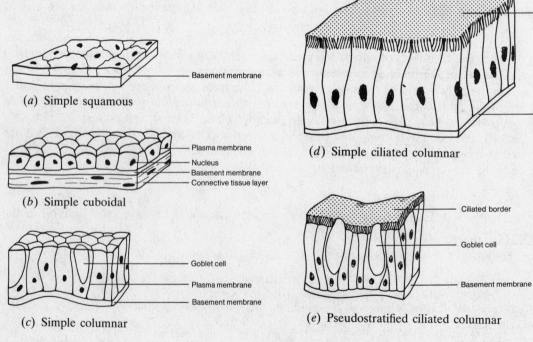

(a) Simple squamous

(b) Simple cuboidal

(c) Simple columnar

(d) Simple ciliated columnar

(e) Pseudostratified ciliated columnar

Fig. 4-1

4.5   Which of the following epithelia contain *goblet cells*? (*a*) simple columnar epithelium, (*b*) simple ciliated columnar epithelium, (*c*) pseudostratified ciliated columnar epithelium.

Specialized unicellular glands, called goblet cells, are dispersed throughout each of the columnar epithelia; they are especially numerous in type (*c*). Goblet cells secrete a lubricative and protective mucus along the exposed surfaces of the tissues.

4.6   Repeat Problem 4.3 for stratified epithelium.

See Table 4-2 and Fig. 4-2.

Table 4-2

Type	Structure and Function	Location
Stratified squamous epithelium (keratinized)	Multilayered, contains *keratin* (Problem 4.7), outer layers flattened and dead; protection	Epidermis
Stratified squamous epithelium (nonkeratinized)	Multilayered, lacks keratin, outer layers moistened and alive; protection and pliability	Linings of oral and nasal cavities, vagina, and anal canal
Stratified cuboidal epithelium	Usually two layers of cube-shaped cells; strengthening of luminal walls	Larger ducts of sweat glands, salivary glands, and pancreas
Transitional epithelium	Numerous layers of rounded, nonkeratinized cells; distension	Luminal walls of ureters and urinary bladder

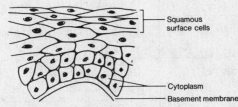

(*a*) Stratified squamous

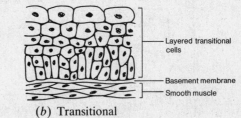

(*b*) Transitional

Fig. 4-2

4.7   Discuss the physical value of *keratinization* and *cornification* of stratified squamous epithelium.

Stratified squamous epithelium that is keratinized forms the epidermis, or outer layer of the skin. The presence of the protein keratin strengthens the tissue, permitting it to withstand abrasion, desiccation, inundation, and bacterial invasion. Cornification of the epidermis occurs as the outer layer of cells dies, dries, and flattens as they are moved toward the surface. This "armoring" too is of adaptive protective value.

4.8 How does transitional epithelium differ from stratified squamous epithelium?

Transitional epithelium is similar to nonkeratinized stratified squamous epithelium, except that the surface cells are large, are round rather than flat, and may have two nuclei. Transitional epithelium is specialized to permit distension of the ureters and urinary bladder and to withstand the toxicity of urine.

OBJECTIVE C *To define* glandular epithelia *and to describe the process of* exocrine gland formation.

Survey During prenatal development, certain epithelial cells invade the underlying connective tissue, forming specialized secretory accumulations called *exocrine glands* which retain a connection to the surface in the form of a duct.

4.9 Name some exocrine glands and identify the body systems where they occur.

Exocrine glands within the integumentary system include *sebaceous glands*, (apocrine) *sweat glands*, and *mammary glands*. Within the digestive system, they include the *salivary* and *pancreatic glands*.

4.10 Classify exocrine glands according to structure and according to mode of discharge of the secretory product.

The unicellular exocrine glands are the goblet cells of Problem 4.5; the multicellular varieties, which are either *simple* or *compound* (see Fig. 4-3), are classified in Tables 4-3 and 4-4.

Table 4-3. Structural Classification

	Type	Function	Examples
Simple	Tubular	Aid in digestion	Intestinal glands
	Branched tubular	Protect, aid in digestion	Uterine glands, gastric glands
	Coiled tubular	Regulate temperature	Certain sweat glands
	Acinar	Additive to spermatozoa	Seminal vesicle of male
	Branched acinar	Skin conditioner	Sebaceous glands of skin
Compound	Tubular	Lubricate urethra of male, assist in digestion	Bulbourethral gland of male reproductive system, liver
	Acinar	Nourishment of infant, aid in digestion	Mammary glands, salivary glands (sublingual and submandibular)
	Tubuloacinar	Aid in digestion	Salivary glands (parotid), pancreas

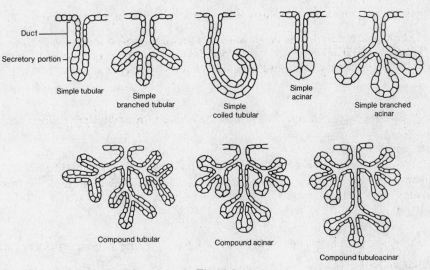

Fig. 4-3

Table 4-4.　Secretory Classification

Type	Discharge and Function	Examples
Merocrine	Anchored cell secretes water; regulates temperature, aids digestion	Salivary and pancreatic glands, certain sweat glands
Apocrine	Portion of secretory cell and secretion are discharged; provides nourishment to infant, assists in regulating temperature	Mammary glands, certain sweat glands
Holocrine	Entire secretory cell with enclosed secretion is discharged; skin conditioner	Sebaceous glands of skin

OBJECTIVE D　*To give the general characteristics, the locations, and the functions of connective tissues.*

Survey　Connective tissues consist of widely separated cells in a bed (*matrix*) of fibrous extracellular material. Derived from embryonic mesoderm, connective tissues are located throughout the body, where they support and bind other tissues, store nutrients, and/or manufacture protective and regulatory materials.

4.11　What are the various kinds of connective tissues, what are their structures and functions, and where are they located?

　　Throughout the embryo is found an undifferentiated connective tissue that can migrate and from which mature, differentiated connective tissues derive. These last fall into the four major categories indicated in Fig. 4-4. Note that one of these, blood tissue, differs from the rest in having a fluid matrix. Further classification is given in Table 4-5.

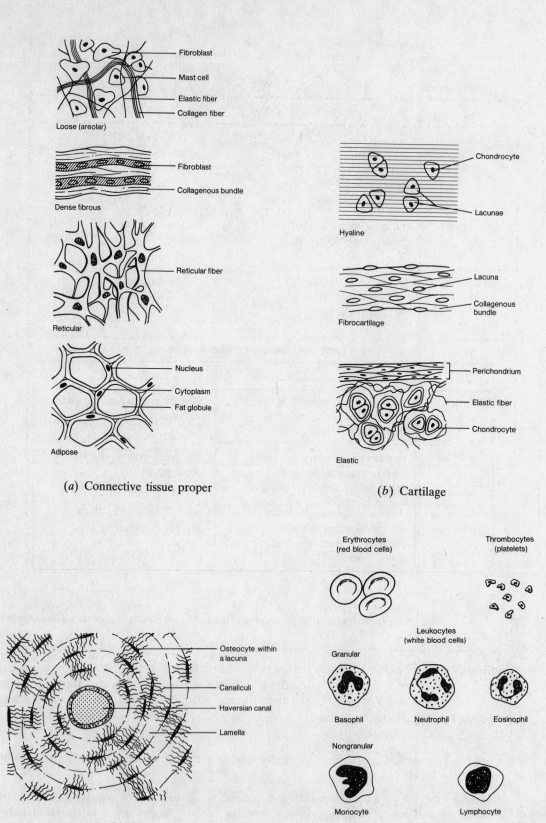

(a) Connective tissue proper

(b) Cartilage

(c) Bone tissue

(d) Blood

Fig. 4-4

Table 4-5

	Type	Structure and Function	Location
Proper	Loose (areolar)	Mostly fibroblast cells with lesser amounts of collagen and elastin cells; binds organs, holds tissue fluids, mediates diffusion	Surrounding nerves and vessels, between muscles, beneath the skin
	Dense fibrous	Densely packed collagen fibers; provides strong, flexible support	Tendons, ligaments, heart valves, sclera of eye, deep skin layers
	Elastic	Predominantly irregularly arranged elastic fibers; supports, provides framework	Large arteries, lower respiratory tract, between vertebrae
	Reticular	Crossed fibers forming supportive network; stores, phagocytic	Lymph nodes, liver, spleen, thymus, bone marrow
	Adipose	Distended cells; protects, stores fat, insulates	Hypodermis (of skin), surface of heart, folds of peritoneum, around kidneys, back of eyeball, surrounding joints
Cartilage	Hyaline	Homogeneous matrix with extremely fine collagenous fibers; flexible support, protection, precursor of bone	Articular surfaces of bones, nose, walls of respiratory passages, fetal skeleton
	Fibro-	Abundant collagenous fibers within matrix; supports, withstands compression	*Symphysis pubis*, inter-vertebral discs, knee joint
	Elastic	Abundant elastic fibers within matrix; provides support and flexibility	Framework of external ear, auditory canal, portions of larynx
Bone	Cancellous (porous)	Spongy organic-inorganic matrix, pores filled with bone marrow; *hematopoiesis* (making blood)	Middle layers of flat bones and ends of long bones
	Compact	*Haversian system* of dense, organic-inorganic rings (*lamellae*) separated by canals; body support and protection	Outer layers of flat bones and shafts of long bones
Blood		Blood cells in a fluid matrix of plasma; homeostasis and nutrient transport	Within vessels of circulatory system and chambers of heart

4.12 Which of the following connective tissues are important in body immunity? (*a*) blood, (*b*) dense fibrous, (*c*) fibrocartilage, (*d*) reticular.

Both the white cells (*leukocytes*) of the blood and the reticular tissue of lymphoid organs protect the body through phagocytosis.

4.13 Why are joint injuries involving cartilage slow to heal?

Cartilage is avascular and must therefore receive nutrients through diffusion from surrounding tissue. For this reason, cartilagenous tissue has a low rate of mitotic activity and, if damaged, heals slowly.

4.14 Distinguish between "fat" and "adipose tissue."

The cells of adipose tissue contain large vacuoles adapted to store lipids, or fats. Feeding an infant an excessive amount during the first year, when adipose cells are forming, causes a greater amount of adipose tissue to develop. A person with a lot of adipose tissue is more susceptible to developing obesity later in life than a person with a lesser amount. Dieting eliminates the fat stored within the tissue but not the tissue itself.

4.15 What do *fibroblasts*, *reticulum cells*, *mast cells*, *chondrocytes*, and *osteocytes* all have in common? How do they differ?

All are specialized cells of connective tissues; they are compared in Table 4-6.

Table 4-6

Cell type	Description	Site and Function
Fibroblast	Large, star-shaped	Throughout connective tissue proper; produces collagenous, elastic, and reticular fibers
Reticulum cell	Branched	Principal cell of reticular connective tissue; produces phagocytes
Mast cell	Resembles a basophil (a granulated white blood cell)	Throughout loose connective tissue surrounding blood vessels; produces *heparin*, an anticoagulant
Chondrocyte	Large, ovoid	Principal cell of cartilage; secretes matrix
Osteocyte	Small, ovoid	Principal cell of bone tissue; secretes durable matrix

4.16 How is the clinical condition *edema* related to connective tissue?

Approximately 11% of the body fluid is found within loose connective tissue; it is called *tissue fluid* or *interstitial fluid*. Sometimes excessive tissue fluid accumulates, causing the swollen condition known as edema. The fluid surplus is generally symptomatic of other conditions.

4.17 What accounts for the hardness of bone tissue?

The hardness of bone is largely due to the (inorganic) calcium phosphate and calcium carbonate salts deposited within the intercellular matrix. Numerous collagenous fibers, also embedded within the matrix, give some flexibility to bone.

OBJECTIVE E   *To describe muscle tissue and to distinguish between three types.*

Survey   Through the property of *contractility*, muscle tissues cause movement of materials through the body, movement of one part of the body with respect to another, and locomotion. Muscle cells, or *muscle fibers*, are elongated in the direction of contraction, and movement is accomplished through the shortening of the fibers in response to a stimulus. Derived from mesoderm, muscle tissue is so specialized for contraction that, once the tissue formation is completed shortly after birth, it cannot replicate. There are three types of muscle tissue in the body: *smooth*, *cardiac*, and *skeletal*.

4.18   What are the structure, function, and location of each type of muscle tissue?

   See Table 4-7 and Fig. 4-5.

Table 4-7

Type	Structure and Function	Location
Smooth	Elongated, spindle-shaped fiber with single nucleus; involuntary movements of internal organs	Walls of hollow internal organs
Cardiac	Branched, striated fiber with single nucleus and intercalated discs; involuntary rhythmic contraction	Heart
Skeletal	Multinucleated, striated, cylindrical fiber that occurs in *fasciculi* (slender bundles); voluntary movement of skeletal parts	Spans joints of skeleton via tendons

(*a*) Smooth

(*b*) Cardiac

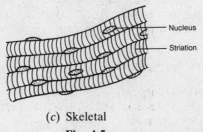

(*c*) Skeletal

Fig. 4-5

4.19 Which of the following are characteristic properties of all muscle tissue? (*a*) irritability, (*b*) contractility, (*c*) extensibility, (*d*) elasticity.

All are characteristic of muscle fibers. A muscle fiber exhibits irritability as it responds to a nerve impulse and contracts, or shortens. Once a stimulus has subsided and the muscle fiber is shortened but relaxed, it may passively stretch back or be extended by contracting fibers of opposing muscles. Each muscle fiber has innate tension, or elasticity, that causes it to have a particular shape as it is relaxed.

4.20 *True or false*:   One very important function of muscle tissue is heat production.

True. Body temperature is remarkably constant. Metabolism within cells releases heat as an end product. Since muscles constitute nearly one-half of the body weight and are in a continuous state of fiber activity, they are major heat sources. The rate of heat production increases immensely as a person exercises strenuously.

OBJECTIVE F   *To describe the basic characteristics and functions of nervous tissue.*

Survey Nervous tissue principally consists of two types of cells: *neurons* and *neuroglia* (literally, "nerve glue"). Neurons, derived from ectoderm, are highly specialized to conduct impulses; neuroglia primarily function to support and assist neurons. The number of neurons is established shortly after birth and thereafter they are incapable of mitosis. Neuroglia are about five times as abundant as neurons; they have mitotic capabilities throughout life.

4.21 How does the structure of a neuron reflect its function?

Ramified *dendrites* (Fig. 4-6) provide a large surface area to receive a stimulus and conduct the impulse to the cell body. The elongated axon conducts the impulse away from the cell body to another neuron or to an organ that responds to the stimulus.

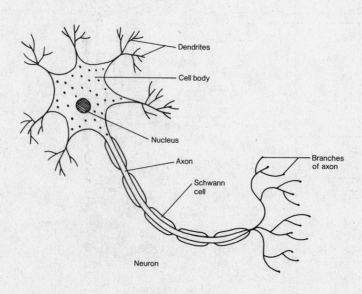

Fig. 4-6

4.22   Describe the association of *Schwann cells* (Fig. 4-6) with certain neurons.

Schwann cells are a type of neuroglial cell that support the axon by coating it with a lipid-protein substance, myelin. This *myelin sheath* also aids in the conduction of nerve impulses and promotes regeneration of a damaged neuron.

4.23   Describe the structure and function of neuroglia.

Besides Schwann cells, and a similar variety called *satellite cells*, there are four kinds of neuroglia, as illustrated in Fig. 4-7. Table 4-8 describes all six types (CNS ≡ central nervous system, PNS ≡ peripheral nervous system).

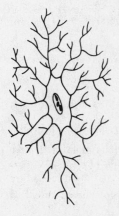         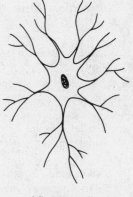

(*a*) Astrocyte   (*b*) Ependyma   (*c*) Oligodendrocyte   (*d*) Microglia

Fig. 4-7

Table 4-8

Type	Structure	Function
Astrocytes	Stellate cells, with numerous processes	Form structural support between capillaries and neurons in CNS; part of *blood–brain barrier* (Chapter 9)
Oligodendrocytes	Similar to astrocytes, but with shorter and fewer processes	Form myelin in CNS; guide development of neurons within CNS
Microglia	Minute cells with few short processes	Phagocytize pathogens and cellular debris within CNS
Ependyma	Columnar cells that may have ciliated free surfaces	Line ventricles and central canal of CNS, where cerebrospinal fluid is circulated by ciliary motion
Satellite cells	Small, flattened cells	Support ganglia within PNS
Schwann cells	Flattened cells arranged in series around axons of dendrites	Form myelin within PNS

Review Questions

Multiple Choice

1. Epithelia are involved in all the following except: (*a*) protection, (*b*) secretion, (*c*) connection, (*d*) absorption, (*e*) excretion.

2. Which of the following is *not* a type of epithelium? (*a*) simple squamous, (*b*) transitional, (*c*) simple ciliated columnar, (*d*) complex stratified, (*e*) pseudostratified ciliated.

3. Classification of epithelia is based on the number of layers of cells and on the: (*a*) shape, (*b*) staining properties, (*c*) size, (*d*) location.

4. The presence of a basement membrane is characteristic of: (*a*) epithelia, (*b*) connective tissues, (*c*) nervous tissue, (*d*) muscle tissues.

5. Simple squamous epithelium is *not* found in: (*a*) lymph vessels, (*b*) pancreatic duct, (*c*) small arterioles, (*d*) air sacs of lung, (*e*) linings of body cavities.

6. Goblet cells are a type of: (*a*) multicellular gland, (*b*) intracellular gland, (*c*) unicellular gland, (*d*) intercellular gland.

7. An example of a holocrine gland would be a: (*a*) sweat gland, (*b*) salivary gland, (*c*) pancreatic gland, (*d*) sebaceous gland.

8. An exocrine gland in which a portion of the secretory cell is discharged with the secretion is termed: (*a*) apocrine, (*b*) merocrine, (*c*) endocrine, (*d*) holocrine.

9. The inability to absorb digested nutrients may be due to damage of which type of epithelia? (*a*) ciliated columnar, (*b*) simple columnar, (*c*) simple squamous, (*d*) simple cuboidal, (*e*) stratified squamous.

10. Which word combination correctly applies to stratified squamous epithelium? (*a*) mesoderm–calcification, (*b*) ectoderm–keratinization, (*c*) mesoderm–ossification, (*d*) endoderm–cornification.

11. Which statement best describes connective tissue? (*a*) derived from endoderm and secretes metabolic substances, (*b*) derived from mesoderm and conducts impulses, (*c*) derived from mesoderm and contains much matrix, (*d*) derived from ectoderm and usually layered.

12. An infection would most likely increase activity in: (*a*) elastic tissue, (*b*) transitional tissue, (*c*) adipose tissue, (*d*) reticular tissue, (*e*) collagenous tissue.

13. Cartilage tissues are generally slow to heal after an injury because: (*a*) cartilage is avascular, (*b*) cartilage does not undergo mitosis, (*c*) the intercellular matrix is semisolid, (*d*) chondrocytes are surrounded by fluids.

14. Which of the following is *not* a specialized type of cell found in connective tissues? (*a*) reticular fiber, (*b*) collagen fiber, (*c*) goblet cell, (*d*) mast cell, (*e*) fibroblast.

15. The function of dense, regularly arranged, connective tissue is: (*a*) elastic recoil, (*b*) binding and support, (*c*) encapsulation of blood vessels, (*d*) articulation.

16. Phagocytosis is a function of which type of connective tissue? (*a*) cartilage, (*b*) areolar, (*c*) elastic, (*d*) reticular, (*e*) adipose.

17. Adipose tissue forms: (*a*) only during fetal development, (*b*) throughout life, (*c*) mainly during fetal development and the first postpartum year, (*d*) mainly at puberty.

18. Intervertebral discs are composed of: (*a*) elastic connective tissue, (*b*) elastic cartilage, (*c*) hyaline cartilage, (*d*) fibrocartilage.

19. Intercalated discs are found in: (*a*) cardiac muscle, (*b*) movable joints, (*c*) vertebral column, (*d*) bone tissue.

20. Tissue fluid would most likely be found in: (*a*) loose connective tissue, (*b*) reticular tissue, (*c*) adipose tissue, (*d*) elastic tissue.

True/False

1. Connective tissues derive only from mesoderm, and function to bind, support, and protect body parts.

2. *Simple ciliated columnar epithelium* helps to move debris through the lower respiratory tract, away from the lungs.

3. Cells of epithelia are lightly packed, mostly avascular, and have small amounts of intercellular matrix.

4. Nervous tissue is located only in the brain and spinal cord.

5. Neurons are capable of mitosis to accommodate increased learning.

6. Most bones in the body begin as fibrocartilage and then ossify to bone.

7. Acinar glands have a flasklike secretory portion.

8. Mast cells that produce the anticoagulant heparin are dispersed throughout loose connective tissue.

9. Red blood cells are the only cellular component of blood tissue.

10. Based on structure and method of secretion, mammary glands are classified as compound acinar and apocrine.

11. Transitional epithelium occurs only in the urinary system.

12. All stratified squamous epithelium is keratinized and cornified.

13. Adipose tissue dies as a person diets, and new cells are formed as weight is gained.

14. Skeletal and cardiac muscle fibers are striated.

15. Neuroglia are specialized cells of nervous tissue that react to stimuli.

Completion

1. _____ is the scientific study of tissues.

2. Flattened, irregularly shaped cells that are tightly bound in a single-layered, mosaic pattern compose _____ tissue.

3. Epithelium consisting of two or more layers is classified as _____ .

4. _____ is the name given to the simple squamous epithelium that lines the inside of the heart and blood vessels.

5. Rhythmic contractions of sheets of _____ muscle tissue in the intestinal wall result in involuntary movement of food materials.

6. _____ is a protein in the skin that strengthens the epidermal tissue.

7. Pancreatic glands are classified as _____ glands because no portion of the gland is discharged with the secretion.

8. Bone tissue that has a spongy meshwork filled with bone marrow is termed _____ bone.

9. _____ is the matrix of blood tissue.

10. Alien matter is engulfed by leukocytes in the blood and in the _____ tissue of lymph nodes.

11. Excessive tissue fluid causes a swelled condition called _____ .

12. All connective tissue and muscle tissue is derived from the embryonic _____ .

13. _____ muscle tissue is composed of multinucleated, striated, cylindrical fibers arranged into fasciculi.

14. The _____ of a neuron receive a stimulus and conduct the nerve impulse to the cell body.

15. The lipid-protein product of Schwann cells forms a cover of _____ around the axon of a neuron.

Matching I

1. Simple squamous epithelium
2. Simple cuboidal
3. Simple columnar
4. Pseudostratified ciliated columnar
5. Stratified squamous
6. Transitional
7. Simple ciliated columnar

(a) Lining of uterine tubes
(b) Capillary walls
(c) Lining of oral cavity
(d) Lining of pancreatic ducts
(e) Lining of digestive tract
(f) Lining of respiratory passageways
(g) Lining of urinary bladder

Matching II

1. Simple acinar gland
2. Compound tubular
3. Unicellular
4. Compound tubuloacinar
5. Simple tubular
6. Compound acinar
7. Simple branched tubular

(a) Goblet cell
(b) Parotid gland
(c) Seminal vesicle
(d) Intestinal gland
(e) Liver
(f) Gastric gland
(g) Mammary gland

Matching III

1. Hyaline cartilage
2. Spongy bone
3. Canaliculi
4. Elastic cartilage
5. Compact bone
6. Fibrocartilage

(a) Pinna (external ear)
(b) Minute canals
(c) Intervertebral joint
(d) Inner bone tissue
(e) Fetal skeleton
(f) Covered by periosteum

Chapter 5

Integumentary System

OBJECTIVE A *To define the* integumentary system.

Survey The skin, or *integument*, and associated structures (hair, glands, and nails) constitute the integumentary system. This system accounts for approximately 7% of the body weight and is a dynamic interface between the body and the external environment.

5.1 Why is the integument considered an organ?

The skin is an organ because it consists of several kinds of tissues that are structurally arranged to function together. It is the largest organ of the body, covering about 7600 cm^2 (3000 in^2) on the average adult, and has thickness ranging generally between 1.0 and 2.0 mm, but up to 6.0 mm on the soles and palms.

5.2 What is the embryonic origin of the integument?

The principal layers of the skin are established by the eleventh week. The epidermis and associated structures are derived from the ectoderm germ layer, and the dermis and hypodermis are derived from the mesoderm germ layer. (See Objective C.)

OBJECTIVE B *To describe the basic functions of the integumentary system.*

Survey The functions of the integumentary system fall under three heads: (1) social recognition and communication; (2) protection of the body from disease and external injury; (3) maintenance of body homeostasis.

5.3 Which body systems interact with the integumentary system in performing its basic functions?

(1) Various emotions are conveyed through facial expressions involving the *muscular system*. Blushing involves vasodilation of cutaneous arteries of the *circulatory system*. (2) The systems that provide *body immunity* interact with the integumentary system. Countless sensory receptors within the integument convey impulses to the *nervous system*. (3) The *circulatory system* interacts extensively with the integumentary system in maintaining homeostasis. Even certain hormones of the *endocrine system* alter the function and appearance of the integument.

OBJECTIVE C *To list the principal layers of the integument and to describe their histology.*

Survey The *epidermis* is the thinner, outermost layer; it is composed of four or five sublayers. The *dermis* underlies the epidermis and consists of two sublayers. The deeper *hypodermis* is a single layer that affixes the integument to the underlying tissues or organs. A cross-sectional view of the integument is given in Fig. 5-1.

5.4 Putting aside the matter of relative depth, how does the epidermis differ from the dermis and hypodermis?

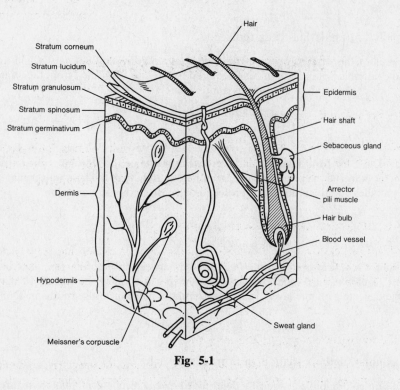

Fig. 5-1

The epidermis, the protective layer of the integument, is composed of stratified squamous epithelium, between 30 and 50 cells thick, which is keratinized and cornified (Problem 4.7). In contrast, the considerably thicker dermis is made up of blood vessels, nerve endings, glands, hair follicles, smooth muscle fibers, and various connective tissue fibers. The hypodermis consists primarily of loose fibrous connective tissue, blood and lymph vessels, and adipose tissue.

OBJECTIVE D *To describe the strata, or structural layers, of the epidermis.*

Survey   In Table 5-1, the strata are listed in the order innermost-to-outermost.

Table 5-1

Layer	Characteristics
Stratum germinativum (basal-cell layer)	A single layer of columnar cells that undergo mitosis; contains *melanocytes* (Problem 5.8)
Stratum spinosum	Several layers of cells with centrally located, large, oval nuclei and spiny processes
Stratum granulosum	One or more layers of granular cells that contain fibers of keratin and have shriveled nuclei
Stratum lucidum	A thin, clear layer found only in the epidermis of the palms and soles
Stratum corneum (horny layer)	Many layers of keratinized, dead cells that are flattened and nonnucleated (cornified)

5.5   How rapidly are epidermal cells replaced?

 The cells of the stratum germinativum are constantly dividing mitotically and moving outward to renew the epidermis. The length of time required for cells to be pushed from the stratum germinativum to the surface of the stratum corneum is approximately six to eight weeks.

5.6   What is a *callus* and why does it form?

 A callus is a localized hyperplasia (overdevelopment) of the stratum corneum due to pressure or friction on the skin and the resulting increase in mitotic activity of the stratum germinativum in that area. A callus provides additional localized protection against mechanical abrasion.

5.7   What is a *blister*? Does it serve a purpose?

 A blister is a vesicle of interstitial fluid located between the stratum germinativum and the stratum spinosum. Developing in response to rapid and intense friction of the surface of the skin, it serves to cushion the delicate germinal layer. In a *blood blister*, a pinch or bruise results in confined and localized hemorrhage.

5.8   What accounts for the variation in normal skin color?

 Normal skin coloration is genetically determined and reflects a combination of three pigments in the skin: *melanin, carotene,* and *hemoglobin*. Melanin is a brown-black pigment formed in cells called *melanocytes* that are found throughout the stratum germinativum and stratum spinosum. The number of melanocytes is virtually the same in all races, but the amount of melanin produced is variable. Carotene is a yellowish pigment found in epidermal cells and fatty parts of the dermis. Hemoglobin of the blood in the vascular dermis and hypodermis gives the skin its pinkish tones.

5.9   What is the functional relationship between melanocytes, melanin, and the tanning process?

 Melanin is a proteinaceous pigment that is protective against the ultraviolet rays present in sunlight. Gradual exposure to sunlight promotes increased production of melanin within melanocytes and hence tanning of the skin. Excessive exposure, however, can result in a *melanoma*, or tumor composed of melanocytes.

5.10   Do albinos lack melanocytes, melanin, or both?

 The skin of a genetically determined albino has the normal complement of melanocytes, but lacks the enzyme tyrosinase that converts the amino acid tyrosine to melanin.

OBJECTIVE E   *To describe the structure and function of the dermis.*

Survey   The upper *papillary layer* of the dermis is in contact with the epidermis. The deeper and thicker *reticular layer* is in contact with the hypodermis. Both layers contain numerous sensory receptors, and are highly vascular to nourish the living portion of the epidermis.

5.11   Which of the following connective-tissue fiber types is not usually found within the dermis? (*a*) reticular, (*b*) elastic, (*c*) white fibrous, (*d*) collagenous.

(*c*):   Elastic fibers are abundant within the papillary layer and provide skin tone; reticular fibers are abundant within the reticular layer and lend a strong meshwork to the skin; collagenous fibers, along with elastic fibers, course in definite directions and are imaged as lines of tension on the surface of the skin (Fig. 5-2).

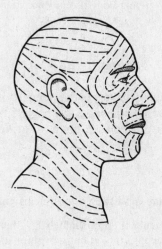

Fig. 5-2

5.12   Define *friction ridges* and explain how they arise.

Friction ridges are print patterns that occur on the palms of the hands and the soles of the feet. They are especially prominent on the skin covering the digits, where they are known as *fingerprints* or *toeprints*. Friction ridges are individualistic and are established prenatally in response to the pull of the elastic fibers of the dermal papillary layer upon the epidermis. As the name implies, friction ridges prevent slippage when grasping objects or in locomotion.

5.13   Describe the innervation of the skin.

Specialized integumentary *effectors* consist of the muscles or glands within the dermis which respond to efferent (motor) impulses transmitted through the autonomic nervous system. Several types of afferent (sensory) *receptors* respond to various tactile (touch), pressure, temperature, tickle, or pain sensations. Certain areas of the body, such as palms, soles, lips, and external genitalia, have a greater concentration of sensory receptors and are therefore more sensitive to touch.

5.14   How is the vascular supply to the skin important in maintaining homeostasis?

Dermal blood vessels figure in regulating body temperature and blood pressure. An autonomic vasoconstriction or vasodilation will, respectively, shunt blood away from the superficial dermal arterioles or permit it to flow more freely. Blood flow in response to thermoregulatory stimuli can vary from 1 to 150 mL/min for each 100 g of skin. Skin color and temperature also depend upon the blood supply. A cold, bluish or grayish skin occurs when the arterioles are constricted and the capillaries dilated; when both are dilated, the skin is warm and ruddy. Vasoconstriction increases the blood pressure.

5.15 How are the two clinical conditions *shock* and *decubitus ulcers* related to dermal blood flow?

Shock is a sudden disturbance of mental equilibrium accompanied by acute peripheral circulatory failure due to marked hypotension. Shock may be caused by loss of blood (from hemorrhage), diffuse systemic vasodilation, and/or inadequate cardiac function. *Decubitus ulcers* (bedsores) may occur in debilitated patients who lie in one position for extended periods of time. Bedsores develop as there is vasocompression in the skin overlying bony prominences, such as on the hip, heel, elbow, or shoulder.

OBJECTIVE F   *To describe the hypodermis.*

Survey   The hypodermis (subcutaneous layer) is the deepest of the three principal integument layers. It is composed primarily of loose fibrous connective tissue and adipose cells, interlaced with blood vessels. Collagenous and elastic fibers reinforce the hypodermis, particularly on the palms and soles.

5.16   *True or false*:   Females have a thicker hypodermis than do males.

True. The hypodermis of adults is approximately 8% thicker in females than in males. The increased thickness is due to greater deposition of lipids within adipose cells and is apparently hormonally influenced.

5.17   What are the functions of the hypodermis?

The hypodermis binds the dermis to underlying organs; it also stores lipids, insulates and cushions the body, and regulates temperature. In mature females, this layer, through its softening of body contour, plays a part in sexual attraction.

OBJECTIVE G   *To describe* hair, nails, sebaceous glands, sudoriferous glands, *and* ceruminous glands.

Survey   Hair, nails, and the three kinds of exocrine glands form from the epidermal skin layer and are therefore of ectodermal derivation. These structures develop as downgrowths of germinal epidermal cells into the vascular dermis, where they receive sustenance and mechanical support.

5.18   What are the functions of hair?

In other mammals, hair has survival value because of its role in body insulation, protective coloration, and social recognition and communication. The primary function of hair in humans is protection, although its effectiveness is limited. Hair on the scalp and eyebrows protects against sunlight, and hair in the nostrils and the eyelashes protects against airborne particles. An important secondary function of hair is as a means of individual recognition and sexual attraction.

5.19   Define: *hair follicle*; *shaft*, *root*, and *bulb* of a hair; *arrector pili muscle*; and the *medulla*, *cortex*, and *cuticle* layers of a hair.

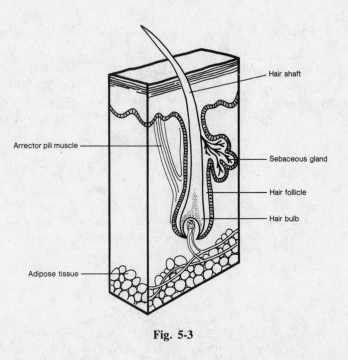

Fig. 5-3

Refer to Fig. 5-3. The *hair follicle* is the germinal epithelial layer that has grown down into the dermis. Mitotic activity of the follicle accounts for growth of the hair. The *shaft* is the dead, visible, projecting portion of a hair; the *root* is the living portion within the follicle; and the *bulb* is the enlarged base of the root that receives nutrients and is surrounded by sensory receptors. Each hair consists of an inner *medulla,* a median *cortex*, and an outer *cuticle* layer. The keratinized cuticle layer appears scaly under a dissecting microscope. Variation in the amount of melanin accounts for different hair colors; a pigment with an iron base (*trichosiderin*) causes red hair; gray or white hair is due to air spaces between the three layers of the shaft. Each hair follicle has an associated *arrector pili muscle* (smooth muscle) that responds involuntarily to thermal or psychological stimuli, causing the hair to be pulled into a more vertical position.

5.20 What are the three distinct kinds of human hair?

Lanugo is fine, silky, fetal hair that appears during the last trimester of development. Its function is unknown, but probably has something to do with the maturation of the hair follicles. *Angora* hair is continuously growing, as on the scalp, or on the face of sexually mature males. *Definitive* hair grows to a certain length and then ceases growing; e.g., eyelashes, eyebrows, and pubic and axillary (armpit) hair.

5.21 Describe the structure and function of nails.

Nails are formed from the hardened, transparent, stratum corneum of the epidermis. A parallel arrangement of keratin fibrils (see Fig. 5-4) accounts for the hardness. Fingernails grow about 1 mm per week, and they serve to protect the digits and aid in grasping small objects.

5.22 Of what value are the sebaceous glands, and why are they of clinical significance?

Sebaceous, or oil, glands are simple, branched glands that develop from the follicular epithelium of hair. They secrete acidic *sebum* onto the shaft of the hair, which is then dissipated to the surface of the skin where it protects, lubricates, and helps to waterproof the stratum corneum. Sebum consists mainly of lipids and some proteins. If the drainage pathway for sebaceous glands becomes blocked, the glands may become infected, resulting in acne. Sex hormones, particularly androgens, regulate the production and secretion of sebum.

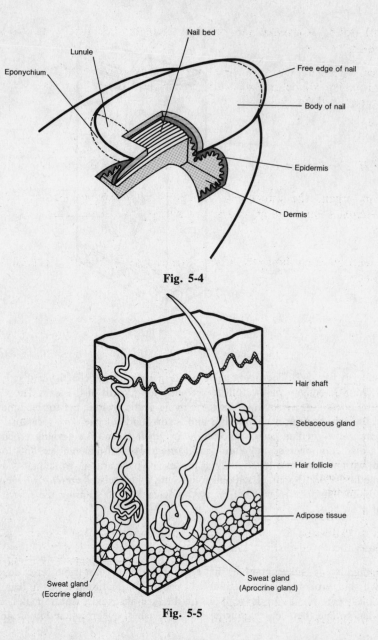

Fig. 5-4

Fig. 5-5

5.23 Differentiate between *eccrine* and *apocrine* sudoriferous (sweat) glands.

As is seen in Fig. 5-5, both eccrine and apocrine sweat glands are coiled, tubular structures that secrete perspiration onto the surface of the skin. *Eccrine sweat glands* are most abundant on the forehead, back, palms, and soles. These glands are formed prenatally and provide evaporative cooling in response to thermal or psychological stimuli. *Relaxed eccrine*, or *insensible*, *perspiration* accounts for 300 to 800 mL of water loss daily, depending upon the external temperature and humidity. *Active perspiration* in response to physical exercise may amount to 5 L of water per day. *Apocrine sweat glands* are much larger; they are restricted to the axillary and pubic regions, in association with hair follicles. They are not functional until puberty, and their odoriferous secretion is thought to act as a sexual attractant.

Perspiration is composed of water, salts, urea, uric acid, and traces of other compounds. Sweating brings about the excretion of certain wastes.

Mammary glands are specialized sudoriferous glands within the breasts. They are potentially functional in the female during her childbearing years, under the stimulus of pituitary and ovarian hormones.

5.24 *True or false*: *Cerumen* (earwax) is normally beneficial but, in some cases, may be detrimental.

True. Cerumen, the secretion of the ceruminous glands of the extenal auditory meatus (ear canal), is an insect repellent and keeps the tympanum (eardrum) pliable. However, excessive amounts may interfere with hearing.

OBJECTIVE H   *To summarize the physiology of the integument.*

Survey   As an organ, the integument functions in *protection*, *synthesis*, *temperature regulation*, *absorption*, *elimination of wastes*, and *sensory reception*.

5.25   Comment on each function of the integument, giving the layer(s) where the function is exercised.

See Table 5-2.

Table 5-2

	Function	Site	Comments
Protection against	Dehydration	Epidermis	Stratification forms dense barrier; sebum provides oily fibers; keratin toughens epidermis; basement membrane seals epidermis
	Mechanical injury	Epidermis	Stratification forms dense barrier; cornification of exposed layer; formation of calluses in response to friction; keratin toughens epidermis
	Pathogens	Epidermis	Stratification forms nearly impenetrable barrier; sebum is acidic (pH 4 to 6.8) and antiseptic, and lipid composition keeps epidermis from cracking; rapid rate of mitosis and shedding of cells from outer layer minimize entry of pathogens
	Ultraviolet light	Epidermis	Stratification forms dense barrier; scalp hair disperses light; melanin within melanocytes absorbs solar radiation
	Blood loss	Epidermis and dermis	Stratification forms dense barrier; process of *wound-healing* (dermal vasoconstriction, blood coagulation, temporary scab, collagenous scar tissue)
	Synthesis	Epidermis and dermis	Keratin, melanin, and carotene synthesized in the epidermis; dermis contains dehydrocholesterol, from which it synthesizes vitamin D in the presence of ultraviolet light
	Temperature regulation	Dermis and hypodermis	Cooling through vasodilation and sweating; warming through vasoconstriction and shivering; insulation provided by lipid content of hypodermis

Table 5-2 (*cont.*)

Function	Site	Comments
Absorption	Epidermis, dermis, and hypodermis	Limited by protective barriers, but some cutaneous absorption of O_2, CO_2, fat-soluble vitamins (A, D, E, and K), certain steroid hormones (cortisol), and certain toxic substances (insecticides)
Elimination of wastes	Epidermis and dermis	Excessive water; salt (NaCl); metabolic wastes (urea, uric acid)
Sensory reception	Epidermis, dermis, and hypodermis	Lower layers of epidermis contain free nerve endings responsive to pain; dermis contains *Meissner's corpuscles* responsive to touch and *Pacinian corpuscles* responsive to pressure; dermis and hypodermis contain *Krause corpuscles* responsive to high and low temperatures and *Ruffini corpuscles* responsive to medium temperatures

Key Clinical Terms

Acne   An inflammatory condition of sebaceous glands. Acne is affected by gonadal hormones and is therefore more common during puberty and adolescence. Pimples and blackheads on the face, chest, or back are the expressions of this condition.

Alopecia   Loss of hair, baldness. Baldness is usually due to genetic factors and cannot be treated. Baldness may accompany old age and is influenced by improper diet and/or poor circulation of blood.

Athlete's foot (*Tinea pedis*)   A fungus disease of the skin of the foot.

Blister   A collection of fluid between the epidermis and dermis, caused by excessive friction or a burn.

Boil   A localized bacterial infection originating in a hair follicle or skin gland; also termed a *furuncle*.

Burn   Lesion of integument caused by heat, chemicals, electricity, or solar exposure. Classified as: *first-degree* (redness or hyperemia in superficial layers of skin), *second-degree* (blisters involving deeper epidermal layers and dermis), or *third-degree* (destruction of areas of integument and damage to underlying tissue).

Callus   A localized buildup of the stratum corneum due to excessive friction.

Carbuncle   Similar to a boil, except involving subcutaneous tissues.

Corn   A small callus of the foot.

Dandruff   Common dandruff is the continual shedding of epidermal cells of the scalp, which can be controlled by normal washing and brushing of the hair. Abnormal dandruff may be due to certain skin diseases, such as *seborrhea* (q.v.) and *psoriasis* (q.v.).

Decubitus ulcer   A bedsore, or exposed ulcer from continual pressure on a localized portion of the skin, restricting the blood supply.

Dermatitis   An inflammation of the skin.

Eczema   A noncontagious inflammatory condition of the skin marked by red, itching, vesicular lesions that may be crusty or scaly.

Gangrene   Necrosis (death) of tissue due to obstruction of blood flow; may be localized or extensive, and perhaps secondarily infected with anaerobic microorganisms.

Melanoma   A cancerous tumor of the melanocytes within the epidermis.

Nevus   A mole or birthmark; congenital pigmentation of a certain area of the skin.

Psoriasis   Inflammatory skin disease usually expressed as circular scaly patches of skin.

Pustule   A small, localized elevation of the skin containing pus.

Seborrhea   A disease characterized by excessive activity of sebaceous glands and accompanied by oily skin and dandruff.

Shingles (*Herpes zoster*)   A virus-caused disease characterized by clusters of blisters along certain nerve tracts.

Urticaria (*hives*)   Skin eruption of reddish wheals, usually with extreme itching; may arise from an allergic reaction or stress.

Wart   Horny projection of epidermal cells; caused by a virus.

Review Questions

Multiple Choice

1. Which of the following word pairs is (are) appropriate? (*a*) integument–gland, (*b*) integument–tissue, (*c*) integument–organ, (*d*) all the above.

2. The integument is derived from: (*a*) ectoderm & endoderm, (*b*) ectoderm & mesoderm, (*c*) mesoderm & endoderm, (*d*) ectoderm & mesoderm & endoderm.

3. The skin accounts for what percentage of the body weight? (*a*) 2%, (*b*) 10%, (*c*) less than 2%, (*d*) 15%, (*e*) 7%.

4. Which of the following is *not* a function of the integument? (*a*) prevention of body dehydration, (*b*) synthesis of vitamin A, (*c*) regulation of body fluids, (*d*) regulation of body temperature.

5. Loss of body fluids through the integument is restricted by: (*a*) keratin, (*b*) the stratum germinativum, (*c*) carotene, (*d*) melanocytes, (*e*) the thickness of the dermis.

6. Which epidermal layer is lacking within the skin of the head and torso? (*a*) stratum spinosum, (*b*) stratum corneum, (*c*) stratum granulosum, (*d*) stratum lucidum, (*e*) stratum germinativum.

7. Which of the following pairings is appropriate? (*a*) stratum germinativum–keratin, (*b*) stratum corneum–melanocytes, (*c*) stratum granulosum–keratin, (*d*) stratum lucidum–blood vessels, (*e*) stratum spinosum–cornified.

8. Fingerprint patterns are established prenatally during development of the: (*a*) stratum corneum, (*b*) dermal papillary layer, (*c*) stratum germinativum, (*d*) dermal reticular layer, (*e*) hypodermis.

9. It is false that the dermis: (*a*) is highly vascular; (*b*) gives rise to sebaceous and sweat glands; (*c*) contains reticular, elastic, and smooth-muscle fibers; (*d*) contains numerous nerve endings.

10. It is false that the epidermis: (*a*) is highly vascular, (*b*) contains melanin and keratin, (*c*) is distinctly stratified, (*d*) gives rise to sebaceous and sweat glands.

11. Which coupling of terms is appropriate? (*a*) mesoderm, stratified squamous epithelium, epidermis; (*b*) epidermis, ectoderm, stratified squamous epithelium; (*c*) hypodermis, ectoderm, adipose tissue; (*e*) dermis, endoderm, vascular tissue.

12. "Rapunzel, Rapunzel, let down your long _____ hair." (*a*) axillary, (*b*) lanugo, (*c*) definitive, (*d*) nasal, (*e*) angora.

13. Which is the proper sequence of epidermal layers pierced as a sliver penetrates the dermis of the hand? (*a*) spinosum, germinativum, granulosum, lucidum, corneum; (*b*) germinativum, spinosum, granulosum, lucidum, corneum; (*c*) corneum, lucidum, granulosum, spinosum, germinativum; (*d*) corneum, lucidum, spinosum, granulosum, germinativum.

14. Cells from the stratum germinativum reach the stratum corneum in approximately: (*a*) 15–20 days, (*b*) 6–8 weeks, (*c*) 8–10 days, (*d*) 12–15 weeks, (*e*) 4–6 months.

15. Which of the following is *not* a type of integumentary sensory receptor? (*a*) Pacinian corpuscle, (*b*) free nerve ending, (*c*) Ruffini corpuscle, (*d*) Krause corpuscle, (*e*) none of the above.

16. Melanin in the skin serves to: (*a*) protect from ultraviolet light, (*b*) prevent infections, (*c*) help regulate body temperature, (*d*) keep the epidermis pliable, (*e*) reduce water loss.

17. Identify the mismatch: (*a*) yellowish coloration in Oriental skin—carotene abundant, (*b*) skin tans in response to sunlight—increased synthesis of melanin, (*c*) skin appears bluish (*cyanosis*)—oxygenated blood, (*d*) lack of skin pigmentation (*albinism*)—heredity, (*e*) dark coloration in Negroid skin—greater synthesis of melanin.

18. The most probable cause of alopecia is: (*a*) protein deficiencies, (*b*) dermal viral infection, (*c*) genetic inheritance, (*d*) stress.

19. Sebaceous glands: (*a*) secrete sebum directly to the surface of the skin, (*b*) are derived from specialized mesoderm, (*c*) are a type of oil-secreting endocrine gland, (*d*) are a compound saccular type, (*e*) are none of the above.

20. Which of the following is *not* a physiological possibility of the integument? (*a*) elimination of certain body salts, urea, and uric acid; (*b*) absorption of fat-soluble vitamins, steroid hormones, and certain toxic chemicals; (*c*) storage of lipids; (*d*) thermoregulation; (*e*) synthesis of proteins and carbohydrates; (*f*) prevention of desiccation and blood loss.

True/False

1. *Integument* means the same as *skin*, neither of which properly includes the hair or glands.

2. Skin is the largest tissue of the body, accounting for approximately 7% of the body weight.

3. Hair, nails, and integumentary glands are of both epidermal and ectodermal derivation.

4. A burn where there is damage to both the epidermis and dermis so that regeneration could occur only from the edges of the wound would be classified as a second-degree burn.

5. The eponychium and lunule are both proximal to the hyponychium of a nail.

6. The skin on the palm of the hand consists of five epidermal layers, two dermal layers, and a single hypodermal layer.

7. Mitotic activity is characteristic of all layers of the epidermis except the dead stratum corneum, which is constantly being shed.

8. Blacks have a significantly greater number of melanocytes within their skin than do individuals of other races.

9. Mammary glands are modified sebaceous glands.

10. Stimulation of the Krause corpuscles within the skin would be perceived as cold and might autonomically induce shivering.

11. Water-soluble substances would be more readily absorbed through the integument than fat-soluble substances.

12. All sudoriferous glands are formed and functional in a newborn.

13. The principal danger of a third-degree burn is excessive body fluid loss and disruption of homeostasis.

14. Alopecia is a disease that results in excessive loss of hair.

15. Warts, shingles, and acne are all virus-caused afflictions of the integument.

Chapter 6

Skeletal System

OBJECTIVE A *To learn the principal functions of the skeletal system.*

Survey Functions of the skeletal system fall into five categories. *Support*—the skeleton forms a rigid framework to which are attached the softer tissues and organs of the body. *Protection*—the skull, vertebral column, rib cage, and pelvic cavity enclose and protect vital organs; sites for blood cell production are protected within hollow cores of certain bones. *Movement*—bones act as levers when attached muscles contract, causing pivotal motion about a joint. *Hemopoiesis*—red bone marrow of an adult produces white and red blood cells and *platelets*. See Problem 6.7. *Mineral storage*—the matrix of bone is composed primarily of calcium and phosphorus, which can be withdrawn in small amounts if needed elsewhere in the body. Lesser amounts of magnesium and sodium are also stored in bone tissue.

6.1 How much of the body's calcium and phosphorus is contained in bones?

About 99% of the calcium, and 90% of the phosphorus, within the body is deposited in bones and teeth. These minerals give bone its rigidity and they account for approximately two-thirds of the weight of bone. In addition, calcium is necessary for muscle contraction, blood clotting, and movement across cell membranes. Phosphorus is required for the activities of DNA and RNA, as well as for ATP utilization.

6.2 In addition to mineralization involving calcium and phosphorus, what other physiological mechanisms determine the stability of bone?

Body organs that have regulatory functions have a direct effect upon the stability of bone. The kidneys, for example, determine blood composition, which in turn affects bone. The digestive system—via proteins and vitamins A, D, and C—and the female reproductive system—via pregnancy—can cause alteration of bone. Enzymatic and metabolic controls (alkaline phosphatase, glycogen, etc.) of the liver affect bone structure. At least five hormones affect bone: pituitary growth hormone stimulates bone growth (*osteogenesis*); thyroid hormone increases both osteogenesis and *osteolysis* (bone destruction); androgens and estrogens of the gonads stimulate bone growth and closure of the growth lines (*epiphyseal plates*); and adrenal cortisol and an excess of thyrocalcitonin lead to *osteoporosis* (bone atrophy through riddling).

OBJECTIVE B *To distinguish between the* axial *and* appendicular *portions of the skeletal system.*

Survey The *axial skeleton* consists of the bones that form the axis of the body and that support and protect the organs of the head, neck, and torso. The *appendicular skeleton* is composed of the bones of the upper and lower extremities and the body girdles that anchor the appendages to the axial skeleton. Tables 6-1 and 6-2 catalog these two parts of the skeleton; some of the bones are located by Fig. 6-1.

6.3 *True or false*: Every human body possesses $80 + 126 = 206$ bones.

False. While there may be 206 bones in the "average" human skeleton, the number differs from person to person depending on age and inheritance. At birth, the skeleton consists of approximately 270 bones. As further bone development (*ossification*) occurs during infancy, the number increases. Following adolescence, however, the number decreases as separate bones gradually fuse (*ankylose*).

Table 6-1.   Axial Skeleton (80 Bones)

Skull		
Cranial Bones	Facial Bones	Jawbone
frontal (1) parietal (2) occipital (1) temporal (2) sphenoid (1) ethmoid (1)	maxilla (2) palatine (2) zygomatic (2) lacrimal (2) nasal (2) vomer (1) inferior nasal concha (2)	mandible (1)

Middle-Ear Ossicles	Hyoid Bone	Vertebral Column	Rib Cage

malleus (2)
incus (2)
stapes (2)
	(1)	cervical vertebra (7)
		thoracic vertebra (12)
		lumbar vertebra (5)
		sacrum (1)
		  (fusion of 5)
		coccyx (1)
		  (fusion of 3–5)
			rib (24)
			sternum (1)

Table 6-2.   Appendicular Skeleton (126 Bones)

Pectoral Girdle	Upper Extremities	Pelvic Girdle	Lower Extremities
scapula (2) clavicle (2)	humerus (2) radius (2) ulna (2) **carpal** (16) **metacarpal** (10) **phalangeal** (28)	os coxa (2) (fusion of 3)	femur (2) tibia (2) fibula (2) patella (2) **tarsal** (14) **metatarsal** (10) **phalangeal** (28)

6.4   What are *sutural* and *sesamoid* bones?

Extra bones within the sutures (Problem 6.12) of the skull are called *sutural*, or *Wormian*, bones. Additional bones that develop in tendons, in response to stress as the tendons repeatedly move across a joint, are called *sesamoid* bones.

OBJECTIVE C   *To classify bones according to shape and to describe their surface features.*

Survey   The bones of the skeleton are divided into four types, on the basis of shape rather than size. ***Long bones*** [Fig. 6-2(*a*)] are longer than wide and function as levers (most of the bones in the appendages). ***Short bones*** [Fig. 6-2(*b*)] are more or less cubical and are found in confined spaces where they transfer forces (wrist and ankle). ***Flat bones*** [Fig. 6-2(*c*)] serve for muscle attachment or protection of underlying organs (cranium and ribs). ***Irregular bones*** [Fig. 6-2(*d*)] are elaborated for muscle attachment or articulation (vertebrae and certain skull bones).

The following terminology is used in the superficial description of bones.

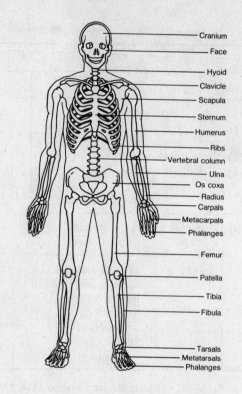

Fig. 6-1

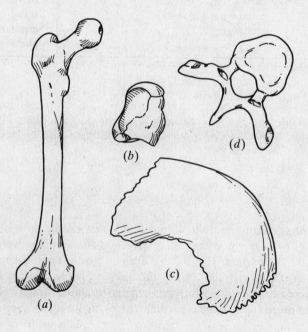

Fig. 6-2

Articulating surfaces

Condyle—a large, rounded, articulating knob (occipital *condyle* of the occipital bone)

Head—a prominent, rounded, articulating, proximal end (*head* of the femur)

Facet—a flattened or shallow articulating surface (costal *facet* of a thoracic vertebra)

Nonarticulating prominences

Process—any marked bony prominence (mastoid *process* of the temporal bone)

Tubercle—a small rounded process (greater *tubercle* of the humerus)

Tuberosity—a large roughened process (radial *tuberosity* of the radius)

Trochanter—a massive process found only on the femur (greater *trochanter* of the femur)

Spine—a sharp, slender process (*spine* of the scapula)

Crest—a narrow, ridgelike projection (iliac *crest* of the os coxa)

Epicondyle—a projection above a condyle (medial *epicondyle* of the femur)

Depressions and openings

Fossa—a shallow "ditch" (mandibular *fossa* of the temporal bone)

Sulcus—a groove that accommodates a vessel, nerve, or tendon (intertubercular *sulcus* of the humerus)

Fissure—a narrow, slitlike opening (superior orbital *fissure* of the sphenoid bone)

Meatus, or canal—a tubelike passageway (external auditory *meatus* of the temporal bone)

Alveolus—a deep pit or socket (maxillary *alveoli* for teeth)

Foramen (pl., foramina) a rounded opening through a bone (*foramen magnum* of the occipital bone)

Sinus—a cavity or hollow space (frontal *sinus* of the frontal bone)

Fovea—a small pit or depression (*fovea capitis* of the femur)

OBJECTIVE D   *To distinguish between* endochondral *and* intramembranous *bone formation*.

Survey   Ossification begins during the fourth week of prenatal development. Bones develop either through *endochondral ossification*—going first through a cartilaginous stage—or through *intramembranous* (*dermal*) *ossification*—forming directly as bone.

6.5   Which bones are endochondral and which membranous?

The majority of bones are found first as hyaline cartilage, which then undergoes endochondral ossification. The bones of the face, those of the roof of the skull, and sesamoid bones are products of intramembranous ossification.

6.6   What are *fontanels*, and why are they important?

During fetal development and infancy, the membranous bones of the top and sides of the cranium are separated by fibrous sutures. There are also six large membranous areas, called *fontanels* ("soft spots"), that permit the skull to undergo changes of shape (*molding*) during parturition (birth); four of these are visible in Fig. 6-3. The fontanels also allow rapid growth of the brain during infancy. Ossification of the fontanels is normally complete by 20–24 months of age.

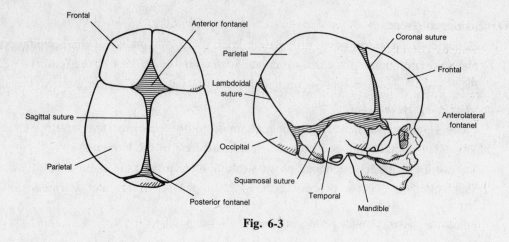

Fig. 6-3

OBJECTIVE E   *To examine the gross structure of a typical long bone.*

Survey   Refer to Fig. 6-4. Within the **diaphysis** (shaft) of a long bone there is a *medullary cavity*, which is lined with a thin layer of connective tissue, called the *endosteum*, and which contains fatty *yellow bone marrow*. On either end of the diaphysis is an *epiphysis*, consisting of cancellous bone surrounded by compact bone (Table 4-5). *Red bone marrow* is found within the pores of the cancellous bone. Separating the diaphysis and epiphysis is an *epiphyseal plate*, a region of mitotic activity responsible for elongation of bone; an *epiphyseal line* replaces the plate when bone growth is completed. A *periosteum* of dense fibrous tissue covers the bone, and is the site of tendon-muscle attachment and diametric (i.e., in width) bone growth.

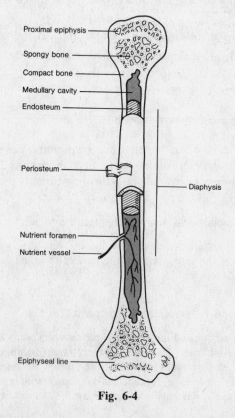

Fig. 6-4

6.7 What is the difference between *hemopoiesis* and *erythropoiesis*?

Hemopoiesis refers to production of all three types of blood cells—erythrocytes (red blood cells), leukocytes (white blood cells), and thrombocytes (blood platelets). *Erythropoiesis* refers specifically to production of erythrocytes. The principal site of hemopoiesis is the red bone marrow of the sternum, *vertebrae*, portions of the *ossa coxae*, and the proximal *epiphyses* of the *femora* and *humeri* [note the italicized plural forms].

6.8 What are *nutrient foramina*?

These are small openings into the diaphysis of a bone that give entry to vessels for the nourishment of the living tissue.

6.9 *True or false*:   Bone growth ceases as a person reaches physical maturity.

Linear bone growth does cease as the epiphyseal lines replace the epiphyseal plates and ossification occurs between the epiphyses and diaphyses. However, diametric bone growth and enlargement of bony processes may occur at any time, to accommodate an increase in body mass (as with a weight lifter).

6.10 Where is *articular cartilage* found?

Articular cartilage, which is thin hyaline cartilage, caps each epiphysis to facilitate joint movement. Technically, *bones* do not articulate together; rather, the *articular cartilage* of one bone articulates with the *articular cartilage* of another.

OBJECTIVE F   *To describe endochondral bone formation.*

Survey   Endochondral ossification begins in a *primary center*, in the shaft of a cartilage model, with hypertrophy of chondrocytes (cartilage cells) and calcification of the cartilage matrix. Then, as suggested in Fig. 6-5, the cartilage model is vascularized, osteogenic cells (Problem 6.11) form a bony collar around the model, and osteoblasts (Problem 6.11) lay down bony matrix around the calcareous spicules. Ossification from primary centers occurs before birth; from *secondary centers* (in the epiphyses), during the first five years.

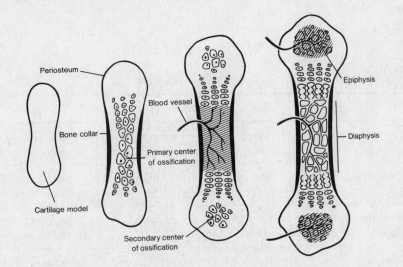

Fig. 6-5

6.11   What are *osteogenic* cells?

Bone cells exist in different forms corresponding to different functional states. *Osteogenic* cells are progenitor cells that give rise to all bone cells. *Osteoblasts* are the principal bone-building cells; they synthesize collagen fiber and bone matrix, and promote mineralization during ossification. Once this is accomplished, osteoblasts are trapped in their own matrix and become *osteocytes*, which maintain the bone tissue. *Osteoclasts* contain lysosomes and phagocytic vacuoles. These bone-destroying cells demineralize bone tissues.

OBJECTIVE G   To identify the cranial and facial bones of the skull, to describe their locations and their structural characteristics, and to name the articulations that affix them together.

Survey   The skull consists of eight cranial bones that articulate firmly with one another to enclose and protect the brain and associated sense organs, and 14 facial bones that form the foundation for the face and hold the teeth. See Table 6-1 and Fig. 6-6.

6.12   Where are the principal sutures of the cranium?

The cranial bones are united by serrated immovable joints called *sutures* (see Fig. 6-3). The frontal bone is joined to the two parietal bones at the *coronal suture*; the parietal bones meet each other at the *sagittal suture*; the occipital bone meets the parietal bones at the *lambdoidal suture*; and a parietal bone joins a temporal bone at a *squamosal suture*.

6.13   List the cavities of the skull.

The *cranial cavity* is the largest cavity of the skull, with a capacity of 1300–1350 cm^3. The *nasal cavity* is formed by both cranial and facial bones. Four sets of *paranasal sinuses* are located within the bones surrounding the nasal area. *Middle* and *inner ear chambers* are located within the temporal bones. The two *orbits* for the eyeballs are formed by both facial and cranial bones. The *oral*, or *buccal*, *cavity* (mouth) is only partially defined by bone (see Problem 6.20).

6.14   Which are the major foramina of the skull, where are they located, and what structures pass through them?

See Table 6-3.

Table 6-3

Foramen	Location	Structures Transmitted
Carotid (canal)	Petrous portion of temporal bone	Internal carotid artery and sympathetic nerves
Greater palatine	Palatine bone of hard palate	Greater palatine nerve and descending palatine vessels
Hypoglossal (foramen/canal)	Anterolateral edge of the occipital condyle	Hypoglossal nerve and branch of ascending pharyngeal artery
Incisive	Anterior region of hard palate, posterior to the incisor teeth	Branches of descending palatine vessels and nasopalatine nerve

Table 6-3   *(cont.)*

Foramen	Location	Structures Transmitted
Inferior orbital	Between maxilla and greater wing of sphenoid	Maxillary branch of trigeminal nerve, zygomatic nerve, and infraorbital vessels
Infraorbital	Inferior to orbit in maxilla	Infraorbital nerve and artery
Jugular	Between petrous portion of temporal (Problem 6.15) and occipital, posterior to carotid canal	Internal jugular vein; vagus, glossopharyngeal, and accessory nerves
. . . lacerum	Between petrous portion of temporal and sphenoid	Branch of ascending pharyngeal artery
Lesser palatine	Posterior to greater palatine foramen in hard palate	Lesser palatine nerves
. . . magnum	Occipital bone	Union of medulla oblongata and spinal cord, meningeal membranes, accessory nerves; vertebral and spinal arteries
Mandibular	Medial surface of ramus (elongated process) of mandible	Inferior alveolar nerve and vessels
Mental	Below the second premolar on the lateral side of mandible	Mental nerve and vessels
Nasolacrimal (canal)	Lacrimal bone	Nasolacrimal (tear) duct
Olfactory	Cribriform (sievelike) plate of the ethmoid	Olfactory nerves
Optic	Back of orbit in lesser wing of sphenoid	Optic nerve and ophthalmic artery
. . . ovale	Greater wing of sphenoid	Mandibular branch of trigeminal nerve
. . . rotundum	Within body of sphenoid	Maxillary branch of trigeminal nerve
. . . spinosum	Posterior angle of sphenoid	Middle meningeal vessels
Stylomastoid	Between styloid and mastoid processes of temporal	Facial nerve and stylomastoid artery
Superior orbital fissure	Between greater and lesser wings of sphenoid	Four cranial nerves (oculomotor, trochlear, ophthalmic branch of trigeminal, and abducens)
Supraorbital	Supraorbital ridge of orbit	Supraorbital nerve and artery
Zygomaticofacial	Anterior surface of zygomatic bone	Zygomaticofacial nerve and vessels

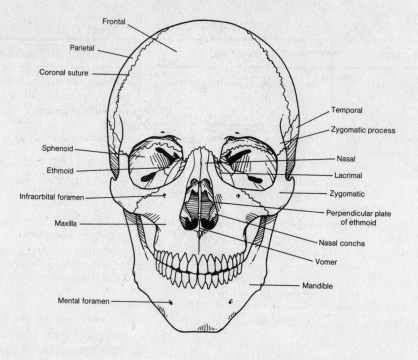

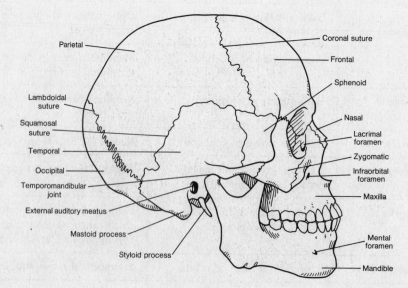

Fig. 6-6

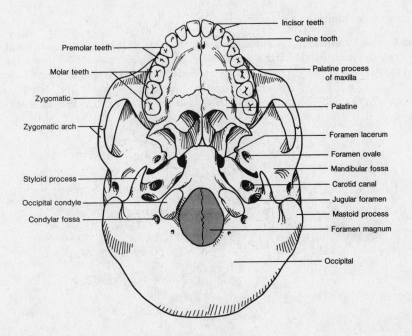

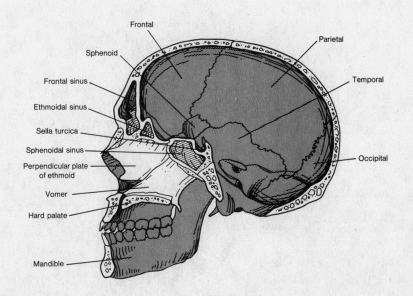

Fig. 6-6 (*cont.*)

6.15   Distinguish four regions of a temporal bone.

The flattened and lateral **squamous portion** forms part of the *zygomatic arch* and has a *mandibular fossa* to receive the condyle of the mandible at the *temporomandibular joint*. The **tympanic portion** contains the *external auditory meatus* (ear canal) and a *styloid process*. The **mastoid portion** consists of the mastoid process, which contains *mastoid* and *stylomastoid foramina*. The dense and inferior **petrous portion** contains the middle and inner ear, as well as the three ear ossicles (malleus, incus, and stapes). See Fig. 6-7.

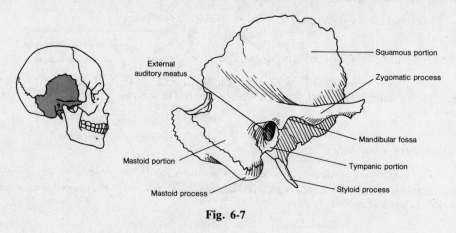

Fig. 6-7

6.16   What structures characterize the occipital bone?

The occipital bone forms the back and much of the base of the skull. It contains the *foramen magnum*, through which the spinal cord attaches to the brain, and the *occipital condyles*, which articulate with the *atlas* (Problem 6.25).

6.17   What do the *perpendicular plate*, *crista galli*, *nasal conchae*, and *cribriform plate* have in common?

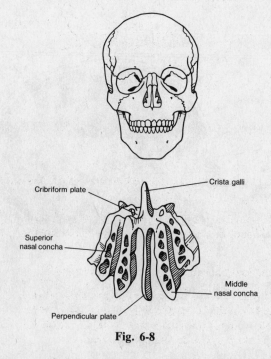

Fig. 6-8

All four structures are located on the ethmoid bone (Fig. 6-8). The perpendicular plate forms part of the *nasal septum*, which divides the nasal cavity into two *nasal fossae*. The crista galli (lit., "cockscomb") attaches to the meninges covering the brain. The epithelium covering the scroll-shaped nasal conchae warms and moistens inhaled air. The perforations in the cribriform plate allow the passage of olfactory nerves.

6.18   What endocrine gland is supported by the ethmoid?

The pituitary gland is lodged in the *sella turcica* of the ethmoid.

6.19   Which of the facial bones contain paranasal sinuses?

The paranasal sinuses are named according to the bones in which they are found; thus, there are the *maxillary*, *frontal*, *sphenoidal*, and *ethmoidal* sinuses. These connect to the nasal cavity and resonate to the voice.

6.20   What is the *hard palate*?

This is the bony partition between the nasal and oral cavities, which is formed through the union of the palatine processes of the maxillae and the palatine bones. The hard palate, along with the fleshy *soft palate*, forms the roof of the mouth.

6.21   Give the diagnostic features of the mandible.

See Fig. 6-9. The mandible has *condyloid processes* for attachment to the skull (at the temporomandibular joint). The *coronoid processes* are for attachment of the temporalis muscles. The *mandibular* and *mental foramina* are for passage of nerves. Sixteen teeth are embedded in the adult mandible.

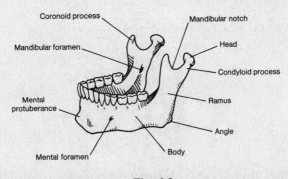

Fig. 6-9

6.22   Why are the ear ossicles, which are contained within the petrous portion of the temporal bones, not considered bones of the skull?

The three small, paired ear ossicles (Fig. 6-10) have a different origin and development than the rest of the skull. These "bonelets" transmit and amplify sound impulses through the middle ear.

6.23   Where is the *hyoid* located and what are its functions?

The U-shaped hyoid (Fig. 6-11) is located in the neck, where it supports the tongue superiorly and the larynx (voice box) inferiorly. In addition, several anterior neck muscles attach to this bone. The hyoid is unique in that it does not attach directly to any other bone. Instead, it is suspended from the styloid process of the skull by the *stylohyoid ligaments*.

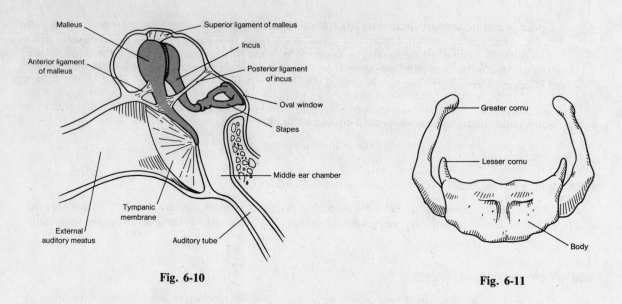

Fig. 6-10

Fig. 6-11

OBJECTIVE H   *To identify the regions of the vertebral column.*

Survey   Table 6-1 and Fig. 6-12 give the distribution of the 24 vertebrae over the ***cervical***, ***thoracic***, and ***lumbar*** regions. The vertebrae are separated by fibrocartilaginous *intervertebral discs*, and have openings between them, the *intervertebral foramina*, for the passage of spinal nerves.

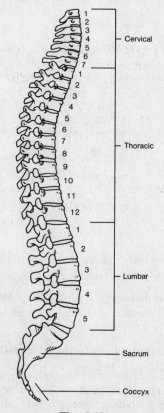

Fig. 6-12

6.24   Could any vertebra be called "structurally typical"?

   While no single vertebra is typical, the structure indicated in Fig. 6-13 generally prevails from one region to another. An anterior *body* makes contact with the intervertebral discs on each end. The *neural arch* on the posterior surface of the body is composed of supporting *pedicles* and arched *laminae*. The hollow space formed by the neural arch is the *vertebral foramen*; it allows passage of the spinal cord. Seven processes arise from the *vertebral arch*: the *spinous* process, two *transverse* processes, two *superior articular* processes, and two *inferior articular* processes.

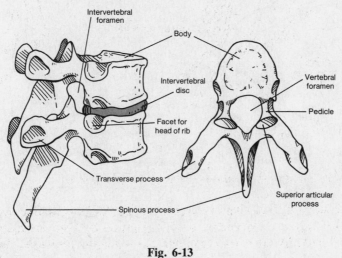

Fig. 6-13

6.25   Do any vertebrae have specific names?

   Only two. The *atlas*, which is the first cervical vertebra, is adapted to articulate with the occipital condyles of the skull, giving support to the head. (The *atlantooccipital joint* permits nodding of the head). The *axis*, or second cervical vertebra, has a peglike *odontoid process* that provides a pivot for rotation with respect to the atlas, as in turning the head to the side.

OBJECTIVE 1   *To describe the parts of the* rib cage *and its functions.*

Survey   The *sternum, costal cartilages,* and *ribs* attached to the thoracic vertebrae form the *rib cage,* or *thoracic cage,* of the thorax. The anteroposteriorly compressed rib cage supports the pectoral girdle and upper extremities, protects and supports the thoracic and upper abdominal viscera, and plays a major role in respiration.

6.26   Discuss the make-up of the sternum.

   The elongated and flattened sternum is a compound bone, consisting of an upper *manubrium,* a center *body,* and a lower *xiphoid process.* On the lateral sides of the sternum are *costal notches* where the costal cartilages attach.

6.27   *True or false:*   Each of the twelve pairs of ribs attaches posteriorly to the thoracic vertebrae and anteriorly to the sternum via costal cartilages.

   False. Only the first seven pairs, the *true ribs,* are anchored to the sternum by individual costal cartilages. Ribs 8, 9, and 10 are attached to the costal cartilage of rib 7; these are termed *false ribs.* The remaining two paired ribs do not attach to the sternum; they are the *floating ribs.*

6.28 What features do ribs have in common?

See Fig. 6-14(*a*). Each of the first ten ribs has a *head* and *tubercle* for articulation with a vertebra. The last two have a head but no tubercle. All ribs have a *neck*, *angle*, and *shaft*.

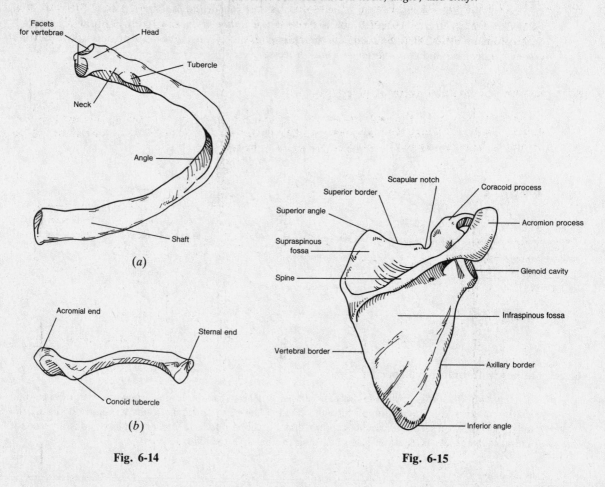

Fig. 6-14

Fig. 6-15

OBJECTIVE J *To describe the structure of the pectoral girdle.*

Survey The two *scapulae* and the two *clavicles* make up the pectoral (shoulder) girdle that attaches to the axial skeleton at the manubrium of the sternum. The pectoral girdle provides attachment for numerous muscles that move the brachium and forearm.

6.29 What is the function of a clavicle?

The S-shaped clavicle [Fig. 6-14(*b*)] binds the shoulder to the axial skeleton and positions the shoulder joint away from the trunk for freedom of movement.

6.30 Identify the following structural features of the scapula: *acromion process*, *coracoid process*, *glenoid cavity*, and *spine*.

See Fig. 6-15. The spine is a diagonal bony ridge that separates the *supraspinous fossa* from the *infraspinous fossa*. The spine broadens towards the shoulder as the acromion process. The glenoid cavity is a shallow depression into which the head of the humerus fits. The coracoid process lies superior and anterior to the glenoid cavity. On the anterior surface of the scapula is a slightly concave area known as the *subscapular fossa*.

OBJECTIVE K   *To identify the bones of the upper extremity and to list the diagnostic features of the bones of the brachium and forearm.*

Survey   The *brachium* is the upper arm; it contains a single bone, the *humerus*. The *radius* and *ulna* are the bones of the forearm, or *antebrachium*. The wrist and hand contain 27 bones, partitioned into **carpus**, **metacarpus**, and **phalanges**.

6.31   Identify the articular surfaces of the humerus.

Refer to Fig. 6-16. The proximal *head* of the humerus articulates with the glenoid cavity of the scapula. On the distal end, the *capitulum* is the lateral rounded condyle that receives the radius, and the *trochlea* is the pulleylike medial condyle that receives the radius.

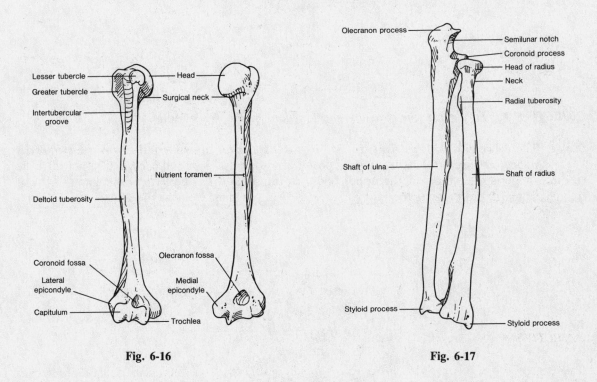

Fig. 6-16   Fig. 6-17

6.32   What do the radius and ulna have in common, and how do they differ?

The medial ulna and the lateral radius both articulate proximally with the humerus and distally with the carpal bones. As shown in Fig. 6-17, both have long shafts, and *styloid processes* for support of the wrist. The ulna has a prominent *olecranon process* (elbow), and the radius has a *radial tuberosity* for muscle attachments.

6.33   Describe the skeletal elements of the *wrist*, *palm*, and *fingers*.

The *carpus*, or wrist, consists of eight bones arranged into two transverse rows of four bones each. The *metacarpus*, or palm of the hand, consists of five bones, and the 14 *phalanges* are the skeletal elements of the fingers. See Fig. 6-18. The joints between the cube-shaped carpal bones permit movement in a confined area, while the elongated metacarpal bones and phalanges act as levers about their freely movable joints.

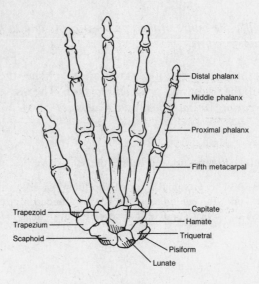

Fig. 6-18

OBJECTIVE L   *To describe the structure and functions of the pelvic girdle.*

Survey   The pelvic girdle, or *pelvis*, is formed by two *ossa coxae* united anteriorly by the *symphysis pubis* (see Fig. 6-19). It is attached posteriorly to the sacrum and coccyx. The pelvic girdle supports the weight of the upper body, as transmitted by the vertebral column. Also, it supports and protects the lower viscera and (in pregnant women) the fetus.

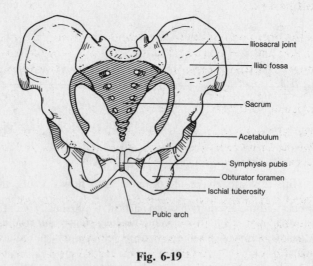

Fig. 6-19

6.34   What three bones comprise the os coxa?

Refer to Fig. 6-20. Each os coxa consists of an *ilium*, an *ischium*, and a *pubis*, which fuse together in the adult. The *acetabulum* is the large circular depression on the lateral side of the os coxa that receives the head of the femur. The large *obturator foramen* is coformed by the ramus of the ischium and the pubis.

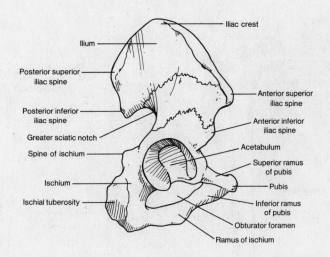

Fig. 6-20

6.35 What structural differences between the pelvis of an adult male and that of an adult female reflect adaptations for pregnancy and childbirth?

In the female: (1) the entire pelvis is tilted forward; (2) the distance between the anterior superior iliac spines (the breadth of the *false pelvis*) is greater; (3) the space posterior to the pelvic brim (the *true pelvis*) is spherical and wider; (4) the symphysis pubis is shallower; (5) the pubic angle is wider and more rounded.

OBJECTIVE M To identify the bones of the lower extremity and to list the diagnostic features of the bones of the thigh and lower leg.

Survey The *femur* is the only bone of the thigh; the *patella* is the (*sesamoid*, "sesame-seed-shaped") bone of the knee; and the *tibia* and *fibula* are the bones of the lower leg. The ankle and foot contain 26 bones, partitioned into **tarsus**, **metatarsus**, and **phalanges**.

6.36 Identify the articular surface of the femur.

See Fig. 6-21. The proximal, rounded *head* of the femur articulates with the acetabulum of the os coxa. The *medial* and *lateral condyles* on the distal end articulate with the tibia. The *patellar surface*, in contact with the patella, is between the condyles anteriorly.

6.37 What do the tibia and fibula have in common, and how do they differ?

As pictured in Fig. 6-22, the tibia is the massive, weight-bearing bone of the lower leg; it articulates proximally with the femur and distally with the *talus* (Problem 6.38). The fibula is the long, narrow bone lateral to the tibia that is more important for muscle attachment than for support. The *malleolus* ("little hammer") of either bone acts as a shield for the ankle.

6.38 Which bones make up the *ankle*, which the *foot*, and which the *toes*?

See Fig. 6-23. The *tarsus*, or ankle, consists of seven bones. Of these, the *talus* articulates with the tibia to form the ankle joint; the *calcaneus* forms the heel of the foot; the *navicular* is anterior to the talus; and the remaining four form a series that articulates with the five *metatarsals*, the bones of the sole of the foot. The 14 *phalanges* are the skeletal elements of the toes.

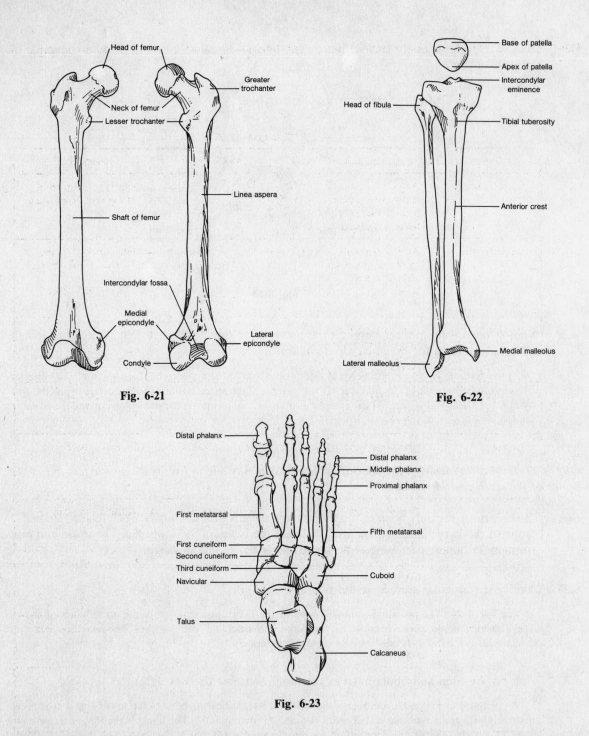

Fig. 6-21

Fig. 6-22

Fig. 6-23

OBJECTIVE N *To describe the kinds of articulations, or* joints, *in the body and the range of movement permitted by each.*

Survey Joints may be classified according to structure or function. In the structural classification, a joint is **fibrous**, **cartilaginous**, or **synovial** (a synovial joint has a joint cavity, and ligaments help to bind the articulating bones). The functional classification distinguishes **synarthroses** (immovable joints), **amphiarthroses** (slightly movable joints), and **diarthroses** (freely movable joints).

6.39   Complete the functional classification by subdividing the diarthroses. Provide an example of
each subcategory.

See Table 6-4.

Table 6-4

Type	Structure	Movements	Examples
Synarthroses	Articulating bones in close contact, bound by a thin layer of fibrous tissue or cartilage	None	Suture between bones of the skull; epiphyseal plate
Amphiarthroses	Articulating bones separated by fibrocartilaginous discs or bound by interosseous ligaments	Slightly movable	Intervertebral joints; symphysis pubis and sacroiliac joint; between tibia-fibula and radius-ulna
Diarthroses	Joint cavity contains *synovial membrane* and *synovial fluid* (Problem 6.40)	Freely movable	
Gliding	Flattened or slightly curved articulating surfaces	Sliding	Intercarpal and intertarsal joints
Hinge	Concave surface of one bone articulates with convex surface of another	Bending motion in one plane	Knee; elbow; joints of phalanges
Pivot	Conical surface of one bone articulates with a depression of another	Rotation about a central axis	Atlantoaxial joint; proximal radioulnar joint
Condyloid	Oval condyle of one bone articulates with elliptical cavity of another	Biaxial movement	Radiocarpal joint
Saddle	Concave-convex surface on either articulating bone	Wide range of movements	Carpometacarpal joint of the thumb
Ball-and-Socket	Rounded convex surface of one bone articulates with cuplike socket of another	Movement in all planes and rotation	Shoulder and hip joints

6.40   Where is synovial fluid produced in diarthroses?

Diarthroses are structurally synovial because they include a joint cavity. The *synovial fluid* is
secreted by a thin *synovial membrane* that lines the cavity (see Fig. 6-24). Certain diarthroses (e.g., knee
joint) have cartilaginous pads, called *menisci*, that cushion and guide the articulating bones. Another
feature are *bursas* (or *bursae*), small sacs filled with synovial fluid that cushion muscles and facilitate the
movement of tendons.

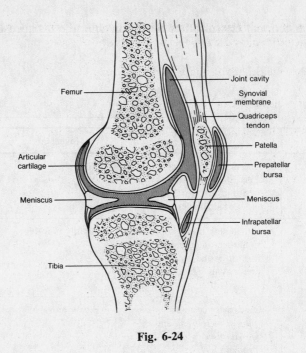

Fig. 6-24

6.41 Give technical terms for the types of movement permitted at diarthrotic joints.

Flexion is a movement which decreases the angle between two bones; *extension* increases the angle. *Abduction* is movement away from the midline of the body; *adduction* is movement toward the midline. *Rotation* is the movement of a bone around its own axis, without lateral displacement. (*Pronation* is the forearm rotation discussed in Problem 1.14; the opposite rotation is called *supination*.) *Circumduction* is a circular, conelike movement of a body segment.

Key Clinical Terms

Arthritis An inflammatory joint disease, usually associated with synovial membrane. In certain types of arthritis, mineral deposits may form.

Bursitis Inflammation of the bursas surrounding a joint.

Dislocation Displacement of one bone away from its natural articulation with another.

Fracture A cracking or breaking of a bone.

Kyphosis (*humpback*) An abnormal posterior convexity of the lower vertebral column.

Lordosis Excessive anteroposterior curvature of the vertebral column, generally in the lumbar region, resulting in a "hollow back" or "saddle back."

Osteoarthritis A localized degeneration of articular cartilage. (Not really an arthritis, since inflammation is not a primary symptom.)

Osteoporosis Atrophy of bone tissue, resulting in marked porosity in skeletal material. Causes include aging, prolonged inactivity, malnutrition, and an unbalanced secretion of hormones.

Scoliosis Excessive lateral deviation of the vertebral column.

Slipped disc Herniation of the *nucleus pulposus* of an intervertebral disc.

Spina bifida Developmental flaw in which the laminae of the vertebrae fail to fuse. The spinal cord may protrude through the opening.

Sprain Straining or tearing of the ligaments and/or tendons of a joint.

Review Questions

Multiple Choice

1. One of the following is *not* a function of the skeletal system: (*a*) production of blood cells, (*b*) storage of minerals, (*c*) storage of carbohydrates, (*d*) protection of vital organs.

2. Mitosis resulting in elongation of bone occurs at the: (*a*) articular cartilage, (*b*) periosteum, (*c*) epiphyseal plate, (*d*) diploe.

3. Which hormone–bone cell combination may result in osteoporosis? (*a*) adrenal cortisol–osteoclast, (*b*) estrogen–osteoblast, (*c*) thyroid hormone–osteoclast, (*d*) thyrocalcitonin–osteoblast.

4. Synovial fluid that lubricates a diarthrotic joint is produced by the: (*a*) menisci, (*b*) synovial membrane, (*c*) bursae, (*d*) articular cartilage, (*e*) mucous membrane.

5. A flattened or shallow articulating surface of a bone is called a: (*a*) tubercle, (*b*) fossa, (*c*) fovea, (*d*) facet.

6. Which type of cartilage is the precursor to endochondral bone? (*a*) costal, (*b*) hyaline, (*c*) fibroelastic, (*d*) articular.

7. Which suture extends from the anterior fontanel to the anterolateral fontanel? (*a*) coronal, (*b*) lambdoidal, (*c*) squamosal, (*d*) longitudinal.

8. A facial bone that is not paired is the: (*a*) maxilla, (*b*) lacrimal, (*c*) vomer, (*d*) nasal, (*e*) palatine.

9. Hemopoiesis would most likely take place in the: (*a*) hyoid, (*b*) vertebra, (*c*) maxilla, (*d*) scapula.

10. Which of the following bones is *not* part of the axial skeleton? (*a*) hyoid, (*b*) sacrum, (*c*) sphenoid, (*d*) clavicle, (*e*) manubrium.

11. The optic foramen is contained in the: (*a*) ethmoid bone, (*b*) occipital bone, (*c*) palatine bone, (*d*) sphenoid bone.

12. An example of a gliding joint is the: (*a*) intercarpal, (*b*) radiocarpal, (*c*) intervertebral, (*d*) phalangeal.

13. Which portion of the temporal bone has a mandibular fossa? (*a*) squamous portion, (*b*) petrous portion, (*c*) tympanic portion, (*d*) articular portion.

14. The superior and middle conchae are bony structures of which bone? (*a*) palatine, (*b*) nasal, (*c*) ethmoid, (*d*) maxilla.

15. Which bone does *not* contain a paranasal sinus? (*a*) frontal, (*b*) ethmoid, (*c*) vomer, (*d*) sphenoid, *e*) maxilla.

16. Teeth are found in the: (*a*) maxillae and mandible; (*b*) palatines and mandible; (*c*) maxillae and palatines; (*d*) maxillae, palatines, mandible.

17. The mastoid process is a structural prominence of the: (*a*) sphenoid, (*b*) parietal, (*c*) occipital, (*d*) temporal.

18. A joint that contains a broad, flat disc of fibrocartilage would be classified as: (*a*) synovial, (*b*) an amphiarthrosis, (*c*) a syndesmosis, (*d*) a diarthrosis.

19. Which bone has a diaphysis and epiphyses, articular cartilages, and a medullary cavity? (*a*) scapula, (*b*) sacrum, (*c*) tibia, (*d*) patella.

20. Remodeling of bone is a function of: (*a*) osteoclasts and osteoblasts, (*b*) osteoblasts and osteocytes, (*c*) chondrocytes and osteocytes, (*d*) chondroblasts and osteoblasts.

21. The cribriform plate is found in the: (*a*) sphenoid, (*b*) maxilla, (*c*) temporal, (*d*) vomer, (*e*) ethmoid.

22. Which is *not* a part of the os coxa? (*a*) acetabulum, (*b*) ischium, (*c*) pubis, (*d*) capitulum, (*e*) obturator foramen.

23. A fractured coracoid process would involve the: (*a*) clavicle, (*b*) scapula, (*c*) ulna, (*d*) rib.

24. The false pelvis is: (*a*) inferior to the true pelvis, (*b*) found in the male only, (*c*) narrower in the male than in the female, (*d*) not really a part of the skeletal system.

25. A fracture of the lateral malleolus would involve the: (*a*) fibula, (*b*) tibia, (*c*) ulna, (*d*) radius.

26. Which bone articulates distally with the talus in the foot? (*a*) navicular, (*b*) first metatarsal, (*c*) calcaneus, (*d*) first cuneiform, (*e*) cuboid.

27. When Joe Student holds out his hand to receive money from his father, the movement may be described as: (*a*) flexion–pronation, (*b*) extension–supination, (*c*) abduction–rotation, (*d*) rejection–depression.

28. An in-bending of the lower vertebral column is called: (*a*) lordosis, (*b*) scoliosis, (*c*) kyphosis, (*d*) spina bifida.

29. Which of the following bones lacks a styloid process? (*a*) sphenoid, (*b*) temporal, (*c*) ulna, (*d*) radius.

30. Surgery to remove a tumor of the pituitary gland would involve the: (*a*) mastoid process, (*b*) pterygoid process, (*c*) styloid process, (*d*) sella turcica.

True/False

1. The tibia and fibula articulate with the femur at the knee joint.

2. The proximal and distal ends of a long bone are referred to as diaphyses.

3. Menisci occur only in certain diarthrotic joints.

4. Supination and pronation are specific kinds of circumductional movements.

5. The yellow bone marrow in certain long bones of an adult produces red and white blood cells and platelets.

6. The matrix of bone is composed primarily of calcium and magnesium, which may be withdrawn in small amounts as needed elsewhere in the body.

7. Thyroid hormone may increase both osteogenesis and osteolysis.

8. A furrow on a bone that accommodates a blood vessel, nerve, or tendon is known as a sulcus.

9. Cervical vertebrae are characterized by the presence of articular facets.

10. The two ossa coxae articulate anteriorly with each other at the symphysis pubis, and posteriorly with the sacrum.

11. The lateral malleolus of the tibia stabilizes the ankle joint.

12. Most of the bones of the skeleton form through intramembranous ossification.

13. There are a total of 56 phalanges in the appendicular skeleton.

14. Articular cartilage and synovial membranes are found only in diarthroses.

15. All joints or articulations in the body permit some degree of movement.

16. Flexion means "contraction of a skeletal muscle."

17. Osteoblasts actually destroy bone tissue in the process of demineralization.

18. A person has seven pairs of true ribs, three pairs of false ribs, and two pairs of floating ribs.

19. An *intertrochanteric* fracture is a crack or break in the femur.

20. Surgery of a meniscus is always performed in the hinge joint of the knee.

Chapter 7

Muscular System

OBJECTIVE A *To recall the classification of muscle tissue.*

Survey Review Problem 4.18.

7.1 Which type of muscle is present in the greatest mass?

Skeletal muscle constitutes a body system by itself; it accounts for some 40% of the body weight. Smooth and cardiac muscle tissues account for approximately 3% of the body weight.

OBJECTIVE B *To give the functions of muscles.*

Survey **Motion.** Contraction of skeletal muscle produces such body movements as walking, writing, and chewing. Movements associated with breathing, digestion, and blood and lymph flow are also produced by muscle (smooth and cardiac).

Heat production. All cells release heat as an end product of metabolism, and a sizable fraction of cells (Problem 7.1) are muscular.

Posture and body support. The muscular system lends form and support to the body and maintains posture in opposition to gravity.

OBJECTIVE C *To distinguish between the various* architectures *of skeletal muscle.*

Survey Skeletal muscle tissue, in association with connective tissue, assumes highly organized patterns whereby the forces of the contracting muscle fibers are united and directed onto the structure to be moved.

7.2 How are skeletal muscles attached to bones?

A skeletal muscle spans a joint, being attached at either end by a *tendon* (Fig. 7-1). The *origin* of a muscle is the more stationary attachment; the *insertion* is the more movable. The body of the muscle proper is called the *belly*. Flattened tendons are called *aponeuroses*.

7.3 Sketch the four principal fiber patterns in skeletal muscle.

See Fig. 7-2. The straplike *parallel (fusiform) muscles* [Fig. 7-2(*a*)] have long excursions (stretch amplitudes) and good endurance, but are not especially strong. The *sartorius* and *rectus abdominis* are examples. The fan-shaped *covergent muscles* [Fig. 7-2(*b*)] have fibers that taper together at the insertion, to maximize contraction. The *deltoid* and *pectoralis major* are examples. *Pennate muscles* [Fig. 7-2(*c*)] have short excursions and are strong, but do not have good endurance. They provide dexterity and include the muscles of the forearm. *Circular muscles* [Fig. 7-2(*d*)] encompass a body orifice, acting as a sphincter when contracted; the *orbicularis oris* and *orbicularis oculi* (Fig. 11-4) are examples.

7.4 How are the fibers bound together in skeletal muscles?

Contracting muscle fibers would not be effective if they worked as isolated units. Each fiber is bound to adjacent fibers by the *endomysium* (Fig. 7-3) to form fasciculi. The fasciculi, in turn, are bound by the *perimysium*. The entire muscle is covered by the *epimysium*.

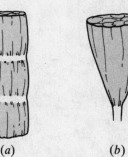

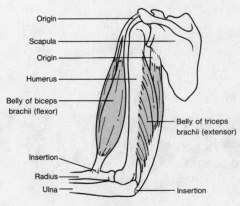

Fig. 7-1

Fig. 7-2

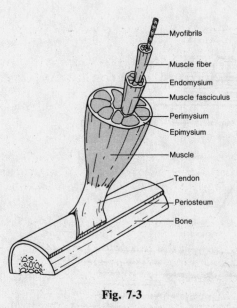

Fig. 7-3

7.5   How do the concepts of *synergism* and *antagonism* apply to skeletal muscles?

Muscles that contract together and are coordinated in effecting a particular movement are *synergistic*. *Antagonistic* muscles perform opposite tasks and are generally located on opposite sides of the limb or portion of the body.

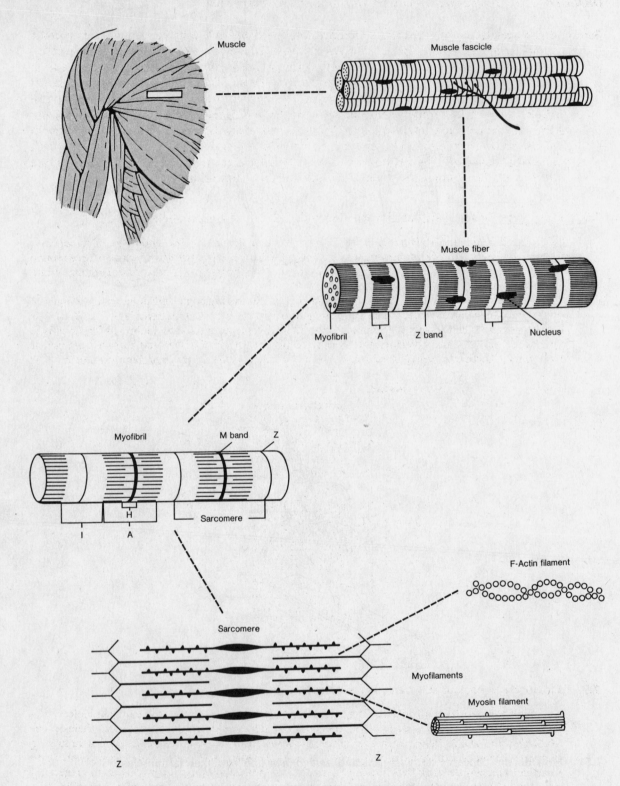

Muscle

Muscle fascicle

Muscle fiber

Myofibril A Z band Nucleus

Myofibril M band Z

H

I A

Sarcomere

F-Actin filament

Sarcomere

Myofilaments

Myosin filament

Z Z

Fig. 7-4

OBJECTIVE D   *To identify the components of a skeletal muscle fiber.*

Survey   Each skeletal muscle fiber is a multinucleated, striated cell containing numerous parallel bundles, or *myofibrils* (Fig. 7-4). Each myofibril is composed of still smaller units, called *myofilaments*, that contain the contractile proteins *actin* and *myosin*. Associated with actin are two additional proteins, *troponin* and *tropomyosin*. The regular spatial organization of the contractile proteins within the myofibrils is responsible for the cross-banding seen in skeletal and cardiac muscle cells: the dark bands are called *A bands* (anisotropic bands) and the lighter bands are referred to as *I bands* (isotropic bands). (Smooth muscle cells contain the same contractile proteins; but, in the absence of a regular spatial arrangement, they lack the cross-banding.) The I bands are bisected by dark *Z lines*, where the *actin filaments* (see Problem 7.7) of adjacent *sarcomeres* (the structural units of myofibrils) join.

7.6   Making use of Fig. 7-5, describe the fine structure of a skeletal muscle fiber.

The cell membrane is given the name *sarcolemma* ("muscle husk"). The cytoplasm, or *sarcoplasm*, is permeated by a network of membranous channels; this *sarcoplasmic reticulum* forms a sleeve around each myofibril. The longitudinal tubules that compose the principal part of the reticulum empty into expanded chambers called *terminal cisternae*.

The *transverse tubules* (T tubules) are not part of the sarcoplasmic reticulum, but are internal extensions of the sarcolemma, which run perpendicular to the sarcoplasmic reticulum. The T tubules pass between adjacent segments of terminal cisternae and penetrate deep into the interior of the muscle fiber. The T tubules also communicate with the outside of the muscle fiber, and contain extracellular fluid in their lumina. A *muscle triad* consists of a T tubule and the cisternae on both sides of it.

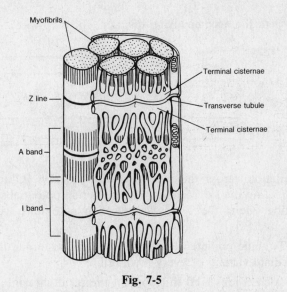

Fig. 7-5

7.7   Describe the protein structures involved in muscle contraction.

Actin filaments (*thin myofilaments*) are composed of actin, tropomyosin, and troponin molecules as shown in Fig. 7-6. *Globular actin* (G-actin) molecules are arranged in double spherical chains called *fibrous actin* (F-actin). Long, threadlike, tropomyosin molecules lie along the surface of F-actin strands, physically covering actin binding sites during the muscle resting state; each tropomyosin covers seven G-actins. Troponin, a small ovoid molecule attached to each tropomyosin, is involved in the regulatory action of calcium ions. ***Myosin filaments*** (*thick myofilaments*) consist of two forms of myosin (Fig. 7-7). *Light meromyosin filaments* (LMM) make up the rodlike backbones of the myosin filaments; *heavy meromyosin filaments* (HMM) form the transverse *crossbridges* that link with the binding sites on the thin myofilaments during contraction. (The thin filaments are intercalated among the thick.)

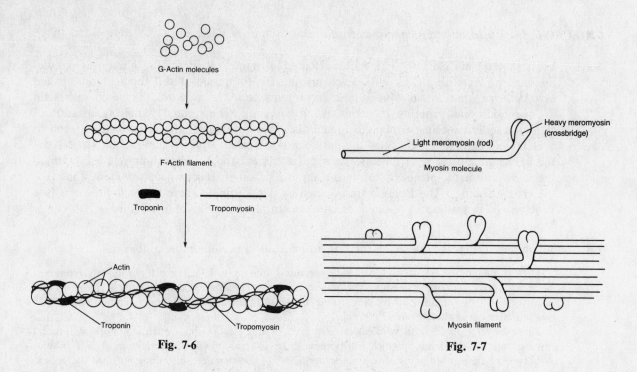

Fig. 7-6   Fig. 7-7

OBJECTIVE E   *To give the sequence of events in muscle contraction.*

Survey   With reference to the accompanying diagram, the steps are:

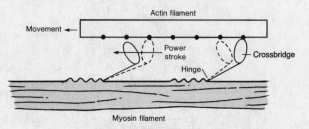

1. Stimulation across the *neuromuscular junction* [Objective F] initiates an *action potential* on the sarcolemma. As this action potential spreads, it is transmitted into the fiber along the T tubules. [Action potentials are treated in Chapter 8; see Objective E.]

2. The T-tubule potential causes the terminal cisternae to release calcium ions in the immediate vicinity of every myofibril.

3. The calcium ions bind to troponin; the resulting conformational change causes the attached tropomysin to move to one side, exposing the actin binding sites.

4. Myosin crossbridges bind to actin. Upon binding, the cocked (energized) HMM undergoes a conformational change, causing the head to tilt (*power stroke*), which pulls the actin filament over the myosin filament.

5. After the power stroke, ATP is able to bind to HMM, causing detachment of the crossbridges from the actin binding sites. ATPase within the HMM cleaves ATP to ADP + energy; the energy is used to recock the HMM. The HMM can then bind with another actin site, if they are still exposed (if calcium is still present), and produce another power stroke.

6. This sliding-with-a-ratchet mechanism—involving numerous actin binding sites and myosin crossbridges—constitutes a single muscle contraction.

7.8   After passage of an action potential, how does the muscle fiber return to the relaxed state?

The continually active calcium pump located in the walls of the sarcoplasmic reticulum forces calcium ions out of the sarcoplasma and back into the terminal cisternae. With binding sites on the action filaments no longer available, the sliding motion ceases and the fiber is relaxed.

OBJECTIVE F   *To describe the* neuromuscular junction.

Survey   At a neuromuscular (*myoneural*) junction (Fig. 7-8), a motor neuron and a muscle fiber meet. The *motor end plate* is the specialized portion of the sarcolemma (of the muscle fiber) surrounding the *axon terminal* (of the motor neuron; Fig. 8-1).

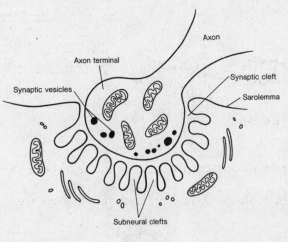

Fig. 7-8

7.9   The membrane of the muscle cell is invaginated (the *synaptic gutter*) at the site of the neuromuscular junction, and at the bottom of the gutter are numerous folds referred to as *subneural clefts*. What is the function of these folds?

The folds in the sarcolemma greatly increase the surface area over which the neurotransmitter (*acetylcholine*; Table 9-3) can produce an action potential.

7.10   List the sequence of events occurring at the neuromuscular junction.

(1) The action potential travels along the motor neuron to the axon terminal (end plate), where it causes an influx of calcium ions. (2) The calcium ions cause neurovesicles to release acetylcholine, which diffuses across the synaptic gutter and combines with specific receptors on the sarcolemma. (3) From these receptors, an action potential radiates over the sarcolemma.

7.11   What disease is caused by abnormal function of the neuromuscular junction?

Myasthenia gravis is the autoimmune disease in which the patient has developed antibodies that bind to and block the receptors for acetylcholine at the synaptic gutter. The numbers of subneural clefts and acetylcholine receptors are also reduced. As a result, transmission of the signal across the neuromuscular junction is significantly reduced, causing muscle weakness.

OBJECTIVE G   *To examine the physiology of a motor unit.*

Survey   A single cranial or spinal motor neuron and the muscle fibers it innervates compose a *motor unit* (Fig. 7-9). Each muscle fiber receives an axon terminal. In large muscles, such as a back or leg muscle, a motor unit may contain 200 to 500 muscle fibers; whereas in small muscles that are concerned with fine, precise movement, there are only about 10 to 15 muscle fibers per motor unit.

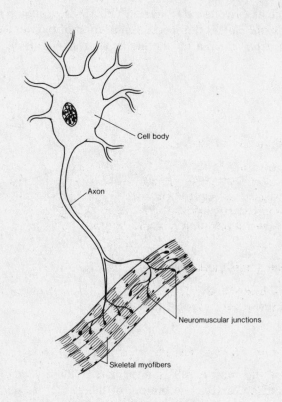

Fig. 7-9

7.12   Describe the response of the individual muscle fibers of a motor unit to an impulsive electrical stimulus (delivered by the motor neuron or externally administered).

Figure 7-10 is the record of a *twitch*, the contraction response of a muscle fiber to electrical stimulation at time zero. Three phases are distinguished: (1) the *latent period*, or time between stimulation and start of contraction; (2) the *contraction period*; and (3) the *relaxation period*.

7.13   Is the duration of the twitch the same for all skeletal muscles?

No. *Slow-twitch* (*red*) *fibers* are found predominantly in the postural muscles and have a twitch duration of about 100 ms (cf. Fig. 7-10). They are capable of contraction over long periods of time and derive energy from oxidative metabolism. *Fast-twitch* (*white*) *fibers* are mainly found in muscles involved in fine, skilled movements (e.g., hand muscles and eye muscles); they have a twitch duration of about 7.5 ms. They are capable of rapid, powerful contractions, but they fatigue quickly. They derive energy from glycolysis.

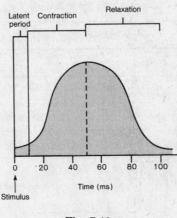

Fig. 7-10

7.14   Explain the phenomenon of *tetanus*.

If electrical impulses are applied to a muscle in rapid succession, one twitch will not completely have ended when the next begins. Therefore, since the muscle is already in a partially contracted state when the second twitch begins, the degree of muscle shortening in the second contraction will be slightly greater than that which occurs with a single twitch. At sufficiently high electrical frequencies, the overlapping twitches sum to one strong, steady contraction (*tetanus*).

7.15   Distinguish beween *isotonic* and *isometric* contractions.

A contraction is *isometric* when the length of the muscle does not change (decrease), and *isotonic* when the muscle shortens but its tension remains the same.

OBJECTIVE H   *To become familiar with the nomenclature for muscles and their actions.*

Survey   See Tables 7-1 and 7-2.

Table 7-1

Named According To:	Examples
Shape	*Rhomboides* (like a rhomboid); *trapezius* (like a trapezoid); or, denoting the number of heads of origin, *biceps* (two heads)
Location	*Pectoralis* (in the chest, or *pectus*); *intercostal* (between ribs); *brachii* (upper arm)
Attachment(s)	*Zygomaticus, temporalis, sternocleidomastoid, tibialis*
Size	*Maximus* (large); *minimus* (small); *longus* (long)
Orientation of fibers	*Rectus* (straight); *transversus* (across)
Relative position	*Lateralis, medialis, external*
Function	*Abductor, flexor, extensor, pronator*

Table 7-2

Action	Definition	Example
Flexion	Decreases a joint angle	Biceps brachii
Extension	Increases a joint angle	Triceps brachii
Abduction	Moves an appendage away from the midline	Deltoid
Adduction	Moves an appendage toward the midline	Adductor longus
Elevation	Raises	Levator scapulae
Depression	Lowers	Depressor labii inferioris
Rotation	Turns a bone around its longitudinal axis	Sternocleidomastoid
Supination	Rotates so that the palm faces forwards	Supinator
Pronation	Rotates so that the palm faces backwards	Pronator teres
Inversion	Turns the sole inward	Tibialis anterior
Eversion	Turns the sole outward	Peroneus tertius

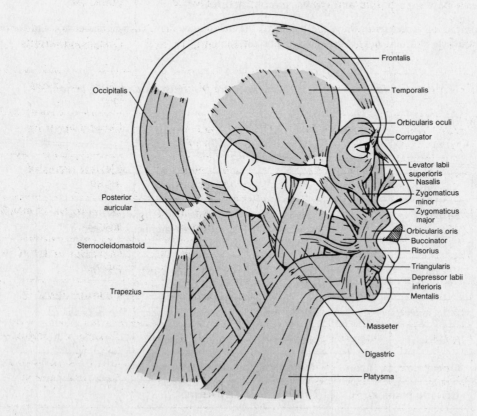

Fig. 7-11

OBJECTIVE 1 *To locate the major muscles of the axial skeleton.*

Survey These comprehend the muscles of facial expression, mastication, neck movement, respiration, the abdominal wall, and the vertebral column.

7.16 Identify by their attachments the muscles of *facial expression* and give their actions.

See Fig. 7-11 and Table 7-3.

Table 7-3

Facial Muscle	Origin(s)	Insertion(s)	Action(s)
Epicranius	Galea aponeurotica, occipital bone	Skin of eyebrow, galea aponeurotica	Wrinkles forehead; moves scalp
Frontalis	Galea aponeurotica	Skin of eyebrow	Wrinkles forehead; elevates eyebrow
Occipitalis	Occipital bone, mastoid process	Galea aponeurotica	Moves scalp backward
Corrugator	Fascia above eyebrow	Root of nose	Draws eyebrows toward midline
Orbicularis oculi	Bones of orbit	Tissue of eyelid	Closes eye
Nasalis	Maxilla, nasal cartilage	Aponeurosis of nose	Compresses nostrils
Orbicularis oris	Fascia surrounding lips	Mucosa of lips	Closes and purses lips
Levator labii superioris	Upper maxilla, zygomatic bone	Orbicularis oris, skin above lips	Elevates upper lip
Zygomaticus	Zygomatic bone	Superior corner of orbicularis oris	Elevates corner of mouth
Risorius	Fascia of cheek	Orbicularis oris at corner of mouth	Draws corner of mouth laterally
Triangularis	Mandible	Inferior corner of orbicularis oris	Depresses corner of mouth
Depressor labii inferioris	Mandible	Orbicularis oris	Depresses lower lip
Mentalis	Mandible (chin)	Orbicularis oris	Elevates and protrudes lower lip
Platysma	Fascia of neck and chest	Inferior border of mandible	Depresses lower lip
Buccinator	Maxilla	Orbicularis oris	Compresses cheek

7.17   Identify by their attachments the muscles of *mastication* and give their actions.

      See Fig. 7-12 and Table 7-4.

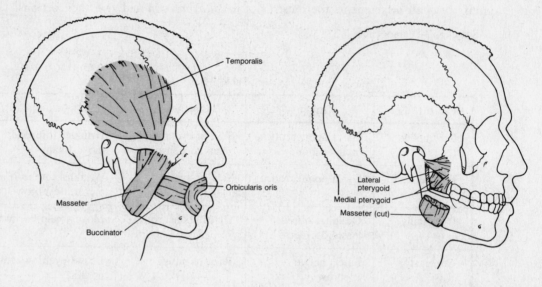

(*a*) Superficial view   (*b*) Deep view

Fig. 7-12

Table 7-4

Chewing Muscle	Origin(s)	Insertion(s)	Action(s)
Temporalis	Temporal fossa	Coronoid process of mandible	Elevates jaw
Masseter	Zygomatic arch	Lateral ramus of mandible	Elevates jaw
Medial pterygoid	Sphenoid bone	Medial ramus of mandible	Elevates jaw; moves jaw laterally
Lateral pterygoid	Sphenoid bone	Anterior side of mandibular condyle	Protracts jaw

7.18   Identify by their attachments the muscles of *neck movement* and give their actions.

      See Fig. 7-13 and Table 7-5.

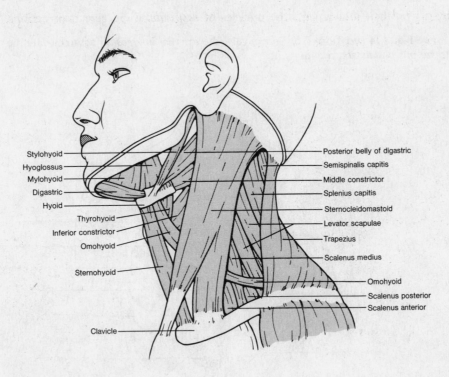

Stylohyoid
Hyoglossus
Mylohyoid
Digastric
Hyoid
Thyrohyoid
Inferior constrictor
Omohyoid
Sternohyoid
Clavicle

Posterior belly of digastric
Semispinalis capitis
Middle constrictor
Splenius capitis
Sternocleidomastoid
Levator scapulae
Trapezius
Scalenus medius
Omohyoid
Scalenus posterior
Scalenus anterior

Fig. 7-13

Table 7-5

Neck Muscle	Origin(s)	Insertion(s)	Action(s)
Sternocleidomastoid	Sternum, clavicle	Mastoid process of temporal bone	Turns head to side; flexes neck
Digastric	Inferior border of mandible, mastoid groove	Hyoid bone	Opens jaw; elevates hyoid
Mylohyoid	Inferior border of mandible	Body of hyoid, median raphe	Elevates hyoid and floor of mouth
Stylohyoid	Styloid process of temporal bone	Body of hyoid	Elevates and retracts tongue
Hyoglossus	Body of hyoid	Side of tongue	Depresses side of tongue
Sternohyoid	Manubrium	Body of hyoid	Depresses hyoid
Sternothyroid	Manubrium	Thyroid cartilage	Depresses thyroid cartilage
Thyrohyoid	Thyroid cartilage	Great cornu of hyoid	Depresses hyoid; elevates thyroid
Omohyoid	Superior border of scapula	Clavicle, body of hyoid	Depresses hyoid

7.19   Identify by their attachments the muscles of *respiration* and give their actions.

See Fig. 7-14 and Table 7-6. Collectively, the *external intercostals* act to elevate the thorax, and the *internal intercostals* act to depress it.

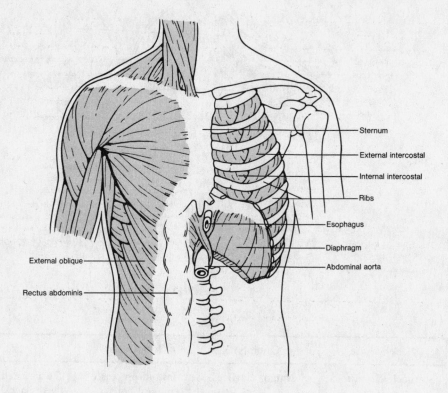

Fig. 7-14

Table 7-6

Breathing Muscle	Origin(s)	Insertion(s)	Action(s)
Diaphragm	Xiphoid process, costal cartilages of last six ribs, lumbar vertebrae	Central tendon	Pulls central tendon inferiorly and increases vertical dimension of thorax
External intercostal	Inferior border of a rib	Superior border of the rib below	Draws ribs together
Internal intercostal	Superior border of a rib	Inferior border of the rib above	Draws ribs together

7.20   Identify by their attachments the muscles of the *abdominal wall* and give their actions.

See Fig. 7-15 and Table 7-7.

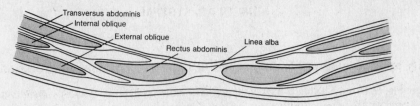

Fig. 7-15

Table 7-7

Abdominal Muscle	Origin(s)	Insertion(s)	Action(s)
External oblique	Lower eight ribs	Iliac crest, linea alba	Compresses abdomen; lateral rotation
Internal oblique	Iliac crest, inguinal ligament, lumbodorsal fascia	Linea alba, costal cartilages of last three or four ribs	Compresses abdomen; lateral rotation
Transversus abdominis	Iliac crest, inguinal ligament, lumbar fascia, costal cartilages of last six ribs	Xiphoid process, linea alba, pubis	Compresses abdomen
Rectus abdominis	Pubic crest, symphysis pubis	Costal cartilages of fifth to seventh ribs, xiphoid process	Flexes vertebral column

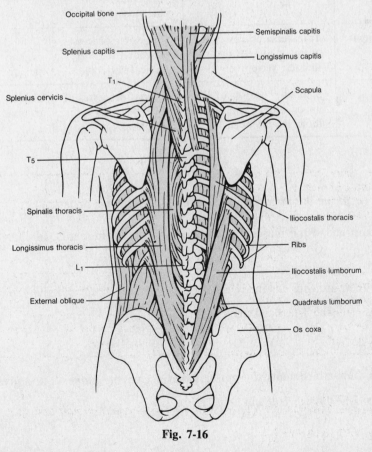

Fig. 7-16

7.21 Identify by their attachments the muscles of the *vertebral column* and give their actions.

See Fig. 7-16 and Table 7-8. The *iliocostalis muscles* (three species), *longissimus muscles* (two species), and *spinalis thoracis muscle* are collectively called the *erector spinae*.

Table 7-8

Spinal Muscle	Origin(s)	Insertion(s)	Action(s)
Quadratus lumborum	Iliac crest, lower three lumbar vertebrae	Twelfth rib, upper four lumbar vertebrae	Extend lumbar region; lateral flexion of vertebral column
Iliocostalis lumborum	Crest of ilium	Lower six ribs	Extend lumbar region
Iliocostalis thoracis	Lower six ribs	Upper six ribs	Extend thoracic region
Iliocostalis cervicis	Angles of three to six ribs	Transverse processes of fourth to sixth cervical vertebrae	Extend cervical region
Longissimus thoracis	Transverse processes of lumbar vertebrae	Transverse processes of all the thoracic vertebrae, lower nine ribs	Extend thoracic region
Longissimus capitis	Transverse processes of upper four or five thoracic vertebrae	Transverse processes of second to sixth cervical vertebrae	Extend cervical region; lateral flexion
Spinalis thoracis	Spinous processes of upper lumbar and lower thoracic vertebrae	Spinous processes of upper thoracic vertebrae	Extend vertebral column

OBJECTIVE J *To locate the major muscles of the appendicular skeleton.*

Survey These comprehend the muscles of the pectoral girdle, humerus, forearm, hand, thigh, and lower leg.

7.22 Identify by their attachments the muscles of the *pectoral girdle* and give their actions.

See Figs. 7-17 and 7-18 and Table 7-9.

7.23 Identify by their attachments the muscles that act on the *humerus* and give their actions.

See Figs. 7-17 through 7-19 and Table 7-10.

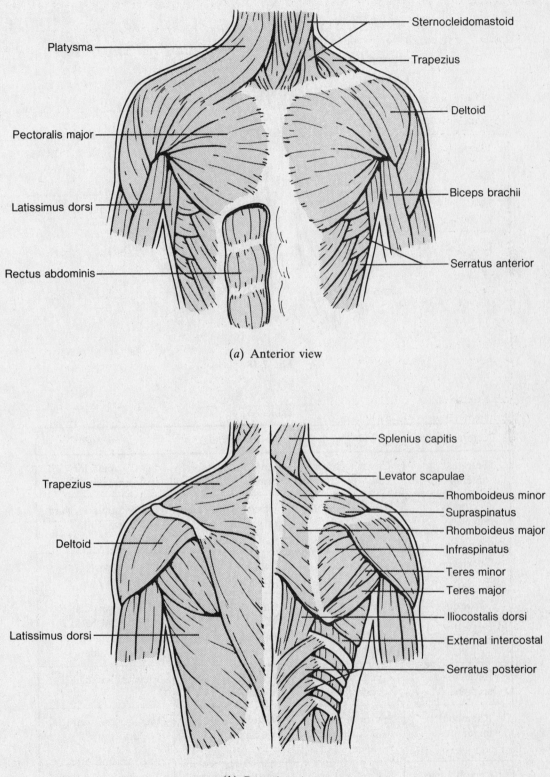

Platysma

Sternocleidomastoid

Trapezius

Pectoralis major

Deltoid

Latissimus dorsi

Biceps brachii

Rectus abdominis

Serratus anterior

(*a*) Anterior view

Trapezius

Splenius capitis

Levator scapulae

Rhomboideus minor

Supraspinatus

Rhomboideus major

Deltoid

Infraspinatus

Teres minor

Teres major

Iliocostalis dorsi

External intercostal

Latissimus dorsi

Serratus posterior

(*b*) Posterior view

Fig. 7-17

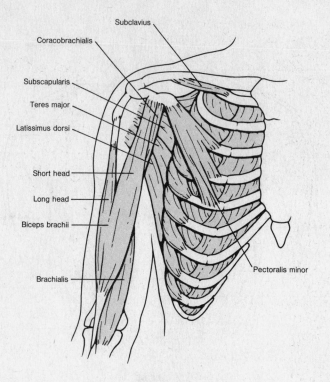

Fig. 7-18

Table 7-9

Pectoral Muscle	Origin(s)	Insertion(s)	Action(s)
Serratus anterior	Upper eight or nine ribs	Anterior vertebral border of scapula	Pulls scapula forward and downward
Pectoralis minor	Sternal ends of third, fourth, and fifth ribs	Coracoid process of scapula	Pulls scapula forward and downward
Subclavius	First rib	Subclavian groove of clavicle	Draws clavicle downward
Trapezius	Occipital bone, spines of cervical and thoracic vertebrae	Clavicle, spine of scapula, acromion process	Elevates scapula; draws head back; adducts scapula; braces shoulder
Levator scapulae	Fourth and fifth cervical vertebrae	Vertebral border of scapula	Elevates scapula
Rhomboideus major	Spines of second to fifth thoracic vertebrae	Vertebral border of scapula	Elevates and retracts scapula
Rhomboideus minor	Seventh cervical and first thoracic vertebrae	Vertebral border of scapula	Elevates and retracts scapula

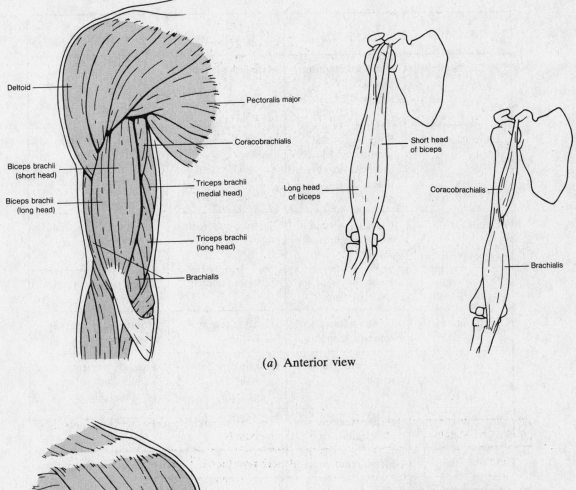

Deltoid

Pectoralis major

Coracobrachialis

Biceps brachii
(short head)

Triceps brachii
(medial head)

Biceps brachii
(long head)

Triceps brachii
(long head)

Brachialis

Short head
of biceps

Long head
of biceps

Coracobrachialis

Brachialis

(a) Anterior view

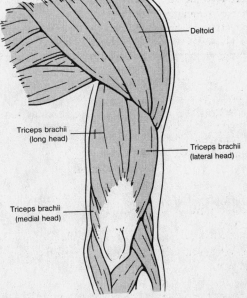

Deltoid

Triceps brachii
(long head)

Triceps brachii
(lateral head)

Triceps brachii
(medial head)

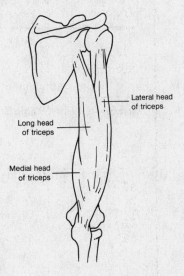

Lateral head
of triceps

Long head
of triceps

Medial head
of triceps

(b) Posterior view

Fig. 7-19

Table 7-10

Muscle	Origin(s)	Insertion(s) on Humerus	Action(s)
Pectoralis major	Clavicle, sternum, costal cartilages of second to sixth ribs	Greater tubercle	Flexes, adducts, and rotates arm medially
Latissimus dorsi	Spines of sacral, lumbar, and lower thoracic vertebrae; iliac crest; lower ribs	Intertubercular groove	Extends, adducts, and rotates humerus medially; retracts shoulder
Deltoid	Clavicle, acromion process, spine of scapula	Deltoid tuberosity	Abducts arm; extends or flexes humerus
Supraspinatus	Fossa (superior to spine of scapula)	Greater tubercle	Abducts and laterally rotates humerus
Infraspinatus	Fossa (inferior to spine of scapula)	Greater tubercle	Rotates arm laterally
Teres major	Inferior angle of scapula	Intertubercular groove	Extends humerus; adducts and rotates arm medially
Teres minor	Axillary border of scapula	Greater tubercle, groove	Rotates arm laterally
Subscapularis	Subscapular fossa	Lesser tubercle	Rotates arm medially
Coracobrachialis	Coracoid process of scapula	Shaft	Flexes and adducts shoulder joint

7.24   Identify by their attachments the muscles that act on the *forearm* and give their actions.

   See Fig. 7-19 and Table 7-11.

7.25   Identify by their attachments the muscles that act on the *wrist*, *hand*, and *digits*, and give their actions.

   See Figs. 7-20 and 7-21 and Table 7-12.

7.26   Identify by their attachments the *anterior* and *posterior* muscles that move the *thigh* and give their actions.

   See Fig. 7-22 and Table 7-13.

Table 7-11

Muscle	Origin(s)	Insertion(s) on Forearm	Action(s)
Biceps brachii	Coracoid process and tuberosity above glenoid fossa of scapula	Radial tuberosity	Flexes and supinates forearm
Brachialis	Anterior shaft of humerus	Coronoid process of ulna	Flexes forearm
Brachioradialis	Supracondyloid ridge of humerus	Proximal to styloid process of radius	Flexes forearm
Triceps brachii	Tuberosity below glenoid fossa, lateral and medial surfaces of humerus	Olecranon process of ulna	Extends forearm

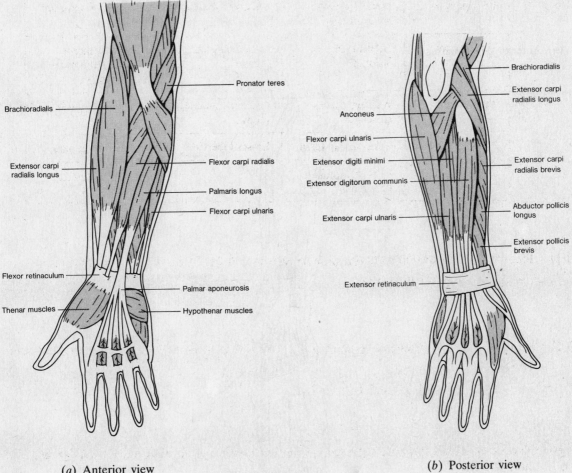

(a) Anterior view   (b) Posterior view

Fig. 7-20

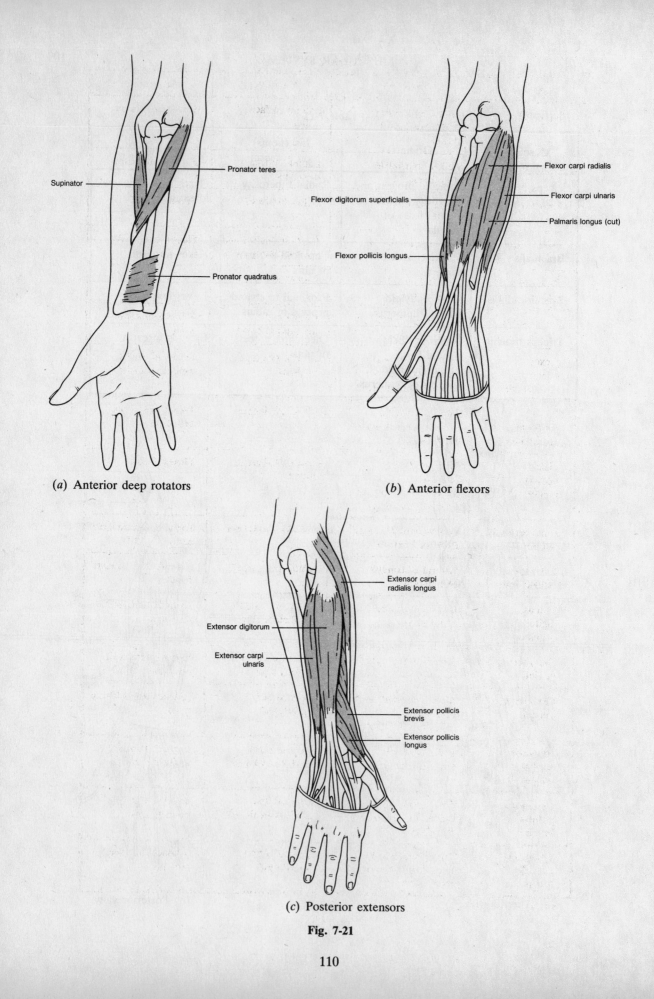

(a) Anterior deep rotators

Supinator

Pronator teres

Pronator quadratus

(b) Anterior flexors

Flexor carpi radialis

Flexor carpi ulnaris

Palmaris longus (cut)

Flexor digitorum superficialis

Flexor pollicis longus

(c) Posterior extensors

Extensor carpi radialis longus

Extensor digitorum

Extensor carpi ulnaris

Extensor pollicis brevis

Extensor pollicis longus

Fig. 7-21

110

Table 7-12

Forearm Muscle	Origin(s)	Insertion(s)	Action(s)
Supinator	Lateral epicondyle of humerus, crest of ulna	Lateral surface of radius	Supinates forearm
Pronator teres	Medial epicondyle of humerus	Lateral surface of radius	Pronates forearm
Pronator quadratus	Distal fourth of ulna	Distal fourth of radius	Pronates hand
Flexor carpi radialis	Medial epicondyle of humerus	Bases of second and third metacarpals	Flexes and abducts hand
Palmaris longus	Medial epicondyle of humerus	Palmar aponeurosis	Flexes hand
Flexor carpi ulnaris	Medial epicondyle, olecranon process	Carpal and metacarpal bones	Flexes and adducts wrist
Flexor digitorum superficialis	Medial epicondyle, olecranon process, anterior border of radius	Middle phalanges of digits	Flexes forearm, wrist, and digits
Flexor digitorum profundus	Proximal two-thirds of ulna, interosseous membrane	Distal phalanges	Flexes wrist, hand, and digits
Flexor pollicis longus	Shaft of radius, interosseous membrane, coronoid process of ulna	Distal phalanx of thumb	Flexes thumb
Extensor carpi radialis longus	Lateral supracondylar ridge of humerus	Second metacarpal	Extends and adducts hand
Extensor carpi radialis brevis	Lateral epicondyle of humerus	Third metacarpal	Extends and adducts hand
Extensor digitorum communis	Lateral epicondyle of humerus	Posterior surfaces of phalanges II–V	Extends wrist and phalanges
Extensor digiti minimi	Lateral epicondyle of humerus	Extensor aponeurosis of fifth digit	Extends fifth digit and wrist
Extensor carpi ulnaris	Medial epicondyle of humerus, olecranon process	Base of fifth metacarpal	Extends and adducts wrist
Extensor pollicis longus	Distal shaft of ulna (lateral side)	Base of distal phalanx of thumb	Extends thumb; abducts hand
Extensor pollicis brevis	Distal shaft of radius, interosseous membrane	Base of first phalanx of thumb	Extends thumb; abducts hand
Abductor pollicis longus	Distal radius and ulna, interosseous membrane	Base of first metacarpal	Abducts first digit and hand

111

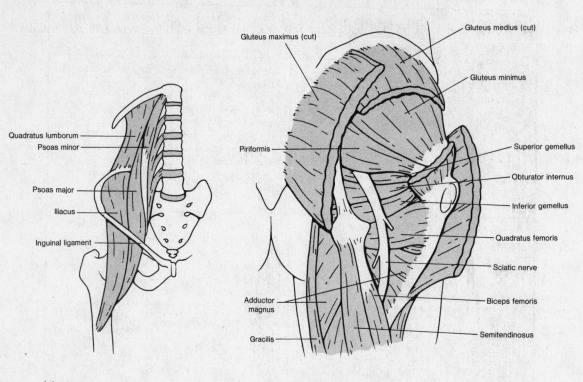

(a) Anterior pelvic muscles   (b) Deep gluteal muscles

Fig. 7-22

Table 7-13

Muscle	Origin(s)	Insertion(s)	Action(s)
Iliacus	Iliac fossa	Lesser trochanter of femur	Flexes and rotates thigh laterally; flexes vertebral column
Psoas major	Transverse processes of all lumbar vertebrae	Lesser trochanter with iliacus	Flexes and rotates thigh laterally; flexes vertebral column
Gluteus maximus	Iliac crest, sacrum, coccyx, aponeurosis of back	Gluteal tuberosity, iliotibial tract	Extends and rotates thigh laterally
Gluteus medius	Lateral surface of ilium	Greater trochanter	Abducts and rotates thigh medially
Gluteus minimus	Lateral surface of lower half of ilium	Greater trochanter	Abducts and rotates thigh medially
Tensor fasciae latae (Fig. 7-24)	Anterior border of ilium, iliac crest	Iliotibial tract	Abducts thigh

7.27 Identify by their attachments the *medial* muscles that move the *thigh* and give their actions. See Fig. 7-23 and Table 7-14.

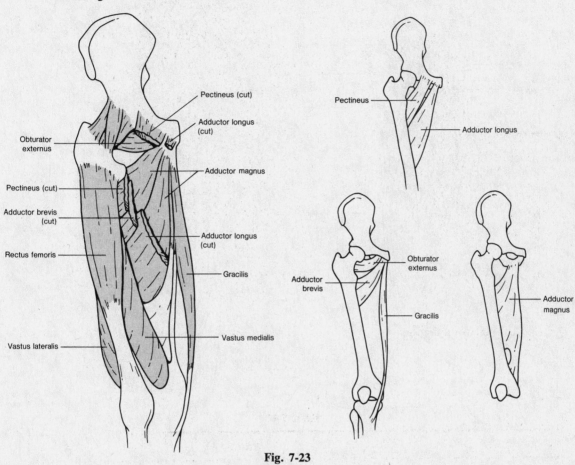

Fig. 7-23

Table 7-14

Muscle	Origin(s)	Insertion(s)	Action(s)
Gracilis	Inferior edge of symphysis pubis	Proximal medial surface of tibia	Adducts thigh; flexes and rotates leg at knee
Pectineus	Pectineal line of pubis	Distal to lesser trochanter of femur	Adducts and flexes thigh
Adductor longus	Pubis (below pubic crest)	Linea aspera of femur	Adducts, flexes, and laterally rotates thigh
Adductor brevis	Inferior ramus of pubis	Linea aspera of femur	Adducts, flexes, and laterally rotates thigh
Adductor magnus	Inferior ramus of ischium, pubis	Linea aspera and medial epicondyle of femur	Adducts, flexes, and laterally rotates thigh

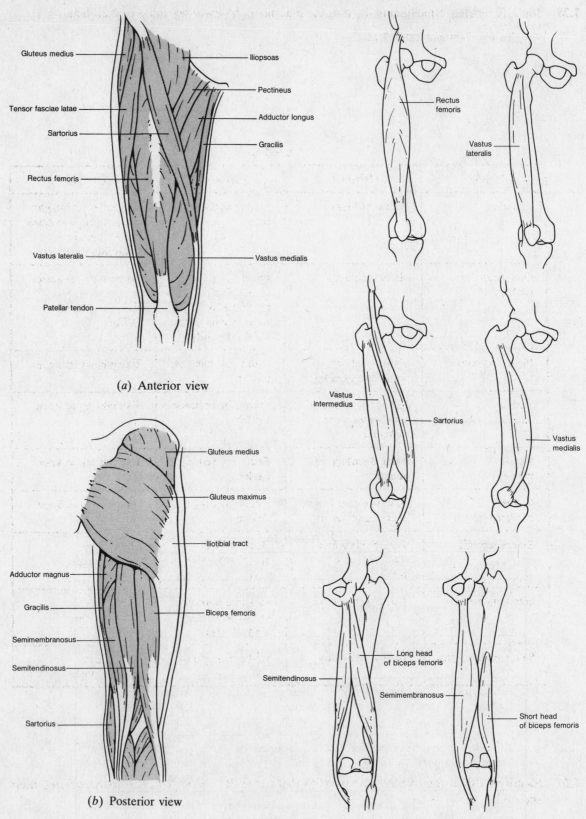

(a) Anterior view

(b) Posterior view

Fig. 7-24

7.28 Identify by their attachments the muscles that move the *lower leg* and give their actions.

See Fig. 7-24 and Table 7-15.

Table 7-15

Muscle	Origin(s)	Insertion(s)	Action(s)
Sartorius	Spine of ilium	Medial surface of tibia	Flexes leg and thigh; abducts thigh; rotates thigh laterally; rotates leg medially
Quadriceps femoris	Spine of ilium	Patella by common tendon (which continues as patellar tendon to tibial tuberosity)	Extends leg at knee
Rectus femoris	Spine of ilium, lip of acetabulum	Patella by common tendon	Extends leg at knee
Vastus lateralis	Greater trochanter and linea aspera of femur	Patella by common tendon	Extends leg at knee
Vastus medialis	Medial surface of femur	Patella by common tendon	Extends leg at knee
Vastus intermedius	Anterior and lateral surfaces of femur	Patella by common tendon	Extends leg at knee
Biceps femoris	Ischial tuberosity, linea aspera of femur	Head of fibula, lateral condyle of tibia	Flexes leg; extends thigh
Semitendinosus	Ischial tuberosity	Proximal portion of medial surface of body of tibia	Flexes leg; extends thigh
Semimembranosus	Ischial tuberosity	Medial condyle of tibia	Flexes leg; extends thigh

7.29 Identify by their attachments the muscles that move the *ankle*, *foot*, and *toes*, and give their actions.

See Figs. 7-25 and 7-26 and Table 7-16.

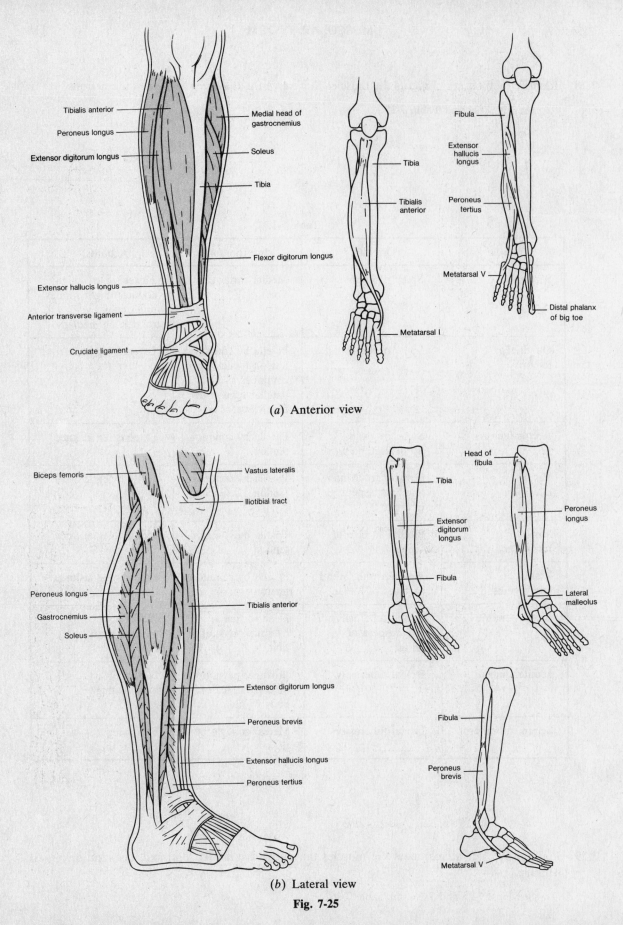

(a) Anterior view

(b) Lateral view

Fig. 7-25

116

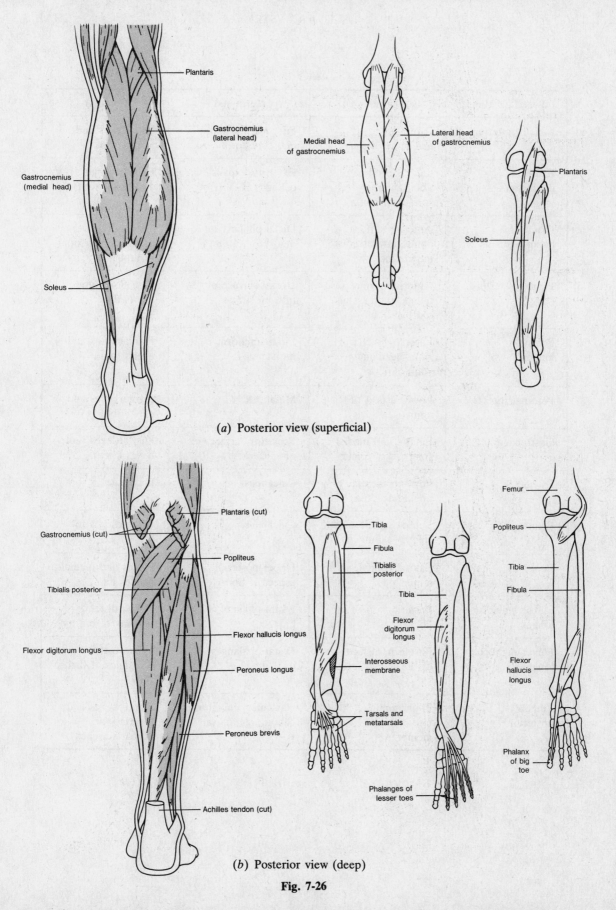

(a) Posterior view (superficial)

(b) Posterior view (deep)

Fig. 7-26

Table 7-16

Lower Leg Muscle	Origin(s)	Insertion(s)	Action(s)
Tibialis anterior	Lateral condyle and body of tibia	First metatarsal, first cuneiform	Dorsiflexes and inverts foot
Extensor digitorum longus	Lateral condyle of tibia, anterior surface of fibula	Extensor expansions of digits II–V	Extends digits II–V; dorsiflexes foot
Extensor hallucis longus	Anterior surface of fibula, interosseous membrane	Distal phalanx of digit I	Extends big toe; assists dorsiflexion of foot
Peroneus tertius	Anterior surface of fibula, interosseous shaft of fibula	Dorsal surface of fifth metatarsal	Dorsiflexes and everts foot
Peroneus longus	Lateral condyle of tibia, head and shaft of fibula	First cuneiform, metatarsal I	Plantarflexes and everts foot
Peroneus brevis	Lower aspect of fibula	Metatarsal V	Plantarflexes and everts foot
Gastrocnemius	Lateral and medial condyles of femur	Posterior surface of calcaneus	Plantarflexes foot; flexes knee
Soleus	Posterior aspects of fibula and tibia	Calcaneus	Plantarflexes foot
Plantaris	Lateral supracondylar ridge of femur	Calcaneus	Plantarflexes foot
Popliteus	Lateral condyle of femur	Upper posterior aspect of tibia	Flexes and medially rotates leg
Flexor hallucis longus	Posterior aspect of fibula	Distal phalanx of big toe	Flexes distal phalanx of big toe
Flexor digitorum longus	Posterior surface of tibia	Distal phalanges of digits II–V	Flexes distal phalanges of digits II–V
Tibialis posterior	Tibia, fibula, interosseous membrane	Navicular, cuneiforms, cuboid, metatarsals II–IV	Plantarflexes and inverts foot; supports arches

Key Clinical Terms

Charley horse   A bruised or torn muscle, with *cramp* (q.v.) and severe pain.

Cramp   A sustained spasmodic contraction of a muscle, usually accompanied by severe localized pain.

Graphospasm   Writer's cramp.

Hernia   Rupture, or protrusion through muscle tissue, of a portion of the underlying viscera. The most common hernias are the *femoral* (viscera passing through the femoral ring), the *inguinal* (viscera protruding through the inguinal canal), and the *umbilical* (viscera protruding through the navel).

Intramuscular injection   Hypodermic injection at certain heavily muscled areas (most commonly the buttock) so that nerves will not be damaged.

Muscular atrophy   A decrease in the size of muscle that had previously reached mature size. It may be caused by disease, disuse, infections, nutritional problems, or aging.

Muscular dystrophy   A genetic abnormality of muscle tissue, characterized by dysfunction and, ultimately, deterioration.

Myasthenia gravis   Refer to Problem 7.11. The disease is progressive, with muscles of facial expression, speech, mastication, and swallowing usually the first affected.

Myopathy   Any disease of the muscles.

Podiatry   The diagnosis and treatment of defects, injuries, and diseases of the feet.

Poliomyelitis   A viral disease that often attacks and destroys the cell bodies of the somatic motor neurons of the skeletal muscles, causing paralysis.

Shin splints   Tenderness and pain on the anterior surface of the lower leg, caused by straining the flexor digitorum longus.

Tetanus (*lockjaw*)   A disease caused by the bacterium *Clostridium tetani*, which produces a toxin causing muscles to go into tetanus (Problem 7.14). The jaw muscles are affected the earliest.

Torticollis (*wryneck*)   Persistent contraction of a sternocleidomastoid muscle, drawing the head to one side and distorting the face. Torticollis may be acquired or congenital.

Review Questions

Multiple Choice

1. Muscle fibers that are under involuntary control, lack cross striations, and have one nucleus at the center of the cell, are referred to as: (*a*) skeletal muscle fibers, (*b*) smooth muscle fibers, (*c*) cardiac muscle fibers, (*d*) none of the above.

2. The numerous folds in the sarcolemma at the neuromuscular junction are called: (*a*) subneural clefts, (*b*) gutter folds, (*c*) synaptic vesicles, (*d*) front plate folds.

3. The neurotransmitter released at the neuromuscular junction is: (*a*) norepinephrine, (*b*) L-dopa, (*c*) glycine, (*d*) acetylcholine.

4. The anisotropic dark bands of muscle fibers are called: (*a*) Z bands, (*b*) I bands, (*c*) A bands, (*d*) D bands.

5. The H bands are composed of: (*a*) thick myosin filaments, (*b*) thin myosin filaments, (*c*) thick actin filaments, (*d*) thin actin filaments.

6. The basic contractile element of striated muscle is the: (*a*) myofibril, (*b*) myosin, (*c*) A band, (*d*) sarcomere.

7. Muscle contraction is produced by shortening of all the following except: (*a*) myofibrils, (*b*) sarcomeres, (*c*) A bands, (*d*) T bands.

8. Muscle contraction is initiated when: (*a*) calcium is removed from troponin, (*b*) actin is removed from troponin, (*c*) actin is made available to troponin, (*d*) calcium is made available to troponin.

9. The source of calcium for the muscle lies in the: (*a*) T tubule, (*b*) lateral sac, (*c*) sarcoplasmic reticulum, (*d*) endoplasmic reticulum.

10. Shortening of sarcomeres is produced by: (*a*) shortening of filaments, (*b*) sliding of thin filaments over thick filaments, (*c*) sliding of thick filaments over thin filaments, (*d*) sliding of both thin and thick filaments.

11. In a relaxed muscle: (*a*) tropomyosin blocks attachment of crossbridges to actin, (*b*) the concentration of sarcoplasmic Ca^{2+} is high, (*c*) tropomyosin is moved out of the way so that crossbridges can attach to actin, (*d*) myosin ATPase is activated.

12. Muscle relaxation occurs: (*a*) as Ca^{2+} is released from the sarcoplasmic reticulum, (*b*) as long as Ca^{2+} is attached to troponin, (*c*) as action potentials are transmitted through the transverse tubules, (*d*) as the sarcoplasmic reticulum actively accumulates Ca^{2+}.

13. As compared to the red fibers, the white fibers: (*a*) contract more slowly, (*b*) develop more power, (*c*) have greater capacity for aerobic respiration, (*d*) have greater resistance to fatigue.

14. A muscle triad consists of: (*a*) a T tubule and a sarcomere, (*b*) a T tubule and two terminal cisternae, (*c*) a T pump and two calcium pumps, (*d*) three myofibrils.

15. A single motor neuron and all the skeletal muscle fibers it innervates compose a: (*a*) motor unit, (*b*) muscle triad, (*c*) sarcounit, (*d*) neuromuscular junction.

16. A muscle that develops tension against some load but does not shorten is undergoing: (*a*) isometric contraction, (*b*) isotonic contraction, (*c*) neither (*a*) nor (*b*), (*d*) both (*a*) and (*b*).

17. The channels that run from the sarcolemma into the interior of a skeletal muscle cell form the: (*a*) sarcoplasmic reticulum, (*b*) myofibers, (*c*) T tubule network, (*d*) tropomyosin.

18. The crossbridges of the myosin filament: (*a*) are made up of troponin molecules, (*b*) are believed to be attached to ATP molecules which are used to drive the power stroke of the head of the crossbridge, (*c*) shorten during the contraction process, (*d*) have a very high affinity for calcium ions released from the cisternae of the sarcoplasmic reticulum.

19. Troponin is a protein which: (*a*) is bound to myosin to form a complex that is normally inhibited in the resting muscle fiber, (*b*) forms the binding site for the myosin crossbridges when they attach to actin, (*c*) has a high affinity for calcium ions, (*d*) contains numerous molecules of ADP.

20. A flexor of the shoulder joint is the: (*a*) supraspinatus, (*b*) trapezius, (*c*) latissimus dorsi, (*d*) pectoralis major, (*e*) teres major.

21. Acetylcholine is stored in synaptic vesicles within: (*a*) motor end plates, (*b*) motor units, (*c*) myofibrils, (*d*) terminal ends of axons.

22. Which of the following muscles does not attach to the humerus? (*a*) teres major, (*b*) supraspinatus, (*c*) biceps brachii, (*d*) brachialis, (*e*) pectoralis major.

23. Which of the following muscles does not insert on the orbicularis oris muscle? (*a*) triangularis, (*b*) zygomaticus, (*c*) risorius, (*d*) platysma, (*e*) levator labii superioris.

24. The erector spinae does not include the: (*a*) iliocostalis, (*b*) longissimus, (*c*) spinalis, (*d*) semispinalis.

25. Which muscles are synergistic with the diaphragm during inspiration? (*a*) external intercostals, (*b*) internal intercostals, (*c*) abdominals, (*d*) all the above.

26. All the following muscles are synergists in flexing the elbow joint, except the: (*a*) biceps brachii, (*b*) brachialis, (*c*) coracobrachialis, (*d*) brachioradialis.

27. Which muscle does not attach to the scapula? (*a*) deltoid, (*b*) latissimus dorsi, (*c*) coracobrachialis, (*d*) teres major, (*e*) rhomboideus major.

28. Which muscle extends and rotates the thigh laterally? (*a*) iliacus, (*b*) gluteus maximus, (*c*) psoas major, (*d*) gluteus medius, (*e*) gluteus minimus.

29. Of the four quadriceps femoris muscles, which contracts over the hip and knee joints? (*a*) rectus femoris, (*b*) vastus medialis, (*c*) vastus intermedius, (*d*) vastus lateralis.

True/False

1. Muscle tissue accounts for approximately 40% of the body weight.

2. Actin is found only in the striated fibers of cardiac and skeletal muscle tissue.

3. Of the various types of muscles, the parallel-fibered muscles have the greatest strength and most endurance.

4. Fasciculi are enclosed in a perimysium covering.

5. The sarcolemma is the region of a myofibril that lies between two consecutive Z lines.

6. An action potential in a muscle fiber is initiated by stimulation across the neuromuscular junction.

7. Smooth, sustained muscle contraction is known as tetanus.

8. White fibers in muscles are surrounded by fewer capillaries than are red fibers, and thus contract more slowly.

9. A muscle triad consists of a sarcoplasmic reticulum, a T tubule, and a terminal cisternum.

10. A motor unit is a single motor neuron plus the muscle fibers which it innervates.

11. Lifting dumbbells is an example of isometric contraction.

12. The zygomaticus muscles cause a smile when contracted.

13. Contraction of the orbicularis oris muscle compresses the lips together.

14. Extension and abduction are interchangeable terms in that both actions result in an appendage's being moved away from the body.

15. The digastric muscles are important in chewing because, when contracted, they open the mouth.

16. The diaphragm and the internal intercostal muscles act synergistically to increase the dimension of the thorax.

17. Flexion of the vertebral column results when the sacrospinalis muscles are contracted.

18. The triceps brachii originates on the humerus and the scapula.

19. The palmaris longus pronates the hand.

20. A pulled groin muscle could involve the gracilis.

21. The sartorius muscle acts only on the hip joint.

22. The quadriceps femoris muscles are antagonistic to the hamstring muscles.

23. All three gluteal muscles insert on the greater trochanter of the femur.

24. The pectoralis minor rotates and adducts the humerus when contracted.

25. The gastrocnemius, soleus, and plantaris function synergistically in plantarflexion of the foot.

Matching

1. Z line
2. Sarcomere
3. A band
4. Sarcoplasmic reticulum
5. Troponin
6. Calcium
7. ATP-myosin complex

(*a*) ATP
(*b*) Flat protein structure to which the thin filaments attach
(*c*) Basic unit of muscle
(*d*) Intramuscular saclike structure (tubules) derived from membranes
(*e*) Structure that binds calcium
(*f*) "Trigger" or regulator of contraction
(*g*) Made up mainly of myosin molecules
(*h*) Functions to release the energy in ATP

Labeling

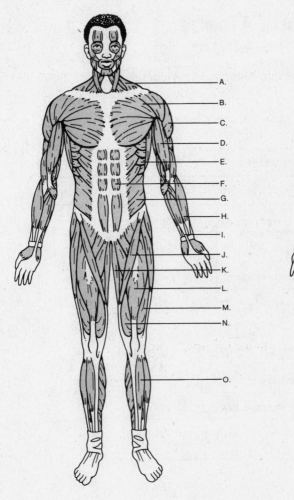

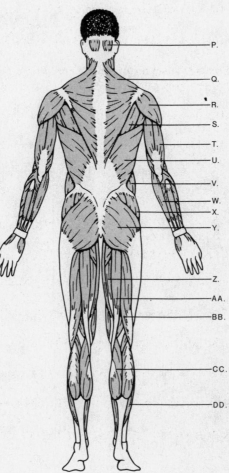

Chapter 8

Nervous Tissue

OBJECTIVE A   *To classify the nervous system into central and peripheral divisions.*

Survey

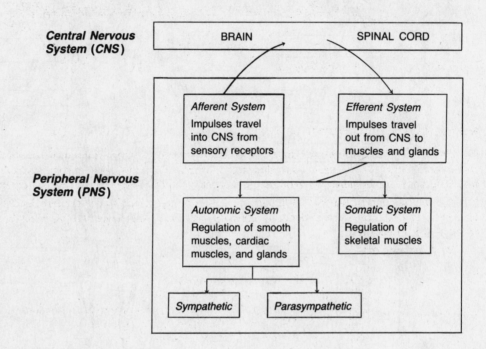

8.1   What are the major functions of the nervous system?

(1) To sense changes within the body and in the external environment; (2) to interpret the changes; (3) to respond to the interpretations by initiating glandular secretion and/or muscle contraction; (4) to assimilate experiences as required in memory, learning, and intelligence; (5) to program instinctual behavior (more important in vertebrates other than man).

8.2   Which component of the efferent nervous system is *voluntary*, and which *involuntary*?

The *somatic system* initiates movement in skeletal muscles; it is under conscious control and therefore voluntary. The *autonomic system* regulates the activities of smooth muscles, cardiac muscles, and visceral glands; it usually operates without conscious control and is considered to be involuntary.

8.3   List several differences between the central and peripheral nervous systems.

(1) A cluster of nerve cell bodies constitutes a *nucleus* (CNS) vs. *ganglion* (PNS). (2) A group of nerve fibers with a common origin and destination constitutes a *nerve tract* (CNS) vs. *nerve* (PNS). (3) The neuroglial cells forming the myelin sheath are *oligodendrocytes* (CNS) vs. *Schwann cells* (PNS). (4) The CNS is divided grossly into *gray matter* and *white matter*, but there is no such division of the PNS.

OBJECTIVE B *To describe the general structure of a neuron.*

Survey Refer to Fig. 8-1. Although neurons vary considerably in size and shape, they are generally made up of a *cell body* (or *soma*), *dendrites*, and an *axon*.

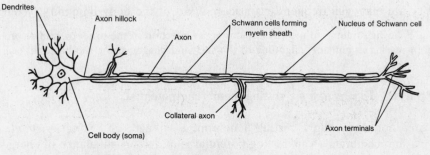

Fig. 8-1

8.4 Give three ways of classifying neurons.

By direction of impulse conduction. *Sensory* (*afferent*) neurons transmit nerve impulses to the spinal cord or brain. *Interneurons* (*internuncial* or *intercalated* neurons) conduct impulses from sensory to motor neurons. *Motor* (*efferent*) neurons conduct impulses away from the spinal cord or brain.

By number of processes. *Multipolar* neurons have one axon and two or more dendrites. *Bipolar* neurons have one axon and one dendrite. *Unipolar* neurons have a single process, which branches into an axon and a dendrite.

By fiber diameter.

Group Aα	12–20 μm	proprioception
Aβ	5–12 μm	pressure, touch
Aγ	3–6 μm	motor-nerve–muscle-spindle junctions
Aδ	2–5 μm	temperature, touch, pain
B	<3 μm	preganglionic autonomic
C	0.3–1.3 μm	postganglionic sympathetic

8.5 Describe the formation and function of the myelin sheath.

Figure 8-2 is a cross-sectional diagram of the process of myelination. In the CNS, certain oligodendrocytes—called *oligodendroglia*—are supposed to produce the myelin sheath; in the PNS, Schwann cells take this role. There are small gaps, the *nodes of Ranvier*, between segments of the myelin sheath. The sheath insulates nerve fibers and thereby inhibits flow of ions between intra- and extracellular fluid compartments.

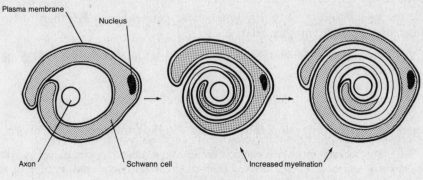

Fig. 8-2

OBJECTIVE C   *To classify neuroglial cells.*

Survey   Refer to Problem 4.23.

8.6   Why are microglia frequently considered part of the body immunity system?

  Following trauma to the CNS or during an infection of the brain or spinal cord, microglia respond by increasing in number, migrating to the site, and phagocytizing the bacterial cells or cellular debris.

OBJECTIVE D   *To describe the* resting membrane potential.

Survey   In a nonconducting ("resting") neuron, a voltage, or *resting potential*, exists across the plasma membrane. This resting potential is due to an imbalance of charged particles (ions) between the extracellular and the intracellular fluids. The mechanisms responsible for the membrane's having a net positive charge on its outer surface and a net negative charge on its inner surface (Fig. 8-3) are the following:

  (i)   A "sodium-potassium pump" transports Na^+ to the outside and K^+ to the inside, with three Na^+ ions moved out for every two K^+ ions moved in.

  (ii)   The plasma membrane is more permeable to K^+ ions than to Na^+ ions, so that the K^+, which is relatively concentrated inside the cell, moves outward faster than the Na^+, relatively concentrated outside the cell, moves inward.

  (iii)   The plasma membrane is essentially impermeable to the large (negatively charged) anions which are present inside the neuron, and therefore fewer negatively charged particles move out than positively charged.

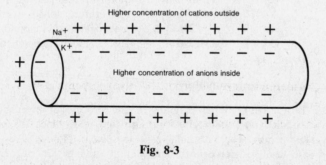

Fig. 8-3

8.7   Since the membrane is 50 to 100 times more permeable to K^+ than to Na^+, do these ions diffuse through different channels?

  Yes: For example, tetradotoxin (a poison obtained from puffer fish) blocks the diffusion of Na^+ but not that of K^+.

8.8   Is energy required to develop and maintain a resting membrane potential?

  Yes: The sodium-potassium pump, like all other cellular active transport systems, requires the expenditure of metabolic energy derived from the hydrolysis of adenosine triphosphate (ATP).

OBJECTIVE E   *To describe the chain of events associated with an* action potential.

Survey   Nerve impulses, which carry information from one point of the body to another, may be described as the progression along the neuron membrane of an abrupt change in the resting potential. This traveling disturbance, called an *action potential*, is schematized in Fig. 8-4.

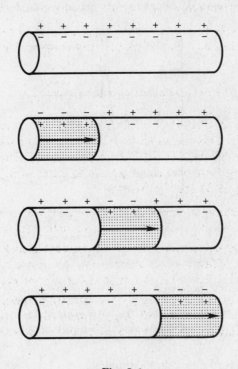

Fig. 8-4

The sequence of events is:

(1)   An adequate *stimulus* (chemical-electrical-mechanical) of the membrane at some point.

(2)   Increased *permeability* of the membrane to sodium at the point of stimulation.

(3)   *Sodium* ions rapidly move *inward* through the membrane.

(4)   As sodium ions move inward, the transmembrane potential reaches zero (the membrane becomes locally *depolarized*).

(5)   Sodium ions continue to move inward and the inside of the membrane becomes positively charged relative to the outside (*reverse polarization*).

(6)   Reverse polarization at the original site of stimulation causes a *local current* that acts as a stimulus to the adjacent region of the membrane.

(7)   At the point originally stimulated there is a decrease in the membrane's permeability to sodium and an increased permeability to potassium.

(8)   Potassium (K^+) ions rapidly move outward, again making the outside of the membrane positive in relation to the inside (*repolarization*).

(9)   Sodium and potassium pumps transport Na^+ back out of, and K^+ back into, the cell. [Now the cycle repeats at (1), relative to the advanced site.]

8.9   What determines whether a stimulus will be strong enough to produce an action potential in a nerve cell?

See Fig. 8-5. The resting membrane potential is about -70 mV; i.e., the inner surface is normally 70 millivolts below the outer surface. A just adequate, or *threshold*, stimulus will sufficiently increase the permeability of the membrane to Na^+ ions to raise the membrane potential to about -55 mV. Once this *threshold potential* is reached, complete depolarization and repolarization will occur and an action potential is generated.

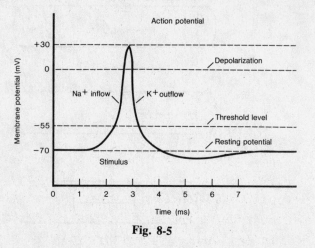

Fig. 8-5

8.10   Is the size of the action potential related to the strength of the (adequate) stimulus?

No: Nerve and muscle cells obey the *all-or-nothing law*, which states that a threshold stimulus evokes a maximal response, and that a subthreshold stimulus evokes no response.

8.11   If a neuron has received a threshold stimulus and is undergoing depolarization and repolarization, how much time must pass before a stimulus can produce a second action potential?

In the interval from the onset of an action potential until repolarization is about 1/3 completed, no stimulus can elicit another response; the "dead phase" is referred to as the *absolute refractory period*. Following the absolute refractory period is an interval during which the neuron will not respond to a normal threshold stimulus, but will respond to a suprathreshold stimulus; this is the *relative refractory period*.

8.12   What factors influence the speed at which impulses are conducted along excitable cell membranes?

Diameter of the conducting fiber.   Conduction velocity is directly proportional to fiber diameter.

Temperature of the cell.   Warmer nerve fibers conduct impulses at higher speeds.

Presence or absence of the myelin sheath.   Myelinated fibers conduct impulses more rapidly than unmyelinated, because the action potential leaps from one node of Ranvier to the next instead of progressing from point to point along the axon. This leaping or jumping of the impulse is called *saltatory conduction*. Saltatory conduction is not only faster but also consumes less energy, because the pumping of sodium and potassium ions need occur only at the nodes.

OBJECTIVE F   *To define* synapse *and* synaptic transmission.

Survey   A *synapse* is the specialized junction through which impulses pass from one neuron to another (*synaptic transmission*). With reference to Fig. 8-6, the steps in the process are:

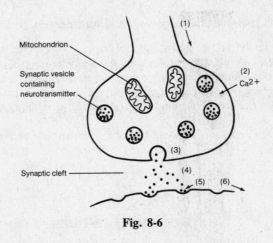

Fig. 8-6

(1) Action potential spreads over axon terminal.

(2) Ca^{2+} influx causes vesicles to move and fuse with presynaptic membrane.

(3) Release of neurotransmitter from vesicles into synaptic cleft.

(4) Neurotransmitter diffuses across synaptic cleft to postsynaptic membrane.

(5) Neurotransmitter combines with specific receptors on postsynaptic membrane.

(6) Permeability of postsynaptic membrane is altered, whereby an impulse is initiated on the second neuron.

(7) Neurotransmitter is removed from synapse—through being enzymatically degraded, taken up in presynaptic terminal, or diffused out of synaptic region.

8.13 Briefly define (a) *synaptic delay*, (b) *synaptic fatigue*, and (c) *one-way conduction*.

(a) There is a delay of about 0.5 ms in the transmission of an impulse from the presynaptic to the postsynaptic neuron. The time is consumed in: (i) release of the neurotransmitter, (ii) diffusion of the neurotransmitter across the cleft, (iii) interaction of the neurotransmitter with receptors on the postsynaptic membrane, and (iv) initiation of the impulse in the postsynaptic neuron.

(b) With repetitive stimulation there is a progressive decline in synaptic transmission due to depletion of the store of neurotransmitter in the synaptic knob.

(c) Most synapses conduct impulses in one direction only, because the neutotransmitter is usually present on only one side of the synapse.

8.14 Neurotransmitters may be *excitatory*, causing the postsynaptic neuron to become active, or *inhibitory*, preventing the postsynaptic neuron from becoming active. Briefly differentiate between excitatory and inhibitory mechanisms.

Excitatory neurotransmitters increase the postsynaptic membrane's permeability to sodium ions. The increased (cf. Problem 8.9) but still subthreshold membrane potential is known as an *excitatory postsynaptic potential* (EPSP), and the membrane is said to be *hypopolarized*. There are two ways in which several EPSP's may combine to reach threshold and elicit an action potential: (1) in *spatial summation*, several presynaptic neurons simultaneously release neurotransmitter to a single postsynaptic neuron; (2) in *temporal summation*, the EPSP's result from the rapid successive discharges of neurotransmitter from the same presynaptic knob.

Inhibitory neurotransmitters increase the postsynaptic membrane's permeability to potassium and chloride ions, resulting in a *hyperpolarized* membrane that exhibits an *inhibitory postsynaptic potential* (IPSP). During the time the membrane is hyperpolarized, the potential is farther below threshold, making it more difficult to generate an action potential.

8.15 List specific drugs that can influence synaptic transmission.

Reserpine can inhibit uptake and storage of the neurotransmitter *norepinephrine* in neurovesicles.
Botulinum toxin can inhibit the release of the neurotransmitter *acetylcholine* from neurovesicles.
Amphetamines can stimulate the release of *norepinephrine* from neurovesicles.
Atropine can block receptors for *acetylcholine* on the postsynaptic membrane.
So-called *cholinergic* drugs can bind to receptors for *acetylcholine*, where they mimic the neurotransmitter.
So-called *anticholinesterase* drugs can inhibit the destruction or metabolism of *acetylcholine*.

Review Questions

Multiple Choice

1. Which kind of neuroglial cells are not found in the CNS? (*a*) astrocytes, (*b*) ependyma, (*c*) microglia, (*d*) satellite cells, (*e*) oligodendrocytes.

2. The neuroglial cells that have functions similar to white blood cells are: (*a*) oligodendrocytes, (*b*) astrocytes, (*c*) microglia, (*d*) ependyma, (*e*) lymphocytes.

3. The speed of a nerve impulse is independent of the: (*a*) diameter of the nerve fiber, (*b*) physiological condition of the nerve, (*c*) presence of myelin, (*d*) length of the nerve fiber, (*e*) presence of Schwann cells.

4. The basic unit of the nervous system is the: (*a*) axon, (*b*) dendrite, (*c*) neuron, (*d*) cell body, (*e*) synapse.

5. Depolarization of the membrane of a nerve cell occurs by the rapid influx of: (*a*) potassium ions, (*b*) chloride ions, (*c*) organic anions, (*d*) sodium ions.

6. A transmitter substance released into the synaptic cleft is: (*a*) cholinesterase, (*b*) acetylcholine, (*c*) ATP, (*d*) RNA, (*e*) all these.

7. At a synapse, impulse conduction normally: (*a*) occurs in both directions, (*b*) occurs in only one direction, (*c*) depends on acetylcholine, (*d*) depends on epinephrine.

8. In a resting neuron, the: (*a*) membrane is electrically permeable, (*b*) outside of the membrane is positively charged, (*c*) outside is negatively charged, (*d*) potential difference across the membrane is zero.

9. Dendrites carry nerve impulses: (*a*) toward the cell body, (*b*) away from the cell body, (*c*) across the body of the nerve cell, (*d*) from one nerve cell to another.

10. The enzyme that destroys acetylcholine is: (*a*) ATPase, (*b*) epinephrine, (*c*) cholinesterase, (*d*) lipase, (*e*) acetylcholinase.

11. The transmitter substance in the presynaptic neuron is contained in the: (*a*) synaptic cleft, (*b*) neuron vesicle, (*c*) synaptic gutter, (*d*) mitochondria.

12. The interior surface of the membrane of a nonconducting neuron differs from the exterior surface in that it is: (*a*) negatively charged and contains less sodium, (*b*) positively charged and contains less sodium, (*c*) negatively charged and contains more sodium, (*d*) positively charged and contains more sodium.

13. The presence of myelin gives a nerve fiber its: (*a*) gray color and regenerative abilities, (*b*) white color and increased rate of impulse transmission, (*c*) white color and decreased rate of impulse transmission, (*d*) gray color and increased rate of impulse transmission.

14. During repolarization of the neuronal membrane: (*a*) sodium ions rapidly move to the inside of the cell, (*b*) sodium ions rapidly move to the outside of the cell, (*c*) potassium ions rapidly move to the outside of the cell, (*d*) potassium ions rapidly move to the inside of the cell.

15. The arrival on a given neuron of a series of impulses from a series of terminal axons, resulting in an action potential, is an example of: (*a*) temporal summation, (*b*) divergence, (*c*) generation potential, (*d*) spatial summation.

16. Coordination via the nervous system differs from that via the endocrine system in that the former: (*a*) is quick, precise, and localized; (*b*) is slower and more pervasive; (*c*) does not require conscious activity; (*d*) has long-lasting effects.

17. The gray matter of the brain consists mainly of neuron cell: (*a*) axons, (*b*) dendrites, (*c*) secretions, (*d*) bodies.

18. The tightly packed coil of the Schwann cell plasma membrane that encircles certain kinds of axons is called: (*a*) myelin sheath, (*b*) neurilemma, (*c*) a node, (*d*) gray matter.

19. The interruptions occurring at regular intervals along a myelin-coated axon are: (*a*) nodes of Ranvier, (*b*) synapses, (*c*) synaptic clefts, (*d*) gap junctions.

20. The junction between the two neurons is called: (*a*) a neurospace, (*b*) an axon, (*c*) a synapse, (*d*) a neural junction.

21. An IPSP is believed to cause: (*a*) an increase in permeability to all cations; (*b*) a selective permeability to calcium, sodium, and potassium; (*c*) an increase in permeability to all anions; (*d*) a selective permeabilty to potassium and chloride.

22. The general depolarization toward threshold of a cell membrane when excitatory synaptic activities predominate is known as: (*a*) facultation, (*b*) differentiation, (*c*) inhibition, (*d*) facilitation.

23. Among neurotransmitters are: (*a*) adenine and guanine, (*b*) thymine and cytosine, (*c*) acetylcholine and norepinephrine, (*d*) none of the above.

24. Clusters of neuron cell bodies found in the CNS are termed: (*a*) nerve clusters, (*b*) ganglia, (*c*) axons, (*d*) nuclei.

25. Which of the following is in the peripheral nervous system? (*a*) oligodendrocytes, (*b*) ependymal cells, (*c*) microglial cells, (*d*) satellite cells.

Matching

1. Multipolar neuron   (*a*) Found only in CNS
2. Afferent neuron   (*b*) One dendrite and one axon
3. Interneuron   (*c*) Single branch connected to cell body
4. Unipolar neuron   (*d*) Carries information toward CNS
5. Bipolar neuron   (*e*) One long axon and many dendrites

True/False

1. There are basically only two different types of cells in the nervous system.

2. The axon is the cytoplasmic neuronal extension conducting impulses toward the cell body.

3. A polarized nerve fiber has an abundance of sodium ions on the outside of the axon membrane.

4. Glial cells sustain the CNS neurons metabolically, support them physically, and regulate ionic concentrations in the extracellular space.

5. Dendrites are usually longer than axons.

6. Every postsynaptic neuron has only one synaptic junction on the surface of its dendrites.

7. A single EPSP is sufficient to cause an action potential.

8. Only EPSP's show temporal and spatial summations.

9. Chemical synapses operate in only one direction.

10. All synapses are inhibitory.

11. A nerve impulse can travel along an axon for an indefinite distance without distortion or loss of strength.

12. The nerve impulse is all-or-nothing.

13. The resting potential in a nerve cell is caused by the high concentration of potassium outside the cell.

14. The permeability of the neuron's plasma membrane to sodium decreases as the membrane is depolarized.

15. The sodium pump operates by diffusion and thus requires no ATP for its operation.

16. Hyperpolarization of the postsynaptic membrane by an excitatory synapse produces an EPSP.

17. The myelin sheath surrounds the dendrites.

18. Transmission across the synaptic junction is by diffusion of sodium.

19. Two transmitter substances in the nervous system are dopamine and acetylcholine.

20. Glial cells have an action potential response.

21. Efferent neurons convey information from receptors in the periphery to the CNS.

22. Somatic efferent nerves innervate skeletal muscle, and the autonomic nerves innervate smooth muscle, cardiac muscle, and glands.

Completion

1. The majority of specialized junctions that receive stimuli from other neurons are located on the _____ and _____ of the neuron.

2. Only 10% of the cells in the nervous system are _____ , and the remainder are _____ cells.

3. _____ cells have one long axon and multiple, short, highly branched dendrites extending from the cell body.

4. The velocity with which an action potential is transmitted down the membrane depends on the fiber _____ and on whether or not the fiber is _____ .

5. On a myelinated neuron, the action potential appears to jump from one node to another. This method of propagation is called _____ .

6. Within the peripheral nervous system, myelin is formed by the _____ .

7. A junction between two neurons where the electrical activity in one neuron influences the excitability of the second is called a _____ .

8. The transmitter substance is stored in small membrane-enclosed _____ in the synaptic knob.

9. When an action potential depolarizes the synaptic knob, small quantities of transmitter substance are released into the _____ .

10. The adding together of two or more EPSP's originating at different places, resulting in depolarization of the membrane, is called _____ .

Chapter 9

Central Nervous System

OBJECTIVE A *To describe the general features of the central nervous system.*

Survey The CNS, consisting of the brain and spinal cord, is covered with *meninges*, is bathed in *cerebrospinal fluid*, and contains *gray* and *white matter*. The CNS functions in body orientation and coordination, assimilation of experiences, and programming of instinctual behavior (cf. Problem 8.1).

9.1 What are the compositions and locations of gray matter and white matter?

The *gray matter* consists of either nerve cell bodies and dendrites or of unmyelinated axons and neuroglia. It forms the outer, convoluted, *cortex* ("bark") of the cerebrum and cerebellum, and also exists as special clusters of nerve cells, called *nuclei*, deep within the white matter. In the spinal cord, the gray matter is deep to white matter. The *white matter*, consisting of aggregations of myelinated axons and associated neurons, forms the *tracts*, or bundled nerve fibers, within the CNS.

9.2 How large is the brain, how many neurons does it contain, and how are the neurons interconnected?

The brain of an adult weighs nearly 1.5 kg (3.3 lb) and is composed of an estimated 10 billion neurons. Neurons communicate with one another by means of innumerable synapses between axons and dendrites, with mediation by neurotransmitter chemicals called *neuropeptides*. These specialized protein messengers are thought to account for specific mental functions.

OBJECTIVE B *To describe the organization of the brain into* forebrain, midbrain, *and* hindbrain, *and to explain how this correlates with the division of the brain into five developmental regions.*

Survey The brain begins its embryonic development as the cephalic end of the neural tube starts to grow rapidly and to differentiate. By the fourth week after conception, three distinct swellings are evident: the *prosencephalon* (forebrain), the *mesencephalon* (midbrain), and the *rhombencephalon* (hindbrain). Further development, during the fifth week, results in the formation of five specific regions: the *telencephalon* and the *diencephalon* derive from the forebrain; the *mesencephalon* remains unchanged; and the *metencephalon* and *myelencephalon* form from the hindbrain. See Fig. 9-1.

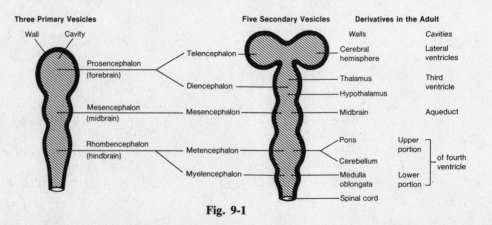

Fig. 9-1

134

9.3 List the principal structures in each of the five regions of the brain and their general functions.

See Table 9-1.

Table 9-1

Region	Structure	Function
Telencephalon	Cerebrum	Controls most sensory and motor activities; reasoning, memory, intelligence, etc.; instinctual and limbic (emotional) functions
Diencephalon	Thalamus	Relay center: all impulses (except olfactory) going into cerebrum synapse here
	Hypothalamus	Regulation of urine formation, body temperature, hunger, heartbeat, etc.; control of secretory activity in anterior pituitary gland; instinctual and limbic functions
	Pituitary gland	Regulation of other endocrine glands
Mesencephalon	Superior colliculus	Visual reflexes
	Inferior colliculus	Auditory reflexes
	Cerebral peduncles	Coordinating reflexes; contain many motor fibers
Metencephalon	Cerebellum	Balance and motor coordination
	Pons	Relay center; contains nuclei (*pontine* nuclei)
Myelencephalon	Medulla oblongata	Relay center; contains many nuclei; visceral autonomic center (e.g., respiration, heart rate, vasoconstriction)

OBJECTIVE C *To decribe the* cerebrum *and the functions of the* cerebral lobes.

Survey The cerebrum consists of five paired lobes comprising two convoluted hemispheres (see Fig. 9-2). The hemispheres are connected by the *corpus callosum*. The cerebrum accounts for about 80% of the brain's mass and is concerned with higher functions, such as perception of sensory impulses; instigation of (voluntary) movement; memory, thought, and reasoning.

9.4 What are the two layers of the cerebrum, and why is it important that the outer layer be convoluted?

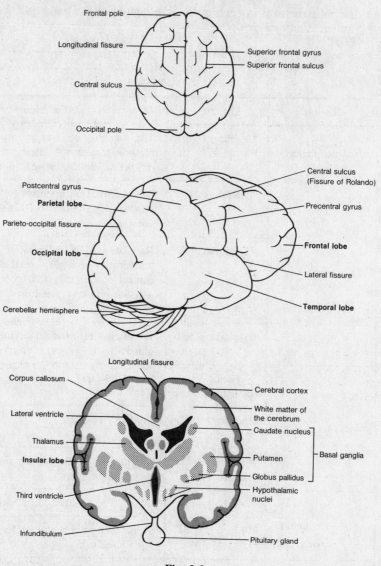

Fig. 9-2

The convoluted surface layer, or *cerebral cortex*, is composed of gray matter 2–4 mm (0.08–0.16 in) in thickness. The elevated folds of the convolutions are the *gyri* (singular, *gyrus*), and the depressed grooves are the *sulci* (singular, *sulcus*). The convolutions greatly increase the surface area of the gray matter and thus the total number of nerve cell bodies.

Beneath the cerebral cortex is the thick *white matter* of the cerebrum, which constitutes the second layer (cf. Problem 9.1).

9.5   What are the specific functions of the cerebral lobes?

See Table 9-2.

9.6   Give the functions of the precentral and postcentral gyri (see Fig. 9-2).

The precentral gyrus of the frontal lobe is an important motor area. The postcentral gyrus of the parietal lobe is designated as the *somatesthetic* area of the brain because it responds to sensory stimuli from cutaneous and muscular receptors throughout the body.

Table 9-2

Lobe	Functions
Frontal	Voluntary motor control of skeletal muscles; personality; higher intellectual processes (e.g., concentration, planning, decision making); verbal communication
Parietal	Somatesthetic interpretation (e.g., cutaneous and muscular sensations); understanding and utterance of speech
Temporal	Interpretation of auditory sensations; auditory and visual memory
Occipital	Integrates movements in focusing the eye; correlates visual images with previous visual experiences and other sensory stimuli; conscious seeing
Insular	Memory; integration of other cerebral activities

9.7   Where is *Broca's area*, and why it is important?

Broca's area, or the *motor speech area*, is a highly specialized portion of the frontal lobe, generally located (in both right-handed and left-handed persons) in the left hemisphere, immediately superior to the lateral sulcus (see Fig. 9-3). Mental activity in Broca's area causes selective stimulation of motor centers elsewhere in the frontal lobe, which, in turn, causes coordinated skeletal-muscle contractions in the pharynx and larynx. At the same time, motor impulses are sent to the respiratory centers to regulate air movement across the vocal cords. The combined muscular stimulation translates thought patterns into speech.

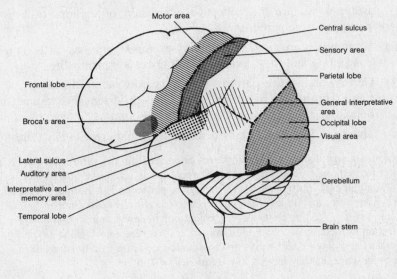

Fig. 9-3

9.8   *True or false*:   The cerebral hemispheres communicate one with the other by nerve impulses passing through fiber tracts.

True. Impulses travel not only between the lobes of a cerebral hemisphere, but between the right and left hemispheres, and to other regions of the brain.

There are three types of fiber tracts within the white matter, which are named according to location and the direction in which they conduct impulses. *Association fibers* are confined to a given hemisphere, where they conduct impulses between neurons in various lobes. *Commissural fibers* (Fig. 9-4) connect the neurons and gyri of one hemisphere with those of the other (e.g., *corpus callosum* and *anterior commissure*). *Projection fibers* (Fig. 9-4) form *descending tracts* that transmit impulses from the cerebrum to other parts of the brain and spinal cord, and *ascending tracts* from the spinal cord and other parts of the brain to the cerebrum.

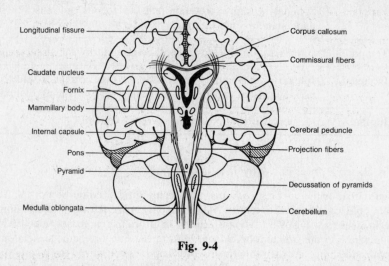

Fig. 9-4

9.9 Comment on the truth or falsity of the following statements concerning *brain waves* as recorded in an *electroencephalogram* (EEG).

(i)   Brain waves are the collective expression of millions of action potentials from neurons of the cerebrum.

(ii) Brain waves are emitted from the developing brain as early as eight weeks following conception and they continue throughout a person's life.

(iii) Certain brain-wave patterns signify healthy mental functions, and deviations from these patterns are of clinical significance in diagnosing trauma, mental depression, hematomas, and various diseases such as tumors, infections, and epilepsy.

(iv) There are four basic kinds of brain-wave patterns: *alpha*, *beta*, *theta*, and *delta*.

All four statements are true. Brain waves originate from the various cerebral lobes and have distinct oscillation frequencies. *Alpha waves* are best recorded in an awake and relaxed person whose eyes are closed. An alpha EEG pattern of 10–12 Hz (cycles per second) is normal for an adult, and a pattern of 4–7 Hz is normal for a child less than eight years old. *Beta waves* accompany visual and mental activity; their frequency is 13–25 Hz. *Theta waves* are common in newborn infants and have a frequency of 5–8 Hz. The detection of theta waves in an adult may indicate severe emotional stress and may perhaps forewarn of a nervous breakdown. *Delta waves* are common in a person who is asleep or in a person who is awake but has brain damage; they have a low frequency of 1–5 Hz.

9.10 What are the *basal ganglia*, and why are they clinically important?

The basal ganglia are specialized paired masses of gray matter located deep within the white matter of the cerebrum. They are made up of the *corpus striatum* and other structures of the mesencephalon. The corpus striatum consists of the *caudate nucleus* and the *lentiform nucleus*; the latter, in turn, consists of the *putamen* and the *globus pallidus*.

Neural diseases such as Parkinson's disease, or physical trauma to the basal ganglia, generally cause a variety of motor dysfunctions, including rigidity, tremor, and rapid and aimless movement.

OBJECTIVE D *To describe the location and structure of the* diencephalon *and to explain the autonomic functions of its chief components, the* thalamus, hypothalamus, epithalamus, *and* pituitary gland (*Fig. 9-5*).

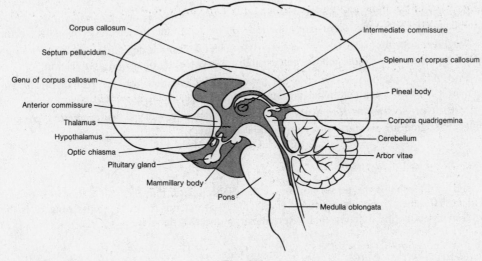

Fig. 9-5

Survey The diencephalon, a major autonomic region of the forebrain, is almost completely surrounded by the cerebral hemispheres of the telencephalon. The *third ventricle* (Problem 9.28) forms a midplane cavity within the diencephalon.

9.11 What is the structure of the thalamus and what are its functions?

The thalamus is actually a paired organ, composed of two ovoid masses of gray matter. Each portion is located immediately below the lateral ventricle of its respective cerebral hemisphere. The thalamus is a relay center for all sensory impulses, except smell, to the cerebral cortex. It also is involved in the initial autonomic response of the body to intensely painful stimuli and is, therefore, partially responsible for the physiological state of *shock* that frequently follows serious trauma.

9.12 Which autonomic function is *not* performed by the hypothalamus? (*a*) heart rate, (*b*) respiration control, (*c*) body-temperature regulation, (*d*) regulation of hunger and thirst, (*e*) sexual response.

(*b*). The hypothalamus consists of several masses of nuclei interconnected to other vital parts of the brain. From the list below, it is seen that although most of the functions of the hypothalamus relate to regulation of visceral activities, it also performs emotional (limbic) and instinctual functions.

Cardiovascular regulation. Impulses from the posterior hypothalamus produce autonomic acceleration of the heartbeat; impulses from the anterior portion produce autonomic deceleration.

Body-temperature regulation. Nuclei in the anterior portion of the hypothalamus monitor the temperature of the surrounding arterial blood. In response to above-normal temperature, the hypothalamus initiates impulses that cause heat loss through sweating and dilation of cutaneous vessels. In response to below-normal temperature, the hypothalamus causes contraction of cutaneous vessels and shivering.

Regulation of water and electrolyte balance. *Osmoreceptors* in the hypothalamus monitor the osmotic concentration of the blood. Viscosity of the blood due to lack of water causes *antidiuretic hormone* (ADH) to be produced and released from the posterior pituitary gland. At the same time, a *thirst center* within the hypothalamus causes the feeling of thirst.

Regulation of gastrointestinal (GI) activity and hunger. In response to sensory impulses from the abdominal viscera, the hypothalamus regulates glandular secretions and GI peristalsis. Glucose, fatty-acid, and amino-acid levels of the blood are monitored by a *feeding center* in the lateral hypothalamus. When sufficient amounts of food have been ingested, a *satiety center* in the midportion of the hypothalamus inhibits the feeding center.

Regulation of sleeping and wakefulness. The *sleep center* and the *wakefulness center* of the hypothalamus function with other parts of the brain to determine the level of conscious alertness.

Sexual response. Specialized *sexual-center* nuclei within the dorsal portion of the hypothalamus respond to sexual stimulation and are responsible for the feeling of sexual gratification.

Emotions. Specific nuclei within the hypothalamus interact with the rest of the *limbic system* (Problem 9.15) in causing emotional responses such as anger, fear, pain, and pleasure.

Control of endocrine functions. The hypothalamus produces neurosecretory chemicals that stimulate the anterior pituitary to release various hormones.

9.13 Describe the *epithalamus*.

The epithalamus is the dorsal portion of the diencephalon that includes a thin roof over the third ventricle. The small, cone-shaped *pineal gland* (Fig. 9-5) extends from the epithalamus; it is thought to have a neuroendocrine function that determines the onset of puberty.

9.14 Locate the *pituitary gland*.

As is shown in Figs. 9-2 and 9-5, the pituitary gland, or *hypophysis*, is attached to the inferior aspect of the diencephalon by the *infundibulum*. Surrounded by a ringed network of blood vessels called the *circle of Willis*, the pituitary is structurally and functionally divided into an anterior portion, the *adenohypophysis*, and a posterior portion, the *neurohypophysis*. The endocrine functions of the pituitary are discussed in Chapter 12.

9.15 Give the principal components of the *limbic system*.

The limbic system is a roughly doughnut-shaped neuronal loop inside the brain, with the thalamic region in the "hole" and the cerebral cortex "outside." Besides involving the hypothalamus, the system includes three structures which are named after their shapes: the *amygdala* ("almond"), the *hippocampus* ("sea horse"), and the *fornix* ("arch"). See Fig. 9-6. The limbic system is believed to generate emotions and, via the hippocampus, to be involved in short-term memory.

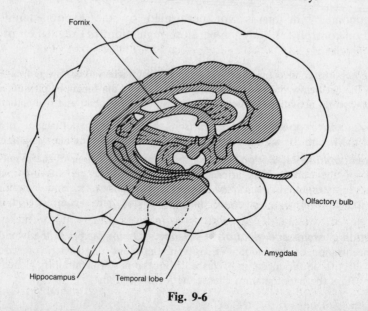

Fig. 9-6

OBJECTIVE E   *To ascertain the location of the* mesencephalon *and the functions of its various structures.*

Survey   The mesencephalon, or midbrain, is a short section of the brain stem between the diencephalon and the pons (Fig. 9-6). It contains the *corpora quadrigemina*, concerned with visual and auditory reflexes, and the *cerebral peduncles*, composed of fiber tracts. It also contains specialized nuclei which help to control posture and movement.

9.16   What are the functions of the *superior* and *inferior colliculi*?

The *corpora quadrigemina* are the four rounded elevations on the dorsal portion of the midbrain. Of these, the two upper eminences, the *superior colliculi*, are concerned with visual reflexes; the two posterior eminences, the *inferior colliculi*, are responsible for auditory reflexes.

9.17   Do the cerebral peduncles contain only motor fibers?

No: they are composed of both motor and sensory fibers, which support the cerebrum and connect it to other regions of the brain.

9.18   What are the functions of the *nuclei* within the midbrain?

The *red nucleus* is gray matter that connects the cerebral hemispheres and the cerebellum; it functions in reflexes concerned with motor coordination and maintaining posture. Another nucleus, the *substantia nigra*, is ventral to the red nucleus and is thought to inhibit involuntary movements.

OBJECTIVE F   *To describe the* metencephalon.

Survey   The metencephalon is the region of the brain stem that contains the *pons*, composed of fiber tracts which relay impulses, and the *cerebellum*, which coordinates skeletal muscle contractions.

9.19   Besides serving as a relay center, what other functions has the pons?

Many of the cranial nerves originate from nuclei located within the pons. Other nuclei of the pons, in the *apneustic* and the *pneumotaxic centers*, cooperate with nuclei in the *rhythmicity area* of the medulla oblongata to regulate the rate of breathing. See Fig. 9-7.

9.20   Comment on the truth or falsity of the following statements concerning the cerebellum. (i) It consists of two hemispheres that are convoluted at the surface. (ii) It functions totally at the subconscious (involuntary) level. (iii) It is the second-largest structure of the brain and is composed of gray matter on the surface, and tracts of white matter, collectively called the *arbor vitae* (Figs. 9-5), deep to the gray matter. (iv) It coordinates skeletal-muscle contractions in response to incoming impulses from proprioceptors within muscles, tendons, joints, and special sense organs.

All four statements are true. The two principal functions of the cerebellum are to coordinate body movement and to maintain balance. In order to perform these functions, the cerebellum is in constant communication with other neurological structures through the *cerebellar peduncles*, which are fiber tracts that extend into and support the cerebellum.

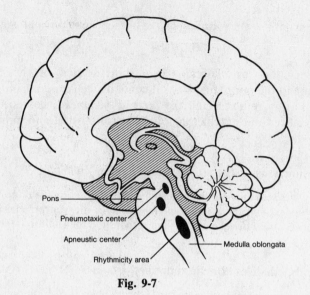

Fig. 9-7

OBJECTIVE G   *To describe the location and structure of the* medulla oblongata *and to give its functions.*

Survey   The medulla oblongata, the preponderant part of the *myelencephalon*, connects to the spinal cord and contains nuclei for cranial nerves and vital autonomic functions. The *reticular formation*, which arouses the cerebrum, is partially located in the myelencephalon.

9.21   Define *decussation*.

The medulla oblongata consists primarily of white matter, in the form of descending and ascending tracts that communicate between the spinal cord and various parts of the brain. Most of these fibers *decussate*, or cross to the other side, through the pyramidal region of the medulla, permitting one side of the brain to receive information from and send information to the opposite side of the body.

9.22   What nuclei are located in the medulla oblongata?

See Fig. 9-8. The gray matter of the medulla oblongata consists of a number of important nuclei for cranial nerves (motor and sensory components), sensory relay to the thalamus, and motor relay from the cerebrum to the cerebellum.

9.23   What are the autonomic functions of the medulla oblongata?

In addition to the nuclei of Problem 9.22, there are three nuclei within the medulla that are autonomic centers for controlling visceral functions. **Cardiac center**: Both *inhibitory fibers* (through the vagus nerves) and *accelerator fibers* (through spinal nerves T1–T5) arise from nuclei of the cardiac center. **Vasomotor center**: Impulses from the vasomotor center cause the smooth muscles of arteriole walls to contract, thus raising the blood pressure. **Respiratory center** (or rhythmicity area): The rate and depth of breathing is controlled by the nuclei of this center, along with those of the pons (Problem 9.19).

9.24   The *reticular formation* is said to house the *reticular activating system* (RAS). Explain.

The reticular formation, a complex network of nuclei and ascending and descending nerve fibers within the brain stem, generates a continuous flow of impulses to rouse the cerebrum, unless the stream is inhibited by other parts of the brain. The RAS is sensitive to chemical changes within, or trauma to, the brain. Severe trauma to the RAS itself may cause a person to become comatose.

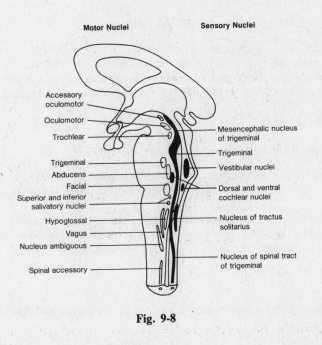

Motor Nuclei   Sensory Nuclei

Accessory oculomotor

Oculomotor

Trochlear

Mesencephalic nucleus of trigeminal

Trigeminal

Trigeminal

Abducens

Vestibular nuclei

Facial

Superior and inferior salivatory nuclei

Dorsal and ventral cochlear nuclei

Hypoglossal

Nucleus of tractus solitarius

Vagus

Nucleus ambiguous

Spinal accessory

Nucleus of spinal tract of trigeminal

Fig. 9-8

OBJECTIVE H   *To describe the protective coverings of the CNS.*

Survey   The CNS is covered by three *meninges* (singular, *meninx*), composed of connective tissue. In order, from outside in, we have the *dura mater*, the *arachnoid membrane*, and the *pia mater* (see Fig. 9-9).

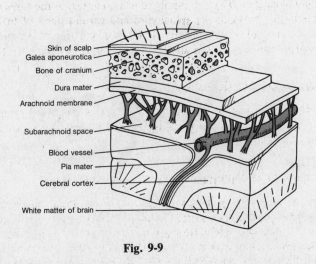

Skin of scalp
Galea aponeurotica
Bone of cranium
Dura mater
Arachnoid membrane
Subarachnoid space
Blood vessel
Pia mater
Cerebral cortex
White matter of brain

Fig. 9-9

9.25   Are the meninges uniform throughout the CNS?

No. The cranial dura mater is divided into a thicker *periosteal layer* and a thinner *meningeal layer*. In certain areas over the brain, the two layers of the cranial dura mater are separated to form enclosed *dural sinuses*, which collect venous blood and drain it to the internal jugular veins of the neck. The spinal dura mater is not double-layered.

9.26   Contrast the *epidural* and the *subarachnoid* spaces.

The spinal dura mater forms a tough, tubular sheath around the spinal cord. The *epidural space* is a fatty, vascular, protective area between the sheath and the vertebral canal.

The *subarachnoid space* is located between the arachnoid membrane and the pia mater. It is maintained by delicate weblike strands (Fig. 9-9) and contains cerebrospinal fluid.

9.27   Discuss the clinical terms *epidural block* and *meningitis*.

An epidural block is an injection of an anesthetic solution into the vicinity of the spinal nerves as they pass through the epidural space. It is administered frequently in the lower lumbar area, to women in labor.

Meningitis is an inflammation of the meninges (usually the arachnoid and pia mater) caused by certain bacteria or viruses. Complications may cause sensory impairment, paralysis, or mental retardation. Untreated meningitis generally results in coma and death.

OBJECTIVE I   *To learn the properties and functions of* cerebrospinal fluid.

Survey   Cerebrospinal fluid (CSF) is a clear, lymphlike fluid formed by active transport of substances from blood plasma in the *choroid plexuses* (Problem 9.29). CSF forms a protective cushion around and within the CNS; it also buoys the brain. As diagrammed in Fig. 9-10, CSF circulates through the ventricles of the brain, the central canal of the spinal cord, and the subarachnoid space around the CNS.

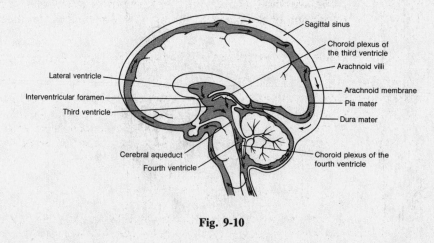

Fig. 9-10

9.28   Describe the *ventricles*.

These are a system of cavities which are connected to one another and to the *central canal* of the spinal cord. As shown in Fig. 9-11, each cerebral hemisphere contains one of the two *lateral ventricles* (*first* and *second ventricles*). The *third ventricle* is located in the diencephalon and is connected to the lateral ventricles by the two *interventricular foramina*. The *fourth ventricle* is located in the brain stem. It is connected to the third ventricle by the *cerebral aqueduct* and meets the central canal inferiorly.

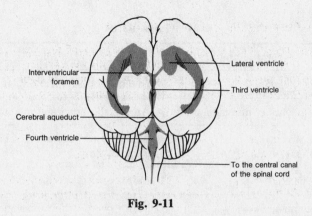

Fig. 9-11

9.29 What are the physical characteristics of CSF?

The 1500-gram brain has a buoyancy weight of some 150 grams, because it is in effect suspended in some 1350 cm^3 of CSF, which has a specific gravity of 1.007. Since the CNS lacks lymphatic circulation, CSF moves cellular wastes into the venous return at its places of drainage into the *arachnoid villi*. CSF is continuously produced (about 800 mL/day) by masses of specialized capillaries, called *choroid plexuses*, located in the roofs of the ventricles. A standing volume of 140 to 200 mL of CSF is maintained at a fluid pressure of about 10 mmHg (1.33 kPa).

OBJECTIVE J *To explain the importance of the* blood–brain barrier *in maintaining homeostasis within the brain.*

Survey The blood–brain barrier (BBB) is a structural arrangement of capillaries, surrounding connective tissue, and astrocytes (Table 4-8), that selectively determines which substances can move from the plasma of the blood to the extracellular fluid of the brain.

9.30 *True or false*: Alcohol passes readily through the BBB because it is a lipid-soluble compound.

True. Fat-soluble compounds pass readily through the BBB, as do H_2O, O_2, CO_2, and glucose. The inorganic ions Na^+, K^+, and Cl^- pass more slowly, so that their concentrations are different in the brain than in the plasma. Other substances, such as macroproteins, lipids, creatine, urea, inulin, certain toxins, and most antibiotics, are restricted in passage. The BBB is an important factor to consider when planning drug therapy for neurological disorders.

9.31 What are the metabolic needs of the brain?

The rate of energy consumption by the brain is very high and constant. The 1.5 kg organ (about 2.5% of body weight) receives approximately 20% of cardiac output at rest, through the paired internal carotid arteries and vertebral arteries. The brain is composed of the most oxygen-dependent tissue of the body, and a failure of cerebral circulation for as little as ten seconds causes unconsciousness.

OBJECTIVE K *To list the common neurotransmitters of the brain, along with their functions.*

Survey Neurotransmitters (see Chapter 8, especially Problem 8.14) are represented by over 200 specific chemicals within the brain. These are secreted by the neurons which synthesize them. The more important are listed in Table 9-3.

Table 9-3

	Neurotransmitter	Function
Excitatory	Acetylcholine	Facilitates transmission of nerve impulses
	Epinephrine, norepinephrine	Arouse the brain and maintain alertness
	Serotonin	Temperature regulation, sensory perception, onset of sleep
	Dopamine	Motor control
Inhibitory	Gamma-aminobutyric acid (GABA)	Motor coordination through inhibiting certain neurons
	Glycine (Problem 2.13)	Inhibits transmission along certain spinal cord tracts
Neuropeptides (short-chain amino acids)	Enkephalins, endorphins	Block transmission and perception of pain
	Substance P	Aids in transmission of impulses from pain receptors

OBJECTIVE L   *To examine the* spinal cord.

Survey   The spinal cord is the portion of the CNS that extends through the neural canal of the vertebral column to the level of the first lumbar vertebra. It is continuous with the brain through the foramen magnum of the skull (Table 6-3). The spinal cord consists of centrally located gray matter, involved in reflexes, and peripheral ascending and descending tracts of white matter, which conduct impulses to and from the brain (cf. Problem 9.8). Thirty-one pairs of spinal nerves arise from the spinal cord; these will be treated in Chapter 10.

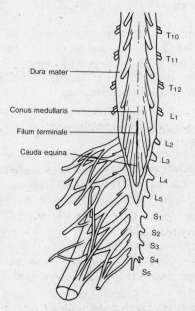

Fig. 9-12

9.32 Discuss the three lumbar structures indicated in Fig. 9-12.

The *conus medullaris* is the terminal portion of the spinal cord. The *filum terminale* is a supportive, fibrous strand of the pia mater. The nerves that collectively radiate from the conus medullaris are known as the *cauda equina* ("horse's tail").

9.33 What is the geometry of the gray matter and the white matter within the spinal cord?

See Fig. 9-13. The deep *gray matter* has, in cross section, a four-horned or letter-H appearance. The *posterior horns* receive the axons of sensory fibers that enter the spinal cord from a spinal nerve; the *anterior horns* contain the dendrites and cell bodies of motor neurons that leave the spinal cord to enter a spinal nerve. At the thoracic and lumbar levels, there are also *lateral horns* containing preganglionic sympathetic neurons whose axons leave via the ventral root. (See Problem 10.20.)

The superficial *white matter* is composed primarily of myelinated fibers. Those of common origin, destination, and function form *tracts*. The tracts are separated by the horns of gray matter into four regions, called the *posterior*, *lateral*, and *anterior funiculi* (Fig. 9-14). Subdivisions of the funiculi, which contain fibers from more than one tract, are called *fasciculi*.

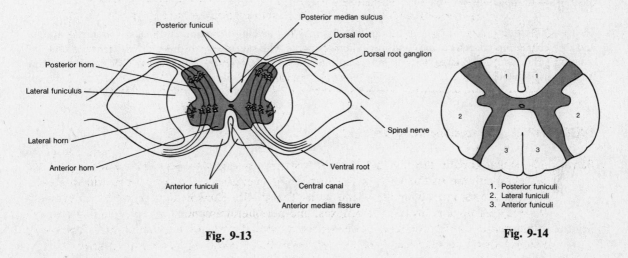

Fig. 9-13

Fig. 9-14

1. Posterior funiculi
2. Lateral funiculi
3. Anterior funiculi

Key Clinical Terms

Cerebral angiography A technique to reveal abnormalities of cerebral blood vessels, such as aneurysms, or a brain tumor that displaces blood vessels. A radiopaque substance is injected into the carotid arteries; then X-ray films are taken of the blood vessels of the brain.

Cerebral concussion A transient state of unconsciousness following head injury and damage to the brain stem.

Cerebral palsy A pathological condition of the brain marked by paralysis, uncoordination, and other dysfunctions of motor and sensory mechanism.

Cerebrovascular disease Any pathological change in cerebral blood vessels. Cerebrovascular diseases include *aneurysms*, *atherosclerosis*, *embolism*, *infarction*, *thrombosis*, *stroke*, and *hemorrhage*.

Chorea A nervous disorder characterized by bizarre, abrupt, involuntary movements. It may be hereditary or a result of rheumatic fever.

Coma   Varying degrees of unconsciousness from any of a number of causes.

Convulsions   Spasmodic contractions of muscles, associated with semiconsciousness or unconsciousness. Convulsions are the result of extreme irritability of the nervous system brought on by such things as brain damage, infection, or prolonged high fever.

Delirium   A state of extreme mental confusion caused by interference with the metabolic processes of the brain. Illusions or hallucinations, and disorientation as to time, place, or person, may all be symptoms.

Electroencephalogram   A record of the electrical impulses of the brain.

Encephalitis   An infectious disease of the central nervous system, with damage to both white and gray matter. It may be caused by a virus or by certain chemicals, such as lead, arsenic, and carbon monoxide.

Epilepsy   A chronic, convulsive disorder characterized by recurrent seizures and impaired consciousness.

Hydrocephalus   Enlarged head due to abnormal accumulation of cerebrospinal fluid, brought on by an obstruction in the circulatory pathway of the fluid.

Multiple sclerosis   A neurological disease that destroys the myelin of neurons. MS causes gradual paralysis and disturbances in speech, vision, and mentation. Patients with advanced MS have difficulty in walking and scanning print, and suffer from body tremors and blurred vision. Currently, the cause of MS is unknown and treatment is limited.

Review Questions

Multiple Choice

1. The white matter of the CNS is always: (*a*) deep to the gray matter, (*b*) unmyelinated, (*c*) arranged into tracts, (*d*) composed of sensory fibers only.

2. The three initial developmental regions of the brain are the: (*a*) telencephalon, prosencephalon, rhombencephalon; (*b*) rhombencephalon, prosencephalon, mesencephalon; (*c*) metencephalon, myelencephalon, prosencephalon; (*d*) prosencephalon, diencephalon, mesencephalon.

3. The third ventricle is located in the: (*a*) cerebrum, (*b*) forebrain, (*c*) hindbrain, (*d*) midbrain, (*e*) cerebellum.

4. Neuropeptides are: (*a*) neurotransmitter chemicals, (*b*) neuroglial cells, (*c*) produced by the choroid plexus, (*d*) nutrients for brain tissue, (*e*) both (*a*) and (*c*).

5. The thalamus is located in the: (*a*) telencephalon, (*b*) mesencephalon, (*c*) diencephalon, (*d*) metencephalon, (*e*) myelencephalon.

6. It is false that the cerebrum: (*a*) comprises about 80% of the brain's mass, (*b*) consists of four paired lobes, (*c*) contains a thin layer of convoluted gray matter, (*d*) is located within the telencephalonic region of the brain.

7. Which is *not* a lobe of the cerebrum? (*a*) parietal, (*b*) insular, (*c*) occipital, (*d*) temporal, (*e*) sphenoidal.

8. Which lobe–function pairing is *incorrect*? (*a*) frontal–sensory interpretation, (*b*) parietal–speech patterns, (*c*) occipital–vision, (*d*) temporal–memory, (*e*) parietal–somatesthetic interpretation.

9. The basal ganglia comprise all of the following except the: (*a*) putamen, (*b*) caudate nucleus, (*c*) globus pallidus, (*d*) infundibulum.

10. Clusters of neuron cell bodies embedded in the white matter of the brain are referred to as: (*a*) nuclei, (*b*) gyri, (*d*) ganglia, (*e*) fasciculi.

11. Tracts of white matter which connect the right and left hemispheres are composed of: (*a*) decussation fibers, (*b*) association fibers, (*c*) commissural fibers, (*d*) projection fibers.

12. Brain waves common to a healthy, sleeping person and a brain-damaged, awake person are called: (*a*) alpha waves, (*b*) beta waves, (*c*) gamma waves, (*d*) theta waves, (*e*) delta waves.

13. Parkinson's disease and other motor disorders are attributed to dysfunction of, or trauma to, the: (*a*) pons, (*b*) basal ganglia, (*c*) parietal lobe, (*d*) thalamus, (*e*) corpus striatum.

14. The inability of a patient to perceive pain might be due to a tumor or trauma of the: (*a*) insula, (*b*) hypothalamus, (*c*) red nucleus, (*d*) thalamus, (*e*) pons.

15. Symptoms of irregular body temperature, intense thirst, and the inability to sleep, might indicate that a patient has dysfunctions of the: (*a*) hypothalamus, (*b*) pons, (*c*) medulla, (*d*) pituitary gland, (*e*) cerebrum.

16. Which property of blood is *not* monitored by the hypothalamus? (*a*) osmotic concentration, (*b*) PCO_2 content, (*c*) fatty-acids content, (*d*) blood glucose, (*e*) amino-acid levels.

17. Which of the following is not involved with motor impulses or motor coordination? (*a*) red nucleus, (*b*) cerebellum, (*c*) basal ganglia, (*d*) precentral gyrus, (*e*) none of the above.

18. The corpora quadrigemina, composed of the superior and inferior colliculi, is located in the: (*a*) telencephalon, (*b*) mesencephalon, (*c*) diencephalon, (*d*) metencephalon, (*e*) constellation Aries.

19. The capillary network which develops in the roofs of the third and fourth ventricles is called the: (*a*) choroid plexus, (*b*) sulcus limitans, (*c*) hyperthalamic plexus, (*d*) cerebral plexus, (*e*) circle of Willis.

20. Which brain structure–autonomic function pairing is *incorrect*? (*a*) pons–respiration, (*b*) corpus callosum–blood pressure, (*c*) medulla oblongata–respiration, (*d*) thalamus–intense pain, (*e*) hypothalamus–body temperature.

21. An abnormal production of antidiuretic hormone (ADH) could result from a dysfunction of the: (*a*) hypothalamus, (*b*) choroid plexus, (*c*) pituitary gland, (*d*) reticular activation system, (*e*) pineal gland.

22. Which statement is *incorrect* concerning the medulla oblongata? (*a*) it is the site of decussation of many fibers; (*b*) it is part of the reticular system; (*c*) it contains specialized nuclei for certain cranial nerves; (*d*) it functions as cardiac, vasomotor, and respiratory centers; (*e*) none of the above.

23. The meninx in contact with the brain and spinal cord is the: (*a*) pia mater, (*b*) dura mater, (*c*) perineural mater, (*d*) arachnoid membrane.

24. Cerebrospinal fluid (CSF) is found within: (*a*) epidural space, subarachnoid space, dural sinuses; (*b*) subarachnoid space, dural sinuses, ventricles; (*c*) central canal, epidural space, subarachnoid space; (*d*) ventricles, central canal, subarachnoid space; (*e*) central canal, epidural space, ventricles.

25. Which statement is *incorrect* concerning cerebrospinal fluid? (*a*) it has a specific gravity of 1.007 and buoys the brain; (*b*) it maintains a volume of 140 to 200 mL and a fluid pressure of 10 mmHg; (*c*) it moves cellular wastes away from nervous tissue; (*d*) it is produced in the choroid plexuses and drains into the circle of Willis; (*e*) none of the above.

26. The presence of theta waves in an adult is an indication of: (*a*) visual activity, (*b*) dreaming, (*c*) brain damage, (*d*) severe emotional stress, (*e*) none of the above.

27. The cerebral aqueduct links the: (*a*) lateral ventricles, (*b*) lateral ventricles and the third ventricle, (*c*) third and fourth ventricles, (*d*) lateral ventricles and the fourth ventricle, (*e*) first and the second ventricles.

28. The spinal cord ends at the level of the: (*a*) coccyx, (*b*) first lumbar vertebra, (*c*) vertebral column, (*d*) sacrum, (*e*) sciatic nerve.

29. The blood–brain barrier restricts passage of: (*a*) lipids, (*b*) Na^+, (*c*) Cl^-, (*d*) H_2O, (*e*) lipid-soluble compounds.

30. Body temperature, sensory perception, and the onset of sleep are partially regulated by the neurotransmitter: (*a*) glycine, (*b*) serotonin, (*c*) acetylcholine, (*d*) dopamine, (*e*) enkephalin.

31. The terminal portion of the spinal cord is known as the: (*a*) cordis terminale, (*b*) conus medullaris, (*c*) cauda equina, (*d*) bulbis caudis, (*e*) filum terminale.

32. Which region of the brain is farthest from the spinal cord? (*a*) mesencephalon, (*b*) telencephalon, (*c*) myelencephalon, (*d*) metencephalon, (*e*) diencephalon.

33. For substances within the blood to reach the neurons within the brain, they must first pass through a cellular membrane derived in part from: (*a*) Schwann cells, (*b*) microglia, (*c*) astrocytes, (*d*) ganglia, (*e*) neurolemma.

34. A patient with symptoms of tremor, halting speech, and an irregular gait might have experienced trauma to the: (*a*) cerebrum, (*b*) pons, (*c*) cerebellum, (*d*) thalamus, (*e*) hypothalamus.   *Ans.*   (*c*)

35. Blockage of **the flow of cere**brospinal fluid may result in: (*a*) meningitis, (*b*) hydrocephalus, (*c*) paraplegia, (*d*) encephalitis, **(*e*) all the above.**

36. Two components of the basal ganglia are the: (*a*) caudate nucleus and lentiform nucleus, (*b*) globus pallidus and infundibulum, (*c*) hypothalamic nucleus and red nucleus, (*d*) insula and putamen.

37. Which is *not* involved in the transmission of perception of pain? (*a*) substance P, (*b*) thalamus, (*c*) enkephalins, (*d*) posterior horns, (*e*) none of the above.

38. A disease of the nervous system in which the myelin sheaths of neurons are altered by the formation of plaques is: (*a*) multiple sclerosis, (*b*) epilepsy, (*c*) cerebral palsy, (*d*) Parkinson's disease, (*e*) neurosyphilis.

39. Which two structures of the brain control respiration? (*a*) pons, hypothalamus; (*b*) cerebrum, hypothalamus; (*c*) pons, medulla oblongata; (*d*) hypothalamus, pituitary gland.

40. Trauma to the superior colliculi would most likely affect: (*a*) speech, (*b*) auditory perception, (*c*) coordination and balance, (*d*) vision, (*e*) perception of pain.

True/False

1. The thalamus is an important relay center in that all sensory impulses (except olfaction) going to the cerebrum synapse there.

2. The longitudinal fissure separates the two cerebral hemispheres, and the central sulcus separates the precentral gyrus from the postcentral gyrus.

3. The convoluted cerebral cortex and the convoluted surface of the cerebellum are the only parts of the brain that contain gray matter.

4. All ventricles of the brain are paired except for the fourth.

5. The posterior horns of the spinal cord contain motor neurons only.

6. Broca's area is the motor speech region of the brain and is generally within the left hemisphere.

7. The gyri and sulci form the convolutions of the cerebrum that greatly increase the surface area of the white matter.

8. Both the hypothalamus and the medulla oblongata mediate vasoconstriction and vasodilation.

9. Association fibers are confined to a single hemisphere and serve to relay impulses to the various cerebral lobes.

10. An alpha brain wave pattern is a healthy sign in a person who is awake but relaxed, and a beta brain wave pattern is a healthy sign in a person who is awake and mentally alert.

11. The hypothalamus is a component of the limbic system that helps determine one's emotions.

12. The pineal gland, the hypothalamus, and the pituitary gland all have neuroendocrine functions.

13. The circle of Willis constitutes the blood–brain barrier, which selectively determines which components of the blood can enter the CNS.

Matching

1.	Broca's area	(a)	Area of decussation
2.	Postcentral gyrus	(b)	Arouses the cerebrum
3.	Medulla oblongata	(c)	Somatesthetic area
4.	Reticular formation	(d)	Auditory reflexes
5.	Choroid plexus	(e)	Inhibits involuntary movements
6.	Substantia nigra	(f)	Motor speech area
7.	Inferior colliculi	(g)	Onset of puberty
8.	Pineal gland	(h)	Monitors osmotic concentration of blood
9.	Pons	(i)	Produces CSF
10.	Hypothalamus	(j)	Apneustic center

Chapter 10

Peripheral and Autonomic Nervous Systems

OBJECTIVE A   *To understand the division of the nervous system into two anatomical categories and two functional categories.*

Survey   Anatomically, the nervous system falls into the CNS (brain and spinal cord) and the **peripheral nervous system** (**PNS**), which consists of *cranial nerves*, arising from the brain, and *spinal nerves*, arising from the spinal cord.

Functionally, the nervous system is divided into the **somatic nervous system**, which innervates *skeletal muscles*, and the **autonomic nervous system** (**ANS**), which innervates *smooth* and *cardiac muscles*, as well as glands. (As most smooth muscle tissue is in the viscera, one sometimes distinguishes the **visceral nervous system** as a major component of the ANS.)

The anatomical and functional categories of course overlap. Thus, the controlling centers of the ANS fall in the CNS (in the brain); whereas the nerve fibers of the ANS, subdivided into *sympathetic* and *parasympathetic* fibers, belong to the PNS.

10.1   Are the nerves of the PNS sensory only, motor only, or mixed?

Most peripheral nerves are composed of both motor and sensory neurons; they are thus mixed nerves. Some cranial nerves, however, are composed either of sensory neurons only (*sensory* or *afferent nerves*) or of motor neurons only (*motor* or *efferent nerves*). Sensory nerves serve the special senses, such as taste, smell, sight, and hearing. Motor nerves conduct impulses to muscles, causing them to contract, or to glands, causing them to secrete.

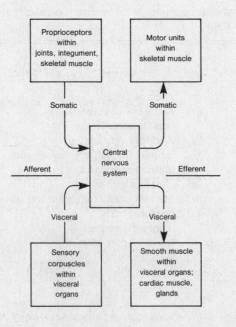

Fig. 10-1

10.2 Draw a diagram of the four-way classification that results when peripheral nerves are typed by area of innervation and by direction of conduction relative to the CNS.

See Fig. 10-1. Somatic afferent nerves are also called *proprioceptive*; and visceral efferent nerves, *autonomic motor*.

10.3 What is the importance of *ganglia* in the PNS (cf. Problem 9.10)?

In the PNS, the nuclei (cell bodies) are clumped together as ganglia. Thus (in the ANS) the ganglia are possible sites of synapse intermediate between end organs and the spinal cord.

10.4 What are *dermatomes* (of the PNS), and how are they important clinically?

A dermatome is the area of the skin innervated by all the cutaneous neurons of a given spinal or cranial nerve. Dermatomes are labeled as in Fig. 10-2.

The pattern of dermatome innervation is of clinical importance when a physician desires to anesthetize a particular portion of the body. Abnormally functioning dermatomes also provide clues about injury to the spinal cord or specific spinal nerves.

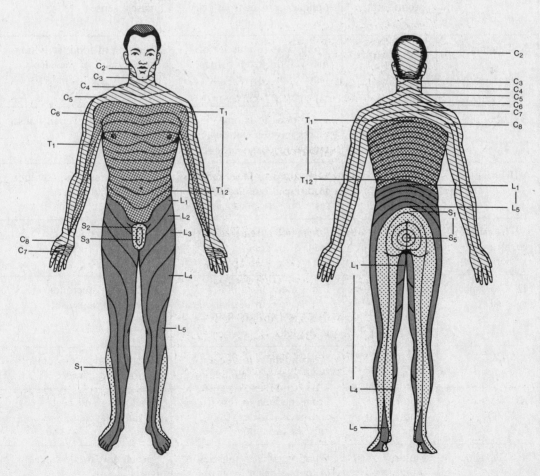

Fig. 10-2

OBJECTIVE B *To identify the twelve pairs of* cranial nerves *and their functions*.

Survey Refer to Fig. 10-3 and Table 10-1. Cranial nerves connect the brain to structures of the head, neck, and trunk. Most are mixed, some are totally sensory, and others are primarily motor. The names of cranial nerves indicate their primary functions or their general distribution; they are also frequently identified by Roman numerals in order of appearance from front to back.

Table 10-1

Nerve	Type	Pathways	Function
I Olfactory	Sensory	From nasal epithelium to olfactory bulb	Smell
II Optic	Sensory	From retina of eye to thalamus	Sight
III Oculomotor	Motor, proprioceptive	From midbrain to four eye muscles; from ciliary body to midbrain	Movement of eyeball and eyelid; focusing; change in pupil size; muscle sense
IV Trochlear	Motor, proprioceptive	From midbrain to superior oblique muscle; from eye muscle to midbrain	Movement of eyeball; muscle sense
V Trigeminal	Mixed	From pons to muscles of mastication; from cornea, facial skin, lips, tongue, and teeth to pons	Chewing of food; sensations from organs of the face
VI Abducens	Motor, proprioceptive	From pons to lateral rectus muscle; from eye muscle to pons	Movement of eyeball; muscle sense
VII Facial	Mixed	From pons to facial muscles; from facial muscles and taste buds to pons	Facial expressions; secretion of saliva and tears; muscle sense; taste
VIII Vestibulo-cochlear or auditory	Sensory	Organs of hearing and balance to pons	Hearing, balance, and posture
IX Glossopha-ryngeal	Mixed	From medulla to muscles of pharynx; from pharyngeal muscles and taste buds to medulla	Swallowing, secretion of saliva; muscle sense; taste
X Vagus	Mixed	From medulla to viscera; from viscera to medulla	Visceral muscle movement; visceral sensations
XI Accessory	Motor, proprioceptive	From medulla to throat and neck muscles; from muscles to medulla	Swallowing and head movements; muscle sense
XII Hypoglossal	Motor, proprioceptive	From medulla to muscles of tongue; from tongue muscles to medulla	Speech and swallowing; muscle sense

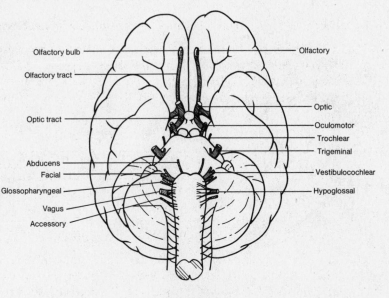

Olfactory bulb
Olfactory tract
Optic tract
Abducens
Facial
Glossopharyngeal
Vagus
Accessory

Olfactory
Optic
Oculomotor
Trochlear
Trigeminal
Vestibulocochlear
Hypoglossal

Fig. 10-3

10.5   Where do the cranial nerves attach to the brain?

The cranial nerves emerge from the inferior surface of the brain and pass through foramina of the skull. The first two pairs of cranial nerves are attached to the forebrain; the remaining ten pairs are attached to the brain stem. Sensory nerves originate in nerve trunks and special sense organs, and terminate at nuclei in the brain; motor nerves originate at nuclei in the brain.

10.6   What have the olfactory, optic, and vestibulocochlear nerves in common?

They are the purely sensory cranial nerves. The *olfactory nerve* consists of bipolar neurons that function as chemoreceptors and relay sensory impulses of smell from mucous membranes of the nasal cavity. Synapses of these neurons are in the *olfactory bulb*; the sensory impulses are then passed through the *olfactory tract* to the olfactory area of the cerebral cortex. The *optic nerve* conducts sensory impulses from the *photoreceptors* (rods and cones) in the retina of the eye to the *optic chiasma*, and on to the *optic tracts* and eventually to the visual cortex of the cerebrum. The *vestibulocochlear nerve* consists of a *vestibular branch* arising from the vestibular organs of equilibrium and balance, and a *cochlear branch* arising from the cochlea of hearing.

10.7   Which cranial nerves innervate the muscles that move the eyeball?

Movements of the eyeball are controlled by six *extrinsic ocular muscles*. The *oculomotor* cranial nerve innervates the *superior*, *inferior*, and *medial recti* muscles and the *inferior oblique*. The *abducens* cranial nerve innervates the *lateral rectus* muscle, and the *trochlear* cranial nerve innervates the *superior oblique* muscle. Inability to look cross-eyed may signal oculomotor nerve damage; problems with lateral eye movements, abducens nerve damage; and trouble looking downward away from the midline, trochlear nerve damage.

10.8   Of the cranial nerves, which is most important to the dentist?

The *trigeminal* cranial nerve [Fig. 10-4(*a*)], which conveys sensory information from the face, nasal area, tongue, teeth, and jaws, and supplies motor innervation to the muscles of mastication. Blockage along the *maxillary* nerve or its branches desensitizes the upper teeth, and blockage along the *mandibular* nerve or its branches desensitizes the lower teeth.

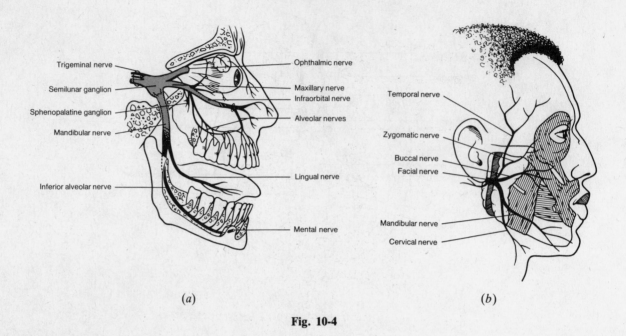

Fig. 10-4

10.9 Describe *tic douloureux*.

Also called trigeminal neuralgia, this disorder of the trigeminal nerve is characterized by severe recurring pain in one side of the face. Because the pain cannot be treated with drugs, denervation may be performed. The patient must then always be careful when chewing, as he might bite his cheek unawares.

10.10 What are the functions of the *facial nerve* [Fig. 10-4(*b*)]?

The facial nerve provides motor innervation to the facial muscles and salivary glands, and conducts sensory impulses from taste buds on the anterior two-thirds of the tongue.

10.11 Define *Bell's palsy*.

Bell's palsy is a temporary functional disorder of a facial nerve. The facial muscles on the affected side will lose tonus, causing them to sag.

10.12 Describe the distribution of the *vagus nerve*.

The vagus ("wandering"; see Fig. 10-5) nerve contains autonomic motor and sensory fibers that innervate visceral organs of the thoracic and abdominal cavities. Sensory fibers of the vagus convey sensations of hunger, abdominal distension, intestinal discomfort, and laryngeal state.

10.13 *True or false*: Because the cranial nerves are inferior to the brain, against the floor of the braincase, they are well protected from trauma.

False. The brain, immersed in and filled with cerebrospinal fluid, is like a water-sodden log: a blow to the top of the head (as in an automobile accident) can cause a heavy rebound of the brain from the floor of the cranium. Neurological examinations routinely involve testing for cranial-nerve dysfunction.

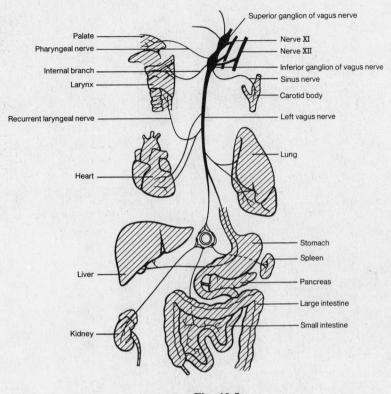

Fig. 10-5

OBJECTIVE C   *To locate and describe the* spinal nerves.

Survey   The 31 pairs of spinal nerves are grouped as: 8 *cervical*, 12 *thoracic*, 5 *lumbar*, 5 *sacral*, and 1 *coccygeal* (see Fig. 10-6). We know (Problem 9.33) that each spinal nerve is a mixed nerve, attached to the spinal cord by a *dorsal* (*posterior*) *root* of sensory fibers and a *ventral* (*anterior*) *root* of motor fibers. Except for C1, the spinal nerves leave the spinal cord and vertebral canal through intervertebral foramina.

10.14   Trace the branching of the spinal nerves.

Refer to Fig. 10-7. Upon emergence through the foramina, the ventral roots (immediately) and the dorsal roots (after swelling into *dorsal root ganglia*, where the cell bodies of the sensory neurons are located) become, respectively, *ventral rami* and *dorsal rami*. These rami further divide, or ramify. Except in the thoracic nerves T2–T12, the ventral rami of different spinal nerves combine and then split again, forming a network known as a *plexus*. There are four plexuses of spinal nerves: the *cervical*, *brachial*, *lumbar*, and *sacral* (Fig. 10-6). The latter two may be referred to jointly as the *lumbosacral plexus*. Nerves that emerge from a plexus no longer carry a spinal designation but are named according to the structure they innervate.

10.15   Which of the following nerves does *not* arise from the brachial plexus? (*a*) axillary, (*b*) phrenic, (*c*) radial, (*d*) ulnar, (*e*) median.

(*b*): the paired *phrenic nerves* arise from the cervical plexus and innervate the diaphragm. The other four, plus the *musculocutaneous nerve*, do arise from the brachial plexus.

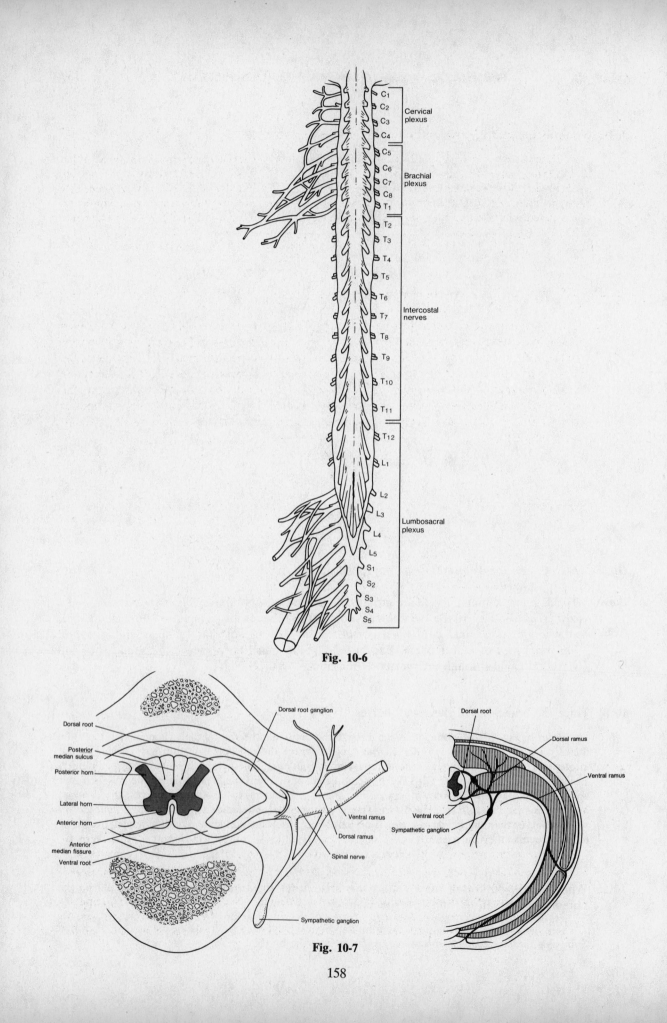

C₁
C₂
C₃
C₄

Cervical
plexus

C₅
C₆
C₇
C₈
T₁

Brachial
plexus

T₂
T₃
T₄
T₅
T₆
T₇
T₈
T₉
T₁₀
T₁₁

Intercostal
nerves

T₁₂
L₁
L₂
L₃
L₄
L₅
S₁
S₂
S₃
S₄
S₅

Lumbosacral
plexus

Fig. 10-6

Dorsal root ganglion

Dorsal root

Posterior
median sulcus

Posterior horn

Lateral horn

Anterior horn

Anterior
median fissure

Ventral root

Ventral ramus

Dorsal ramus

Spinal nerve

Sympathetic ganglion

Dorsal root

Dorsal ramus

Ventral ramus

Ventral root

Sympathetic ganglion

Fig. 10-7

158

10.16 Where is the *sciatic nerve* and how may it be traumatized?

> The sciatic nerve (which branches into *tibial* and common *peroneal* nerves) arises from L4–S3 of the sacral plexus, passes through the pelvis, and extends down the posterior aspect of the leg within the *sciatic sheath*. It is the largest nerve in the body. A posterior dislocation of the hip joint will generally injure the sciatic nerve. A herniated disc, pressure from the uterus during pregnancy, or an improperly administered injection into the buttock may damage the roots leading to the sciatic nerve or the nerve itself.

OBJECTIVE D   *To be able to trace a* spinal reflex arc.

Survey   In the *polysynaptic* pathway diagramed in Fig. 10-8, there are seven components: a **receptor** in the peripheral end of a sensory neuron; the **sensory (afferent) neuron**, which carries impulses from the receptor to the spinal cord; a **synapse**, within the spinal cord, between the sensory neuron and an interneuron; the **interneuron (association neuron)**, within the spinal cord; a **synapse**, within the spinal cord, between the interneuron and a motor neuron; the **motor (efferent) neuron**, which transmits impulses from the anterior horn of the gray matter to an effector organ; the **effector** (muscle tissue or organ), which responds to the efferent nerve impulses.

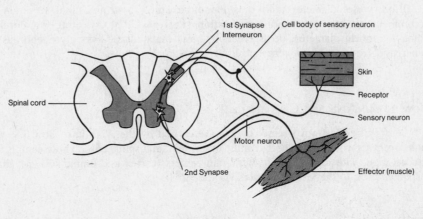

Fig. 10-8

10.17 Give an example of (*a*) a reflex arc without interneurons (*monosynaptic*), (*b*) a polysynaptic reflex arc that involves more than one interneuron.

(*a*)   In the *knee-jerk reflex*, tapping the patellar tendon causes the quadriceps femoris (Table 7-15) to become stretched, which provokes impulses from intrafusal spindle receptors. The impulses are conducted along the sensory neuron to the spinal cord, where the sensory neuron synapses directly with the motor neuron, which stimulates contraction of extrafusal fibers and thus the whole muscle.

(*b*)   In a *withdrawal reflex*, a painful stimulus applied to the body, such as a sharp or hot object, initiates impulses in a sensory receptor. The impulses are transmitted to the spinal cord via an afferent neuron, which stimulates two (or more) interneurons. One interneuron generates impulses in an efferent neuron, which initiates a response such as foot or hand withdrawal; the second interneuron conducts impulses toward the brain, whereby the person is made aware of the painful event.

OBJECTIVE E *To distinguish further between the ANS and the somatic system.*

Survey

Autonomic	Somatic
Functions automatically; i.e., primarily below the conscious level	Conscious or voluntary regulation
Fibers synapse once (at a ganglion) after they leave the CNS:	Fibers do not synapse after they leave the CNS:

Effector cells can be either stimulated or inhibited	Skeletal muscle cells are only stimulated

10.18 Which specific physiological functions are regulated by the autonomic system?

Blood-vessel diameter (and thus blood pressure), gastrointestinal secretion, size of eye pupil, accommodation for near vision, micturition (Problem 20.3), sweating, glomerular filtration rate in kidneys, bronchi diameter, penis erection, basal metabolism, liver glycogenolysis, body temperature, pancreas secretion. (Not a complete list.)

10.19 How do the ANS and CNS interact?

Sensory signals from visceral organs are carried by the ANS into the CNS, where they influence mainly centers within the hypothalamus, brain stem, and spinal cord. These centers integrate the sensory visceral input with input from higher brain centers (cortex and limbic system), then send appropriate responses back to the visceral organs via the ANS.

OBJECTIVE F *To compare the sympathetic and parasympathetic divisions of the ANS as to* (1) *origin in the CNS,* (2) *location of ganglia, and* (3) *neurotransmitter substances.*

Survey

	SYMPATHETIC	*PARASYMPATHETIC*
(1)	Thoracolumbar nerves	Craniosacral nerves
(2)	Far from visceral effector organs	Near or within visceral effector organs
(3)	In ganglia, acetylcholine; in effector organs, norepinephrine	In ganglia, acetylcholine; in effector organs, acetylcholine

10.20 Describe the disposition of ganglia in both sympathetic and parasympathetic divisions.

    Sympathetic division. There are two types of sympathetic ganglia. (i) *Sympathetic chain ganglia*, or *paravertebral ganglia*, are interconnected by neuron fibers to form two chains lateral to the spinal cord. There are 22 ganglia in each chain (3 cervical, 11 thoracic, 4 lumbar, 4 sacral). As diagramed in the upper part of Fig. 10-9(*a*), preganglionic neurons leave the spinal cord, pass through ventral roots into spinal nerves, and then pass from the spinal nerves via *white rami communicantes* (sing. *ramus communicans*) into the sympathetic chains. There most of them synapse (on postganglionic neurons) in the chain ganglia. Some of the postganglionic neurons run back into spinal nerves via *gray rami communicantes*, while the rest pass directly out to the viscera. (ii) *Collateral*, or *prevertebral*, *ganglia* are found outside the sympathetic chain, in the vicinity of the viscera and arteries. As diagramed in the lower part of Fig. 10-9(*a*), some (preganglionic) neurons synapse (on postganglionic neurons) in collateral ganglia (*celiac, superior mesenteric, inferior mesenteric*).

    Parasympathetic division. All parasympathetic ganglia are called *terminal ganglia*, because they are located close to or in an organ that is innervated. Two cases of parasympathetic innervation are schematized in Fig. 10-9(*b*).

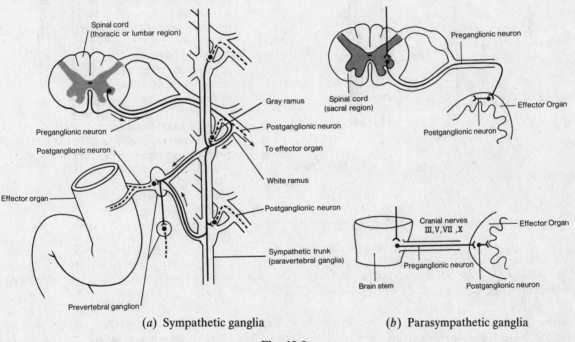

(*a*) Sympathetic ganglia   (*b*) Parasympathetic ganglia

Fig. 10-9

10.21 Explain why parasympathetic fibers and sympathetic fibers are referred to as *cholinergic* and *adrenergic*, respectively.

    The reason is that acetyl*choline* and (nor)epinephrine ((nor)*adrenaline*) are the respective chemical transmitters released at effector organs in the parasympathetic and sympathetic divisions.
    In three exceptional cases—those of sweat glands, smooth muscles of certain blood vessels, and the adrenal medulla—the innervating sympathetic fibers are cholinergic.

10.22 What are the types of acetylcholine receptors (cholinergic) in the ANS?

    Muscarinic receptors are located on all effector cells innervated by postganglionic neurons of the parasympathetic division and on those effector cells innervated by postganglionic, cholinergic neurons of the sympathetic division (see Problem 10.21). *Nicotinic receptors* are located at the ganglia in both sympathetic and parasympathetic divisions.

10.23 What are the types of norepinephrine receptors (adrenergic) in the ANS?

There are two main types, called *alpha receptors* and *beta receptors*, each divided into two subtypes. See Table 10-2.

Norepinephrine stimulates mainly alpha receptors; epinephrine stimulates both alpha and beta receptors, approximately equally. *Isoproteranol*, a synthetic catacholamine, stimulates mainly beta receptors.

Table 10-2

Receptor Subtype	Location	Effects of Stimulation
α_1	Smooth muscle	Vasoconstriction, uterine contraction, dilation of pupil, intestinal sphincter contraction, pilomotor (arrector pili) contraction
α_2	Axon terminals of postganglionic adrenergic neurons	*Negative feedback*: norepinephrine acts to inhibit its own further release
β_1	Heart	Changes in rate and force of heart contraction
β_2	Smooth muscle	Vasodilation, uterine relaxation, intestinal relaxation, bronchodilation, glycogenolysis

OBJECTIVE G   *To be able to predict the effects of sympathetic versus parasympathetic stimulation on specific organs.*

Survey   The heart, as well as most smooth muscles and visceral organs of the body, is innervated by both sympathetic and parasympathetic fibers. One division stimulates, while the other one inhibits. The two divisions are usually activated reciprocally; i.e., as the activity of one is enhanced, the activity of the other is diminished. To predict the effects of each division on a specific organ, use the following rule of thumb:

Sympathetic stimulation activates the body in states of stress, fear, and rage (the "fight-or-flight" reaction), and during strenuous physical activity.

Parasympathetic stimulation maintains body functions under quiet, day-to-day living conditions; it decreases heart rate and promotes digestion and absorption of food.

10.24 List the organs which are innervated by the ANS and indicate the effects of sympathetic and parasympathetic stimulation on each organ.

See Table 10-3; part of the information appears in Table 10-2.

Table 10-3

Organ		Sympathetic (Adrenergic or Cholinergic) Stimulation	Parasympathetic (Cholinergic) Stimulation
Heart		Accelerates rate and increases strength of contraction	Slows rate and decreases strength of contraction, inhibits activity
Vascular Smooth Muscle	Skin	Vasoconstriction (adrenergic); vasodilation, blushing (cholinergic)	None
	Skeletal muscles	Vasoconstriction (adrenergic); vasodilation (cholinergic)	None
	Heart	Vasodilation	Vasodilation
	Lungs	Mild constriction	None
	Viscera	Vasoconstriction (adrenergic to cerebrum, abdominal viscera); vasodilation (cholinergic to external genitalia)	Vasodilation (cerebrum, abdominal viscera, external genitalia)
Other Smooth Muscle	Hair (arrector pili)	Contraction and erection of hair, "goose pimples"	None
	Bronchi	Dilation	Constriction; stimulates secretion
	Digestive tract	Decreases activity (e.g., peristalsis) and tone	Increases activity and tone
	Gallbladder and ducts	Inhibits	Stimulates
	Anal sphincter	Constricts	Dilates
	Urinary bladder	Relaxes and inhibits emptying	Contracts and stimulates emptying
	Ciliary (eye)	None	Contracts (allows thickening of lens to accommodate vision)
	Iris (eye)	Dilation of pupil	Constriction of pupil
Glands	Sweat	Accelerates secretion (cholinergic)	None
	Nasal, lacrimal, salivary, gastric, intestinal, pancreatic	Vasoconstriction and inhibited secretion	Vasodilation and stimulated secretion
	Pancreatic islets	Decreases secretion of insulin	Stimulates secretion
	Liver	Stimulates breakdown of glycogen with release of glucose into blood	None
	Adrenal medulla	Accelerates secretion of norepinephrine and epinephrine (which increase heart rate, blood pressure, blood sugar)	None

10.25 Give four classes of drugs that are used clinically to stimulate or inhibit autonomic functions.

Adrenergic receptor-activating drugs include *epinephrine, norepinephrine, isoproteranol, ephedrine,* and *amphetamine.* Used to: dilate bronchial tubes, treat cardiac arrest, dilate pupil and decrease intraocular pressure, delay absorption of local anesthetics, elevate mood of patient.

Adrenergic receptor-blocking drugs include *phentolamine, phenoxybenzamine, prazosin* (alpha blockers); *propranolal, timolal, nadolal* (beta blockers). Used to: (alpha blockers) lower blood pressure in cases of pheochromocytoma; (beta blockers) lower blood pressure, reduce frequency of anginal episodes, treat heart arrhythmias, reduce intraocular pressure in cases of glaucoma.

Cholinergic receptor stimulants include *acetylcholine* and its mimics *methacholine, carbachol, bethanecol.* Used to: stimulate intestines and urinary bladder postoperatively, lower intraocular pressure in glaucoma, dilate peripheral blood vessels, terminate curarization, treat myasthenia gravis.

Cholinergic receptor antagonists include *atropine, scopolamine, dicyclomine* (antimuscarinic agents). Used to: treat Parkinson's disease, dilate pupil, control motion sickness, treat peptic ulcers and hypermobility of the gut, decrease salivary and bronchial secretion (preoperative use of atropine).

Review Questions

Multiple Choice

1. The chemical transmitter between sympathetic postganglionic fibers and the effector organs is: (*a*) norepinephrine, (*b*) acetylcholine, (*c*) adrenaline, (*d*) epinephrine.

2. Most body organs are innervated by the: (*a*) parasympathetic division, (*b*) sympathetic division, (*c*) two divisions, (*d*) CNS.

3. Parasympathetic fibers arise from which set of cranial nerves? (*a*) 3rd, 5th, 8th, 10th; (*b*) 4th, 5th, 9th, 10th; (*c*) 3rd, 7th, 9th, 10th; (*d*) 5th, 9th, 10th, 12th.

4. A preganglionic fiber entering the sympathetic chain *cannot*: (*a*) synapse with postganglionic neurons at the first ganglion it meets, (*b*) travel down the sympathetic chain before synapsing with postganglionic neurons, (*c*) end in the sympathetic chain without having synapsed, (*d*) pass through the sympathetic chain without having synapsed.

5. The cell bodies of the preganglionic neurons of the sympathetic division are located within the: (*a*) cervical and sacral regions of the spinal cord, (*b*) white matter of the spinal cord, (*c*) lateral horns of the spinal cord gray matter, (*d*) brain and sacral region.

6. The autonomic nervous system is responsible for which function(s)? (*a*) motor, (*b*) sensory, (*c*) motor and sensory, (*d*) none of the above.

7. The white ramus of each spinal nerve has attached to it: (*a*) a prevertebral ganglion, (*b*) a chain ganglion, (*c*) a posterior root ganglion, (*d*) the celiac ganglion.

8. Which describes the effect of the sympathetic system on the eye pupil and the intestinal tract? (*a*) dilates/inhibits, (*b*) dilates/stimulates, (*c*) constricts/inhibits, (*d*) constricts/stimulates.

9. Which of the following would *not* result from sympathetic stimulation? (*a*) glucogenolysis, (*b*) contraction of the spleen, (*c*) secretion of catecholamines from the adrenal medulla, (*d*) profuse secretion of the salivary glands.

10. One reason for the division of the ANS is that: (a) sympathetic signals are transmitted from the spinal cord to the periphery through two successive neurons, in contrast to one neuron for parasympathetic signals; (b) sympathetic fibers alone innervate organs in the abdominal cavity; (c) sympathetic fibers alone arise from the spinal cord; (d) the effects of the two divisions on the organs are usually antagonistic to each other.

11. The sympathetic system does *not*: (a) spring from thoracolumbar levels, (b) summon energy during an emergency, (c) stimulate bile secretion from the gallbladder, (d) dilate the bronchial tubes.

12. The largest of the paravertebral ganglia, which carries sympathetic impulses to the eye, salivary glands, and mucosa of the mouth, is the: (a) superior cervical, (b) celiac, (c) superior mesenteric, (d) stellate.

13. Of the following statements about the parasympathetic division of the ANS:

   (i)　all its neurons release acetylcholine as their primary neurotransmitter substance;

   (ii)　the cell bodies of its postganglionic neurons lie in or near the organ innervated;

   (iii)　the cell bodies of its preganglionic neurons lie in the cervical and sacral spinal cord;

one would say that: (a) all are true, (b) none is true, (c) (i) and (ii) are true, (d) (ii) and (iii) are true (e) (iii) is true.

14. Of the following statements about the sympathetic division of the ANS:

   (i)　all its neurons release norepinephrine as their primary neurotransmitter substance;

   (ii)　all the cell bodies of its postganglionic neurons lie in or near the organ innervated;

   (iii)　the cell bodies of its preganglionic neurons lie in the thoracic and lumbar cord;

one would say that: (a) (i) is true, (b) (ii) is true, (c) (iii) is true, (d) (i) and (iii) are true, (e) all are true.

15. Autoreceptors of the sympathetic division that are involved in negative feedback are the: (a) α_1, (b) α_2, (c) β_1, (d) β_2.

16. Beta receptors are stimulated by: (a) methoxamine, (b) acetylcholine, (c) isoproteranol, (d) atropine.

17. The receptors for acetylcholine at the ganglia of both the sympathetic and parasympathetic divisions are: (a) muscarinic receptors, (b) blocked by atropine, (c) nicotinic receptors, (d) stimulated by isoproteranol.

18. Which type of receptor is found in the heart? (a) alpha, (b) beta, (c) nicotinic, (d) GABA.

19. Which class of drugs may be used to treat bronchial asthma? (a) cholinergic, (b) anticholinesterase, (c) adrenergic, (d) adrenergic-blocking agents.

20. Cholinergic-blocking agents have as an unwanted side effect: (a) increased gastric secretion, (b) spasms of the intestinal tract, (c) diarrhea, (d) dry mouth, (e) bradycardia.

21. A patient scheduled for surgery confides in his nurse the night before that he is "terribly scared." Which of the following indicate(s) increased sympathetic activity in this patient: (a) patient complains that his mouth feels dry; (b) patient's gown is moist with perspiration; (c) his skin looks pale; (d) the pupils of his eyes are widely dilated; (e) all the above.

22. Which of the following is *not* a function of the ANS? (a) innervation of all visceral organs, (b) transmission of afferent and efferent impulses, (c) regulation and control of vital activities, (d) conscious control of motor activities.

23. Concerning (1) the heart, (2) glands, (3) smooth muscle, (4) certain skeletal muscles, we can say that the ANS innervates: (*a*) (1), (2), (3); (*b*) (1), (3), (4); (*c*) (2), (3); (*d*) (1), (2), (3), (4).

24. Atropine (which blocks muscarinic receptors) is liable to cause: (*a*) weakness of cardiac muscles, (*b*) an increase in resting heart rate, (*c*) an excessive flow of saliva, (*d*) overactivity of the small intestine.

25. A *ganglion* is an aggregate of nerve cell bodies: (*a*) inside the brain or spinal cord, (*b*) outside the brain and spinal cord, (*c*) only in the spinal cord, (*d*) only in the brain.

26. A cranial nerve that affects eye movement is the: (*a*) optic (II), (*b*) trigeminal (V), (*c*) trochlear (IV), (*d*) hypoglossal (XII).

27. The cranial nerve with the greatest distribution is the: (*a*) trigeminal (V), (*b*) vagus (X), (*c*) abducens (VI), (*d*) none of these.

28. Taste sensation is mediated by cranial nerves: (*a*) trigeminal (V), facial (VII); (*b*) trochlear (IV), abducens (VI); (*c*) facial (VII), glossopharyngeal (IX); (*d*) trigeminal (V), glossopharyngeal (IX).

29. A patient with a contusion over the parotid region has the following symptoms: paralysis of facial muscles, one eye can't be shut, corner of the mouth droops. Which cranial nerve is damaged? (*a*) abducens (VI), (*b*) facial (VII), (*c*) glossopharyngeal (IX), (*d*) accessory (XI), (*e*) hypoglossal (XII).

30. The knee jerk in response to a tap over the patellar tendon: (*a*) is a conditioned reflex, (*b*) is a polysynaptic reflex, (*c*) has its reflex center in the spinal cord, (*d*) is mediated by a three-neuron reflex arc.

31. Which pairing of nerve and organ innervated is *incorrect*? (*a*) phrenic–diaphragm, (*b*) vagus–abdominal viscera, (*c*) glossopharyngeal–taste buds, (*d*) abducens–facial muscles, (*e*) sciatic–lower extremity.

32. Inability to walk a straight line may indicate damage to which cranial nerve? (*a*) vestibulocochlear, (*b*) trochlear, (*c*) facial, (*d*) hypoglossal, (*e*) accessory.

33. The rectus eye muscle capable of causing the eyeball to turn laterally in a horizontal plane is innervated by which cranial nerve? (*a*) optic, (*b*) abducens, (*c*) facial, (*d*) oculomotor, (*e*) trochlear.

34. Which of the following is *not* a plexus of the spinal nerves? (*a*) cervical, (*b*) sacral, (*c*) choroid, (*d*) brachial, (*e*) lumbar.

35. Which of the following cranial nerves is *not* a mixed nerve? (*a*) oculomotor, (*b*) glossopharyngeal, (*c*) trigeminal, (*d*) vagus, (*e*) vestibulocochlear.

36. For each statement below, determine whether it applies to (1) the sympathetic division, (2) the parasympathetic division, (3) neither division, (4) both divisions. (*a*) a neurotransmitter is mimicked by catecholamine; (*b*) it is involved with visceral reflexes; (*c*) it is generally under voluntary control; (*d*) the preganglionic neurons are myelinated; (*e*) the effector organs possess alpha and beta receptors; (*f*) it increases intestinal contractions; (*g*) it increases blood flow to skeletal muscle; (*h*) the ganglia are in or near the organ innervated.

True/False

1. Cranial nerves innervate structures of the head and neck, and nothing else.

2. The extrinsic ocular muscles are innervated by three different cranial nerves.

3. All spinal nerves are mixed.

4. The parasympathetic division of the ANS functions in meeting stressful and emergency conditions.

5. An inability to shrug the shoulders may indicate a dysfunction of the facial nerve.

6. Erection of the penis is primarily a parasympathetic response.

7. The posterior (dorsal) root of a spinal nerve consists of sensory neurons only.

8. Compression of the brachial plexus could result in paralysis of the hand.

9. The optic nerve controls movement of the eye.

10. There are 12 pairs of cranial nerves, 31 pairs of spinal nerves, and four plexuses of the spinal column included in the peripheral nervous system.

Chapter 11

Sensory Organs

OBJECTIVE A *To identify the receptors and the neural pathways for the sense of taste.*

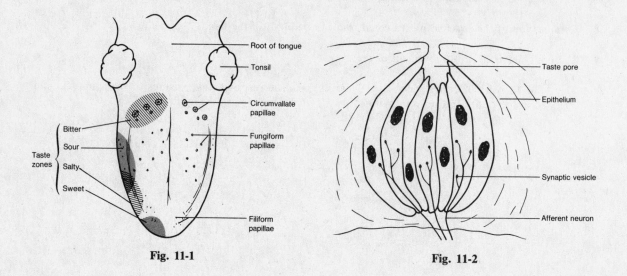

Fig. 11-1   Fig. 11-2

Survey   Receptors for taste (*gustation*) are located in taste buds on the tongue (Fig. 11-1), hard palate, soft palate, and epiglottis, and in the pharynx. A taste bud contains a cluster of 40 to 60 taste cells, as well as some supporting cells (Fig. 11-2). Each taste cell is innervated by an afferent neuron. The four primary taste sensations are *sweet* (evoked by sugars, glycols, aldehydes), *sour* (acids), *bitter* (alkaloids), and *salty* (anions of ionizable salts).

11.1   Which of the four types of lingual papillae (see Fig. 11-1) lacks taste buds on its surface?

Filiform papillae contain only mechanoreceptors and are not involved in taste.

11.2   Do taste receptors undergo adaptation?

Yes: with continuous exposure to a taste stimulus there is a decrease in afferent nerve transmission.

11.3   Which nerves conduct taste impulses to the brain?

Facial nerve (VII)—from anterior two-thirds of tongue; glossopharyngeal nerve (IX)—from posterior third of tongue; vagus nerve (X)—from pharyngeal areas.

11.4   Which areas of the brain receive impulses from the taste receptors?

Taste information is transmitted to the brain stem (*nucleus solitarius*), then to the thalamus (*nucleus ventralis posteromedialis*), and finally to the sensory cortex (postcentral gyrus on the lateral convexity).

OBJECTIVE B　*To identify the receptors and neutral pathway for the sense of smell.*

Survey　Receptors for smell (*olfaction*) are located in each nasal cavity, on the superior nasal concha mucosa (Fig. 11-3). Like taste receptors, smell receptors are sensitive to chemicals in solution. For smell, however, the chemicals are originally airborne and become dissolved in the mucous layer lining the nasal cavity.

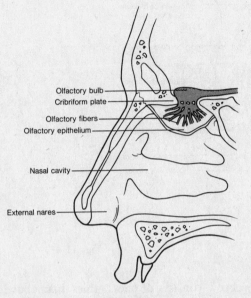

Fig. 11-3

11.5　What are the characteristics of an *odorant*, a substance that stimulates smell receptors?

The substance must be volatile (to reach the smell receptors), water-soluble (to penetrate the watery mucous layer covering the receptors), and lipid-soluble (to penetrate the plasma membranes of the smell receptors).

11.6　Does adaptation of smell receptors occur?

Yes: smell receptors adapt very rapidly to continued exposure to an odorant (50% adaptation within the first second).

11.7　Do all of the volatile chemicals in the nose stimulate smell receptors?

No: only about 2 or 3% of the air that is inhaled comes in contact with the smell receptors, because of their location above the main airstream. Olfaction can be greatly increased by forceful sniffing, which draws the volatile chemical into contact with smell receptors.

11.8　Which of the cranial nerves innervate olfactory mucosa?

The olfactory nerve (cranial nerve I) transmits most impulses related to smell; however, some irritating chemicals stimulate the trigeminal nerve (cranial nerve V), as well as the olfactory nerve. Sensory impulses are conveyed along the olfactory tract to the olfactory portions of the cortex (*prepiriform cortex*, *subcallosal gyrus*, and *olfactory tubercle*).

OBJECTIVE C   *To describe the accessory structures of the eye.*

Survey   The following structures (see also Fig. 11-4) all subserve the *eyeball*, the actual organ of vision.

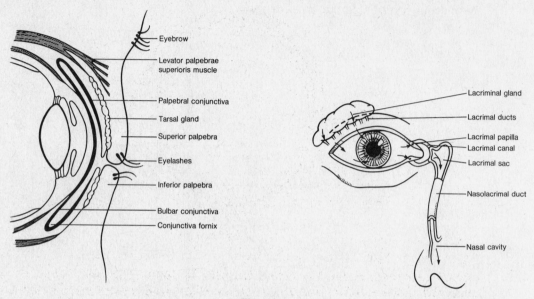

Fig. 11-4

Eyebrow. The eyebrow consists of short, thick hair above the eye which helps to prevent perspiration and other substances from entering the eye and to shade the eye from the sun.

Eyelids. The two eyelids (*palpebrae*) cover and protect the eye from desiccation, foreign matter, and the sun. Each eyelid is covered with skin and is composed of muscle fibers, dense fibrous connective tissue (*tarsal plate*), sebaceous glands (*tarsal* or *Meibomian glands*), and sweat glands (*ciliary glands*). The numerous *eyelashes* attached to the eyelids protect the eye from airborne objects.

Conjunctiva. A thin mucous membrane that lines the anterior surface of the eyeball (*bulbar conjunctiva*) and the interior surface of the eyelid (*palpebral conjunctiva*).

Lacrimal apparatus. The lacrimal apparatus consists of the *lacrimal gland*, which secretes tears (*lacrimal fluid*), and the *lacrimal canals* that drain the tears into the *lacrimal sac*. The tears lubricate the conjunctiva and contain an enzyme, called *lysozyme*, which is bactericidal.

Eye muscles. *Superior rectus* (rotates eye upward and toward midline), *inferior rectus* (rotates eye downward and toward midline), *medial rectus* (rotates eye toward midline), *lateral rectus* (rotates eye away from midline), *superior oblique* (rotates eye downward and away from midline), *inferior oblique* (rotates eye upward and away from midline), *levator palpebrae superioris* (elevates eyelid), *orbicularis oculi* (constricts eyelid).

11.9   Why do you have to blow your nose when you cry?

Tears drain from the eye into the lacrimal canals, through the lacrimal sac, and through the nasolacrimal duct, which empties into the nasal cavity.

OBJECTIVE D   *To describe the structures of the eyeball.*

Survey   The following structures are illustrated in Fig. 11-5.

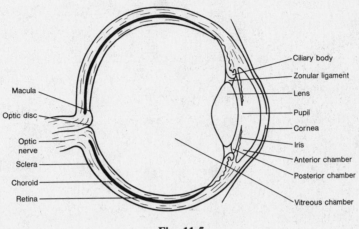

Macula

Optic disc

Optic nerve

Sclera

Choroid

Retina

Ciliary body

Zonular ligament

Lens

Pupil

Cornea

Iris

Anterior chamber

Posterior chamber

Vitreous chamber

Fig. 11-5

*Fibrous tunic (**outer layer**).* The fibrous tunic has two parts. The *sclera* (white of the eye) is composed of dense, tightly bound elastic and collagenous fibers that support and protect the eyeball and provide a base for attachment of eye muscles. The *cornea* forms the anterior surface of the eyeball; it is composed of avascular, dense connective tissue, is both transparent and convex, and refracts incoming light.

*Vascular tunic (**middle layer**).* The vascular tunic has three parts. The *choroid* is a thin, pigmented, highly vascular layer of connective tissue that absorbs light, preventing it from being reflected back through the retina. The *ciliary body* is the thickened, anterior portion of the vascular tunic; it contains smooth muscle that helps regulate the shape of the lens. The *iris*, which forms the most anterior portion of the vascular tunic, consists of pigment (which gives the eye its color) and smooth muscle fibers that are arranged in a circular and radial pattern. Contraction of the smooth muscle fibers regulates the diameter of the *pupil* (opening in the center of the iris).

*Internal tunic (**inner layer, or retina**).* This is the receptor component of the eye, containing two types of photosensitive cells. *Cones* function at high light intensities and are responsible for daytime color vision and acuity (sharpness). *Rods* function at low light intensities and are responsible for night (black-and-white) vision. In addition, the retina has *bipolar cells*, which synapse with the rods and cones, and *ganglion cells*, which synapse with the bipolar cells. The axons of the ganglion cells course along the retina to the optic disc and form the *optic nerve*. Finally, there are pigment cells, containing melanin, which absorb stray light.

Anterior and posterior chambers. The anterior chamber is located between the cornea and iris; the posterior chamber, between the iris and lens. These two chambers are connected through the pupil and are filled with fluid called *aqueous humor*.

Vitreous chamber. The chamber between the lens and the retina; it is filled with a transparent jellylike substance called *vitreous humor*.

Lens. A yellowish, transparent, biconvex body composed of tightly arranged proteins. It is located between the posterior and vitreous chambers and is supported by zonular (*suspensory*) *ligaments* of the ciliary body. The lens focuses light rays for near and far vision.

11.10 Identify the ten layers of the retina.

See Fig. 11-6.

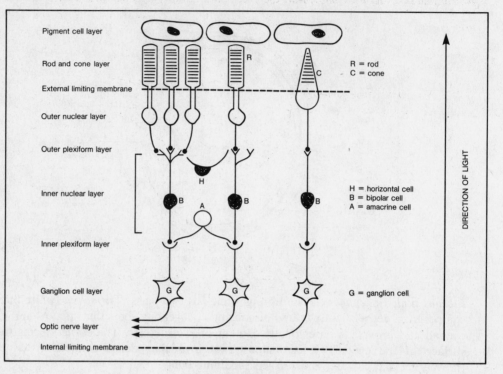

Fig. 11-6

11.11 Do all cone cells respond to the entire visible spectrum?

No: the cones fall into three classes, with absorption peaks corresponding to the three primary colors, blue, green, and orange-red (see Fig. 11-7).

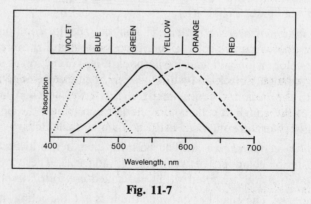

Fig. 11-7

11.12 What is the cause of color blindness?

Color blindness results from the inability of the cones to produce one or more of the photopigments. *Monochromats* have no photopigments and see no colors. *Protonopes* lack red-sensitive pigment. *Deuteranopes* lack green-sensitive pigment. *Tritanopes* lack blue-sensitive pigment.

11.13 What are the mechanisms involved in the stimulation of photoreceptors?

Excitation of rods follows the *rhodopsin cycle*, schematized in Fig. 11-8. With rhodopsin replaced by *iodopsin*, and scotopsin by *photopsin*, the cycle describes the excitation of cones.

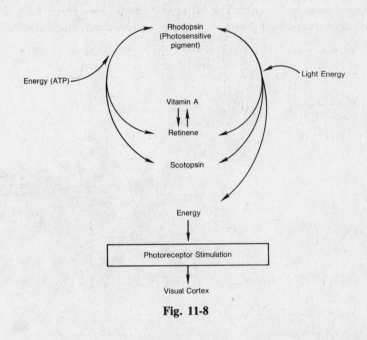

Fig. 11-8

11.14 Do all the photoreceptor fibers that arise in the right eye cross over to the left side of the brain?

No: the fibers *partially* cross over at the optic chiasm. Neurons from the right half of each eye project to the right lateral geniculate, and neurons from the left half of each eye pass to the left lateral geniculate. Thus, the neurons from the temporal portion of each eye pass through the optic chiasm without crossing, whereas the neurons from the nasal half of each eye decussate in the optic chiasm.

OBJECTIVE E　　*To describe how the normal eye views objects at various distances.*

Survey　　For an image to be focused on (neither in front of nor behind) the retina of an eyeball of given length, the more distant the object is, the flatter must be the lens. Adjustments in lens shape, accomplished by the ciliary muscles, are called *accommodation*. When these muscles are contracted, the zonular ligaments (Fig. 11-5) are slackened and the lens is made thicker.

11.15 List the common defects of vision and the treatment for each.

In *myopia* (nearsightedness), the eyeball is too long for the refractive power of the lens, and far objects are focused at a point in front of the retina. The eye can focus on very near objects. *Treatment*: concave lenses. In *hyperopia* (farsightedness), the eyeball is too short for the lens, and near objects are focused behind the retina. Distant objects are focused correctly. *Treatment*: convex lenses. In *astigmatism*, the lens or cornea does not have a smoothly spherical surface, so that some portions of an image are in focus, while other portions are blurred. *Treatment*: lenses of compensating nonuniform curvature.

OBJECTIVE F   *To give the structural components of the* external ear *and their functions.*

Survey   See Fig. 11-9. The external ear directs sound toward the auditory apparatus proper. Its structures are: the *pinna* or *auricle*, the funnellike outer component that collects sounds into the *external auditory meatus*. The latter is a canal (2.5 cm, or 1 in., in length) that runs to the thin, cone-shaped *tympanic membrane*, or *eardrum*, which conducts sound waves to the middle ear.

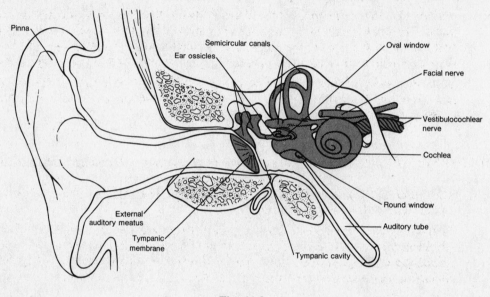

Fig. 11-9

11.16 What are the major physical parameters used to describe a sound wave?

There are two: *amplitude* and *frequency*. As suggested by Fig. 11-10, the amplitude is the "height" of the wave; the *power* or *intensity* of the wave is proportional to the square of its amplitude. Intensity translates psychologically into *loudness*, and is measured (on a logarithmic scale) in *decibels* (dB).

Frequency is the number of oscillations ("back-and-forth," in the case of sound) the wave makes in a unit of time. Frequency translates into *pitch*, and is measured in *hertz* (Hz), where 1 Hz = 1 cycle per second.

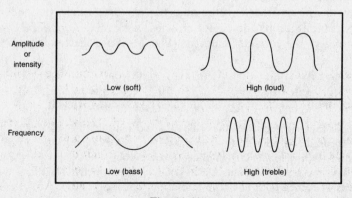

Fig. 11-10

OBJECTIVE G *To give the structural components of the* middle ear, *along with their functions.*

Survey The middle ear, or tympanic cavity, is the air-filled space medial to the eardrum (see Fig. 11-9). Its structures are as follows.

Auditory ossicles (cf. Table 6-1), composed of the *malleus* ("hammer"), attached to the eardrum; the *incus* ("anvil"); and the *stapes* ("stirrup"), attached to the *oval window*, which is a membrane-covered opening into the inner ear. These three small bones form a sound-conducting bridge between external and inner ear.

Muscles. The *tensor tympani* is the skeletal muscle that inserts on the medial surface of the malleus and is innervated by the trigeminal nerve (V). The *stapedius* is the skeletal muscle that inserts on the neck of the stapes and is innervated by the facial nerve (VII). These two muscles function to reduce the pressure of loud sounds before it can injure the inner ear.

Eustachian tube (**auditory tube**). The eustachian tube connects the middle ear to the throat and allows air to pass between the middle ear and the outside, thus maintaining equal air pressure on the two sides of the eardrum.

OBJECTIVE H *To give the structural components of the* inner ear, *along with their functions.*

Survey The inner ear contains not only the organs of hearing, but those of orientation and equilibrium. It has two main parts.

Bony labyrinth (Fig. 11-11). This is a network of cavities in the petrous temporal bone (Problem 6.15). These include three *semicircular canals*, each of which swells into a globular *ampulla*; a central *vestibule*; and a spiral ($2\frac{1}{2}$ turns) *cochlea* ("snail"). The cochlea has three chambers: the *scala tympani*, the *scala vestibuli*, and the *cochlear duct*.

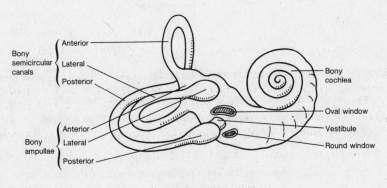

Fig. 11-11

Membranous labyrinth (Fig. 11-12). This intercommunicating system of membranous ducts is seated in the bony labyrinth, and its parts are conamed with those of the bony labyrinth. Thus we have the (membranous) semicircular canals and their ampullae; these possess receptors sensitive to rotary motions of the head. The vesibule is made up of two connected sacs—the *utricle* and *saccule*—which possess receptors sensitive to gravity and linear motions of the head. In the cochlear duct is found the *organ of Corti* (Fig. 11-13), a "transducer" that converts sound (mechanical) impulses into nerve (electrical) impulses. The membranous labyrinth is filled with a fluid called *endolymph* and is surrounded by a fluid called *perilymph*.

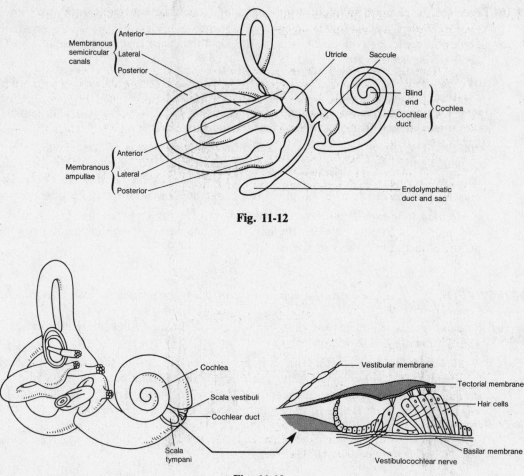

Fig. 11-12

Fig. 11-13

11.17 List the sequence of events involved in hearing.

1. Sound waves are funneled by the pinna into the external auditory meatus.
2. The waves strike the tympanic membrane and cause it to vibrate.
3. Vibrations of the tympanic membrane are passed through the malleus, incus, and stapes.
4. Oval·window is pushed back and forth by the stapes.
5. Vibrations of the oval window set up pressure waves in the perilymph.
6. Pressure waves are propagated through the scala vestibuli and scala tympani to the endolymph of the cochlear duct.
7. Receptor cells of organ of Corti are stimulated, leading to the generation of impulses in the vestibulocochlear nerve; these pass into the brain (pons).

11.18 Identify the common hearing disorders.

Tinnitus is a ringing sensation in the ears, caused by abnormal stimulation of either the inner ear or the vestibulocochlear nerve. *Deafness*, the lack of the sense of hearing, is of two types. *Conduction deafness* is caused by a defect of the external or middle ear, which inhibits sound transmission. An example would be wax accumulation (*impacted cerumen*) in the external auditory canal, preventing sound waves from reaching the tympanic membrane. *Nerve deafness* is caused by defects of either the inner ear or the vestibulocochlear nerve. *Otitis media* is an acute infection of the middle ear.

11.19 Explain how changes in body motion (that involve the head) are monitored by hair-cell receptors within the *vestibular apparatus* (the three semicircular canals, the utricle, and the saccule).

Whenever the head is moved—or, more precisely, is *accelerated*—in a certain direction, the hair cells of the vestibular apparatus move with the head. However, because of inertia, the endolymph within the vestibular apparatus tends to keep its original position in space; thus, it pushes in the opposite direction against the hair-cell receptors, stimulating them. The information generated by the receptors, in the form of action potentials, is transmitted into the CNS, where it helps regulate postural reflexes and equilibrium.

Receptors in the roughly spherical utricle and saccule detect linear acceleration in any given direction. Receptors in the semicircular canals detect rotational acceleration—also in any direction, since the semicircular canals are disposed in perpendicular planes.

Review Questions

Multiple Choice

1. The structure that is directly in contact with the tympanic membrane is the: (*a*) stapes, (*b*) incus, (*c*) malleus, (*d*) semicircular canal.

2. In the central region of the retina there is a yellowish spot, the *macula lutea*, with a depression in its center that produces the sharpest vision. This depressed area is called the: (*a*) optic disc, (*b*) rod and cone receptor site, (*c*) fovea centralis, (*d*) rhodopsin.

3. The modality of taste which is sensed over the tip of the tongue is: (*a*) sweet, (*b*) sour, (*c*) salty, (*d*) bitter.

4. Contraction of muscles of the ciliary body causes: (*a*) dilation of the pupil, (*b*) constriction of the pupil, (*c*) the lens to become more round and convex, (*d*) the lens to become more flat.

5. Which membrane separates the external auditory meatus from the middle ear? (*a*) auditory, (*b*) tympanic, (*c*) vestibular, (*d*) auricular.

6. Aqueous humor produced by the ciliary body is secreted into the posterior chamber and enters the anterior chamber through the: (*a*) canal of Schlemm, (*b*) vitreous body, (*c*) suspensory ligaments, (*d*) pupil.

7. Loss of equilibrium is referred to as: (*a*) vertigo, (*b*) vestibular nystagmus, (*c*) Ménière's disease, (*d*) tinnitus.

8. Advance protection against nerve damage within the cochlea is provided by the: (*a*) tympanic membrane, (*b*) oval window, (*c*) bony labyrinth, (*d*) stapedius muscle.

9. When the eyeball is too long and an image is focused in front of the retina, the condition is termed: (*a*) presbyopia, (*b*) hyperopia, (*c*) myopia, (*d*) astigmatism.

10. Which of the following is lateral to the eye? (*a*) lacrimal gland, (*b*) lacrimal sac, (*c*) cornea, (*d*) lacrimal caruncle.

11. The automatic adjustment of lens curvature by ciliary muscle contraction is known as: (*a*) adaptation, (*b*) accommodation, (*c*) convergence, (*d*) all the above.

12. The hair cells of the organ of Corti are supported by the: (*a*) basilar membrane, (*b*) vestibule, (*c*) tectorial membrane, (*d*) utricle.

13. The structure that drains the aqueous humor from the eye is the: (*a*) tear duct, (*b*) tarsal duct, (*c*) nasolacrimal duct, (*d*) canal of Schlemm.

14. Which of the following does not refract light? (*a*) pupil, (*b*) lens, (*c*) cornea, (*d*) vitreous humor.

15. Which structure belongs to the vestibular apparatus? (*a*) saccule, (*b*) semicircular canal, (*c*) utricle, (*d*) all the above.

16. Immobilization of the stapes results in: (*a*) bone deafness, (*b*) conduction deafness, (*c*) sensory deafness, (*d*) nerve deafness.

17. The funnellike outer portion of the ear is referred to as the: (*a*) external auditory meatus, (*b*) pinna, (*c*) auditory ossicle, (*d*) external apparatus.

True/False

1. When a person is emotionally upset, the tear glands are likely to secrete excessive fluids. This response involves motor impulses carried to the lacrimal glands on parasympathetic nerve fibers.

2. The motor units of the extrinsic eye muscles contain the largest number of muscle fibers of any muscles in the body. Because of this, the eyes can be moved with great precision.

3. Taste buds occur on the surface of the tongue, but are also found in smaller numbers in the roof of the mouth and the walls of the pharynx.

4. The major function of the middle ear is to couple the air in the outer ear to the liquid-filled chambers of the inner ear.

5. The pitch of a sound is directly related to the wave frequency.

6. The sensory cells of hearing are located on the basilar membrane.

7. The photoreceptive rods and cones are sensitive to color and to black and white, respectively.

8. With continued exposure to a taste stimulus comes a decrease in nerve transmission from the affected taste receptors.

9. Tears contain the enzyme *amylase*.

10. The lateral rectus muscles rotate the eyes away from the midline.

11. The anterior chamber is located between the cornea and the iris and is filled with vitreous humor.

12. In nearsightedness, the eyeball is too short for the lens and near objects are focused behind the retina.

13. The bone in the middle ear that is attached to the oval window is the malleus.

14. Vibrations of the oval window set up compressional waves in the perilymph.

15. Conduction deafness may be caused by wax accumulation in the external ear.

16. The saccule, semicircular canals, and cochlea constitute the vestibular apparatus.

17. Night vision is a function of the cone cells of the eye.

Chapter 12

Endocrine System

OBJECTIVE A *To distinguish between* endocrine *and* exocrine *glands.*

Survey ***Endocrine glands*** secrete *hormones* into the extracellular fluid around the secretory cells. The hormones then pass into capillaries and are transported in the blood. ***Exocrine glands*** produce secretions that are transported *through ducts* to the "outside" of the body (considered as a hollow cylinder).

12.1 Furnish some examples of exocrine glands.

Salivary glands (ducts transport saliva to the mouth); *sweat glands* (ducts transport sweat to the body surface); *sebaceous glands* (hair shafts act as ducts and sebum is dissipated to the surface of the skin). See Fig. 5-1.

12.2 What is a *mixed gland*?

A mixed gland has both endocrine and exocrine functions. For example, the pancreas, in its endocrine function, secretes insulin and glucagon into the blood. In its exocrine function, it secretes pancreatic juice (from acinar cells; Problem 4.10) into pancreatic ducts leading to the intestinal wall.

12.3 Give some configurations of endocrine gland cells.

The thyroid exhibits ***follicles*** [Fig. 12-1(*a*)]; the parathyroids and the adrenal cortex exhibit ***cords*** [Fig. 12-1(*b*)]; the pancreas shows ***clumps*** [Fig. 12-1(*c*)], the so-called *islets of Langerhans*.

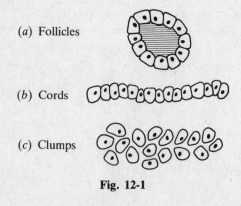

(*a*) Follicles

(*b*) Cords

(*c*) Clumps

Fig. 12-1

OBJECTIVE B *To know the characteristics of a* hormone.

Survey A *hormone* is a chemical that: (1) is synthesized by living cells, (2) is effective in minute quantities, (3) is secreted into the bloodstream and transported by the circulatory system, (4) acts on a target organ that is distant from the site of hormone synthesis, and (5) is a physiological regulator (speeds up or slows down some biological function).

12.4   Distinguish two chemical types of hormones.

>   ***Hormones synthesized from amino acids.***   Subtypes are: *amine hormones* (epinephrine, sero-
> tonin, norepinephrine, melatonin); *small peptides* (antidiuretic hormone, oxytocin, hypothalamic "re-
> leasing" hormones); *large peptides* (gastrin—17 amino acids,   glucagon—32 amino acids,   ACTH—39
> amino acids); and *proteins* (insulin, parathyroid hormone, growth hormone).

>   ***Steroids.***   These include: *glucocorticoids*, *estrogens* and *progesterone* (female sex hormones),
> *mineralocorticoids*, *androgens* (male sex hormones), and *prostaglandins*.

12.5   How do hormones bring about their observed effects?

>   Hormones are specific as to which cells they affect and the cellular changes they elicit. Steroid
> hormones, which are lipid-soluble, cross the cell membrane and interact with receptor molecules in the
> nucleus. Amino-acid hormones, which are insoluble in lipids and cannot passively cross the cell
> membrane, interact with receptor molecules on the surface of the cell.

12.6   Give a model for the negative-feedback mechanism that regulates the production of many
hormones.

>   Gland A secretes hormone A, for which organ X is the target. Gland B, in organ X, is stimulated by
> hormone A to release hormone B, which inhibits synthesis/secretion in its target, gland A.
>   The importance of such a feedback loop to body homeostasis is obvious.

OBJECTIVE C   *To name, locate, and list the secretions of the principal endocrine glands.*

Survey   See Fig. 1-6; there are normally four parathyroids embedded in the thyroid. The most
important of the hormones secreted are listed in Table 12-1 (page 182). Note the functional
divisions of the pituitary and the adrenal glands.

12.7   Which of the endocrine organs develop from two different germ layers?

>   *Pituitary* (anterior from ectoderm, posterior from neuroectoderm) and *adrenals* (cortex from
> mesoderm, medulla from neuroectoderm).

12.8   Which of the endocrine glands secrete steroid hormones?

>   From Problem 12.4 and Table 12-1: the *testes*, the *ovaries*, and the *adrenal glands* (cortices).

OBJECTIVE D   *To describe the relationship between the* hypothalamus *and the* pituitary, *and the
regulation of pituitary secretions.*

Survey   The hypothalamus and pituitary function together as an integrated unit. Oxytocin and ADH
are actually made in the hypothalamus, in neurosecretory neurons. The two hormones are
transported in the axons of the neurosecretory cells to the *posterior pituitary*, which releases
them. The *anterior pituitary* is also controlled by substances made in the hypothalamus,
which are carried to that lobe in small blood vessels. These substances, which stimulate or
inhibit the release of the anterior pituitary hormones, are called *releasing factors* or
inhibitory factors. (Once they have been chemically identified, they are called *releasing
hormones* or *inhibitory hormones*.)

Table 12-1

Gland		Hormones
Pituitary	Adenohypophysis (anterior pituitary)	Growth h. (GH) Thyroid-stimulating h. (TSH) Adrenocorticotropic h. (ACTH) Prolactin Follicle-stimulating h. (FSH) Luteinizing h. (LH)
	Neurohypophysis (posterior pituitary)	Antidiuretic h. (ADH) Oxytocin
Thyroid		Thyroxine (T_4) Triiodothyronine (T_3) Calcitonin
Parathyroids		Parathyroid h. (PTH)
Adrenal glands	Adrenal cortex	Cortisol Corticosterone } *glucocorticoids* Aldosterone Deoxycorticosterone } *mineralocorticoids*
	Adrenal medulla	Epinephrine Norepinephrine
Pancreas		Insulin Glucagon
Testes		Testosterone (an *androgen*)
Ovaries		Estradiol (an *estrogen*) Progesterone

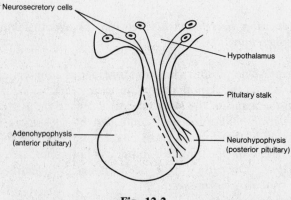

Fig. 12-2

12.9   Describe the two lobes of the pituitary and their embryological origins.

> Refer to Fig. 12-2. The anterior lobe formed from an invagination (*Rathke's pouch*) of the pharyngeal epithelium; thus the epithelioid nature of its cells. The posterior lobe formed from an outgrowth of the hypothalamus and contains axons from the neurosecretory cells of the hypothalamus, along with glia-like cells (*pituicytes*).

12.10 Secretory cells of the adenohypophysis are divided into three groups, according to their staining properties. Name them.

> *Acidophils* take up acidic dyes; they secrete growth hormone (GH) and prolactin. *Basophils* take up basic dyes; they secrete TSH, FSH, and LH. *Chromophobes* are stain-resistant; they secrete *corticotropes* (e.g., ACTH).

OBJECTIVE E   *To learn the specific effects of each of the pituitary hormones.*

Survey   Table 12-2 follows the order of Table 12-1.

Table 12-2

Pituitary Hormone	Target Tissue	Effects
GH (or *somatotropin*)	Skeleton, soft tissues	Accelerates rate of body growth, stimulates uptake of amino acids into cells and protein synthesis; carbohydrate and fat breakdown
TSH (or *thyrotropin*)	Thyroid gland	Growth and development of thyroid gland, uptake of iodine, synthesis and release of thyroid hormones
ACTH	Adrenal cortex	Growth and development of adrenal cortex, stimulates secretion of glucocorticoids (Table 12-1)
Prolactin	Mammary glands	Development of mammary glands, stimulates milk production
FSH	Ovaries and testes	*Female*: stimulates growth of ovarian follicles; *male*: stimulates spermatogenesis
LH	Ovaries and testes	*Female*: stimulates maturation of follicle cells, promotes ovulation, development of corpus luteum, stimulates corpus luteum to secrete estrogens and progesterone; *male*: stimulates interstitial cells to secrete testosterone
ADH (or *vasopressin*)	Kidney tubules	Facilitates water reabsorption in the distal convoluted tubules and collecting ducts
Oxytocin	Uterus and mammary glands	Stimulates contraction of the uterine muscles; stimulates the secretion of milk from the breast

12.11 What regulates the secretion of prolactin?

Prolactin increases progressively during pregnancy, under the influence of an increased production of *somatomammatropin* from the placenta. During lactation, stimulation of the nipple by sucking initiates a neuroendocrine reflex that results in increased prolactin secretion. This stimulates milk production for the next episode of nursing.

12.12 What are the mechanisms by which growth hormone stimulates the growth of body cells?

Protein synthesis is a major prerequisite for tissue growth because proteins make up a large portion of cellular structure and (as enzymes) regulate all cellular function. GH promotes protein synthesis by: (i) stimulating amino-acid uptake by cells; (ii) increasing synthesis of tRNA (Problem 3.14), the limiting factor in protein synthesis; (iii) increasing the number and aggregation of ribosomes (Table 3-1).

12.13 What are some of the factors that stimulate GH secretion?

Hypoglycemia: a 50% reduction in blood glucose will lead to a five-fold increase in GH secretion. *Muscular activity*: walking 30 minutes will cause an increase in GH levels. Increased *amino acids* in the blood. *Stress* (catecholamines).

12.14 What are the mechanisms that stimulate oxytocin and ADH release?

Oxytocin. Stretching of the uterus late in pregnancy initiates impulses to the hypothalamus, which signals the posterior pituitary to release oxytocin. (The oxytocin then stimulates strong uterine contraction.) *Sucking of a nipple* initiates impulses via the hypothalamus that signal the posterior pituitary to release oxytocin. (The oxytocin stimulates contractions in the myoepithelial cells surrounding the alveolus of the mammary gland, thus causing the "let-down" or secretion of milk.)

ADH. Both a *decrease in body water* (dehydration) and an *increase in plasma osmolarity* stimulate ADH secretion. (ADH causes increased water reabsorption in the kidney tubules, and so—negative feedback—water is returned to the body fluids and the plasma osmotic pressure is decreased to normal levels.)

12.15 Why is oxytocin sometimes administered to a woman after parturition?

Oxytocin causes the uterus to shrink, and the uterine contractions initiated by the hormone squeeze the blood vessels, thus minimizing the danger of hemorrhage.

OBJECTIVE F   *To describe the anatomy and physiology of the* thyroid gland.

Survey   An anterior view of the thyroid is presented in Fig. 12-3; a posterior, in Fig. 12-5. The process of secretion from the thyroid follicles (under the stimulation of TSH) is diagramed in Fig. 12-4.

1. *Iodide* is actively transported from plasma into thyroid follicle cells.

2. Iodide and *thyroglobulin* are secreted into the lumina.

3. Iodide is oxidized to iodine and attached to tyrosines in the thyroglobulin, forming mono- and diiodotyrosines (MIT, DIT). Coupling of MIT with DIT forms *triiodotyrosine* (T_3, or triiodothyronine); coupling of two DIT forms *tetraiodotyrosine* (T_4, or thyroxine).

4. Under the influence of TSH, the colloid is taken up by endocytosis (Problem 3.6) into the thyroid follicle cells.

5. T_4 and T_3 are removed from thyroglobulin and secreted.

6. T_4 and T_3 are transported in the blood, in association with plasma proteins: *thyroid-binding globulin* (TBG), *thyroxine-binding prealbumin* (TBPA), and *albumin*.

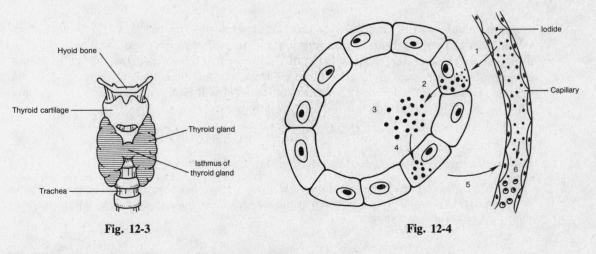

Fig. 12-3   Fig. 12-4

12.16 Describe the actions of thyroid hormones T_4 and T_3.

They (i) accelerate metabolic rate and oxygen consumption in all body tissues; (ii) increase body temperature; (iii) affect growth and development in early life; (iv) accelerate glucose absorption; and (v) enhance the effects of the sympathetic nervous system.

12.17 What are the effects of iodine deficiency on thyroid function?

When dietary intake of iodine is low (below 10 μg/day), T_4 and T_3 synthesis becomes inadequate, and secretion declines. As plasma levels of T_4 and T_3 fall, the negative feedback mechanism causes an increased release of TSH from the anterior pituitary. The excessive TSH causes the thyroid to hypertrophy, producing a goiter that may become very large. (Exposure to cold can also bring about increased secretion of TSH.)

OBJECTIVE G   *To identify the origin and actions of* parathyroid hormone (PTH).

Survey   ***Origin.*** PTH is released from the small, flattened *parathyroid glands*, which are embedded in the posterior surfaces of the lateral lobes of the thyroid gland (Fig. 12-5). ***Actions.*** PTH (1) stimulates the formation and activity of osteoclasts (Problem 6.11), which render bone

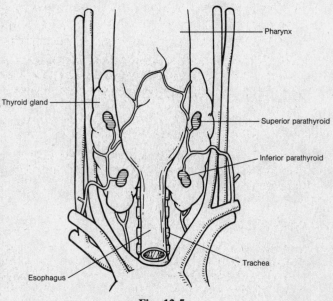

Fig. 12-5

minerals soluble and release calcium from the bones into the blood; (2) acts on kidney tubule cells to increase calcium reabsorption and therefore to decrease calcium loss in the urine; and (3) increases the synthesis of *1,25-dihydroxycholecalciferol*, which increases calcium absorption from the gastrointestinal tract. As all three lead to increased plasma calcium levels, we may be sure that secretion of PTH will be evoked by a drop in plasma Ca^{2+} (or Mg^{2+}) concentration.

12.18 What are the two cell types found in parathyroid glands?

The cells that secrete PTH are called *principal* or *chief* cells; they have a clear cytoplasm. *Oxyntic* ("acid-secreting") cells, which have granules in their cytoplasm, are dispersed throughout parathyroid glands; their function is not known.

12.19 Why is it important that plasma calcium levels be maintained?

Calcium participates in essentially all known biological functions. Among these are: transmission of nerve impulses, muscle contraction, cell division, coagulation of blood, release of neurotransmitters, secretory processes of endocrine and exocrine glands, and enzyme function.

OBJECTIVE H   *To learn the physiological actions of the hormones secreted from the* adrenal cortex *and* adrenal medulla.

Survey   The *adrenal* (*suprarenal*) *glands* are embedded in adipose tissue at the superior ends of the kidneys. Each adrenal gland is triangular in shape [Fig. 12-6(*a*)] and consists of an outer *cortex* and an inner *medulla*. The cortex is composed of three layers, or *zones*, as indicated in Fig. 12-6(*b*). The steroid hormones (Table 12-1) secreted by the cortical zones and the medulla act as follows.

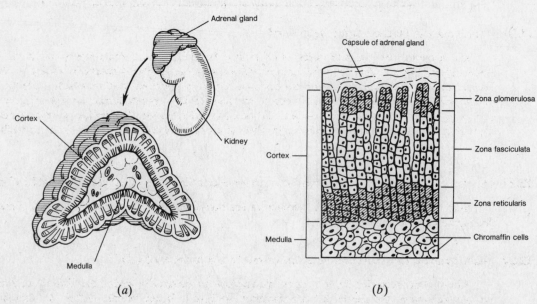

(*a*)   (*b*)

Fig. 12-6

Glucocorticoids (1) regulate carbohydrate and lipid metabolism, stimulate synthesis of glucose from noncarbohydrates (*gluconeogenesis*), increase blood glucose and liver glycogen storage, accelerate the breakdown of proteins; (2) in large doses inhibit inflammatory responses (capillaries fail to dilate, less edema occurs, decreased numbers of white blood cells migrate into the inflamed area); (3) promote vasoconstriction; (4) help the body to resist stress.

Mineralocorticoids regulate the concentration of extracellular electrolytes (cations), especially sodium and potassium.

Amine hormones, from the medulla, are compared as to effects in Table 12-3.

Table 12-3

Epinephrine	Norepinephrine
Elevates blood pressure through increased cardiac output and peripheral vasoconstriction	Elevates blood pressure through generalized vasoconstriction
Accelerates respiratory rate and dilates respiratory passageways	Similar effect, but less marked
Increases efficiency of muscular contraction	Similar effect, but less marked
Increases rate of glycogen breakdown into glucose, so level of blood glucose rises	Similar effect, but less marked
Increases conversion of fats to fatty acids, so level of blood fatty acids rises	Similar effect, but less marked
Increases release of ACTH and TSH from the adenohypophysis	No effect

12.20 What controls glucocorticoid secretion?

The secretion of glucocorticoids is controlled by ACTH from the anterior pituitary. This is evidenced by the fact that hypophysectomy (excision of the pituitary) results in atropy of the zona fasciculata and zona reticularis, and cessation of cortisol production. Factors which influence ACTH release are: (i) negative feedback of cortisol on the pituitary, hypothalamus, or higher brain centers (i.e., high levels of cortisol inhibit ACTH release); (ii) various forms of stress (fever, vigorous exercise, hypoglycemia, injury) increase release; (iii) there are episodic bursts of ACTH release throughout the day.

12.21 Do the adrenal glands secrete any steroid hormones besides those listed in Table 12-1?

Yes: the cortex releases small amounts of sex hormones. It is thought that these supplement the productions of the gonads.

12.22 What controls the secretion of aldosterone (a mineralocorticoid)?

Aldosterone secretion by the zona glomerulosa is principally under the control of the *renin-angiotensin system*, the plasma K^+ concentration, and, to a limited extent, ACTH. A "flowchart" for aldosterone production is given in Fig. 12-7.

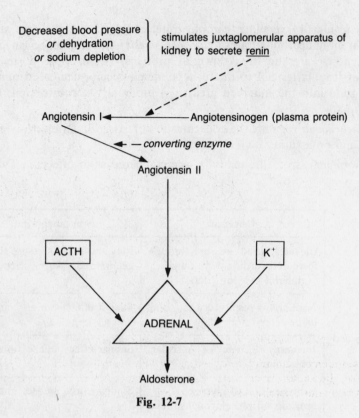

Fig. 12-7

12.23 What factors stimulate the adrenal medulla to secrete epinephrine (adrenaline) and norepinephrine (noradrenaline)?

Adrenal medullary secretion is prompted by sympathetic impulses during stress and in emergency situations in which the body is prepared for "fight or flight."

OBJECTIVE 1   *To identify the* pancreatic hormones *and to explain their physiological effects.*

Survey   Figure 12-8 pictures the pancreas, the endocrine portion of which consists of scattered clusters of cells called *islets of Langerhans*. In a typical islet:

Glucagon is secreted by *alpha cells*, which constitute 20% of the islet, are located mainly on the periphery, and are innervated by cholinergic fibers.

Insulin (from "islet") is secreted by *beta cells*, which constitute 75% of the islet, are located mainly at the center, and are innervated by adrenergic fibers.

Somatostatin is secreted by *delta cells*, which constitute 5% of the islet and are scattered through it.

Physiological effects. *Insulin* stimulates movement of glucose across the plasma membrane; stimulates glycolysis; and lowers blood glucose levels. *Glucagon* stimulates glycogenolysis and maintains blood glucose levels during fasting or starvation. *Somatostatin* stimulates incorporation of sulfur into cartilage; stimulates collagen formation; and has insulinlike properties.

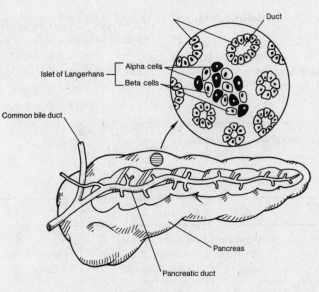

Fig. 12-8

12.24 What are causes of *diabetes mellitus* (insulin deficiency)?

Predisposition to diabetes is inherited; moreover, over 20% of the relatives of diabetic patients have an abnormal glucose tolerance curve. Other possible causes are: environmental chemicals, infectious agents (mumps virus), autoimmune events, nutrition, psychological stress.

12.25 What are the two types of diabetes mellitus?

Insulin-dependent, or *juvenile*, diabetes requires insulin injections; is often severe and complicated by *ketoacidosis* (acetone breath); is usually contracted in youth, but may occur at any age. *Noninsulin-dependent*, or *maturity-onset*, diabetes does not require insulin injections; is mild, ketoacidosis being rare; is often associated with obesity and usually improves with weight loss; may be treated with oral hypoglycemic drugs to stimulate insulin release from beta cells.

12.26 Give the symptoms of diabetes mellitus.

(1) *Glycosuria*, or glucose in the urine; (2) *polyuria*, or increased urine volume; (3) *polydypsia*, or increased drinking; (4) *hyperglycemia*, or high blood glucose levels; (5) weakness; (6) loss of weight; (7) ketoacidosis; (8) vascular abnormalities.

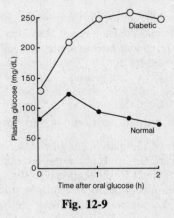

Fig. 12-9

12.27 What test is used to determine whether or not a patient has diabetes mellitus?

The *oral glucose tolerance curve* (Fig. 12-9) is useful in testing for diabetes. A glucose dose (2 g glucose/kg body weight) is given to the fasting patient. Diabetes is present if, just before the dose, the glucose level exceeded 115 mg/dL blood, or if the levels 1, 1.5, and 2 h after the dose exceed 185, 165, and 140 mg/dL blood, respectively.

OBJECTIVE J *To scan the less important sites of endocrine function; i.e.*, pineal gland, thymus, gastric *and* duodenal mucosa, *and* placenta.

Survey See Table 12-4.

Table 12-4

Organ	Location	Endocrine Function
Thymus	Upper mediastinum, in front of the aorta and behind the manubrium of the sternum	Secretes the hormone *thymosin*, which stimulates T-lymphocyte activity
Pineal gland	Small cone-shaped gland located in the roof of the third ventricle, near the corpora quadrigemina	Secretes the hormone *melatonin*, which affects the secretion of gonadotropins and ACTH from the anterior pituitary
Gastric mucosa	Epithelial cells lining the stomach; *G cells* are located in the lateral walls of the glands in the central portion	G cells secrete *gastrin*, which stimulates gastric juice secretion and gastric motility
Duodenal mucosa	Epithelial cells in the upper part of the small intestine	Secretes *secretin*, which stimulates secretion of pancreatic juice rich in bicarbonate, and *cholecystokinin*, which stimulates secretion of pancreatic juice rich in enzymes
Placenta	In the pregnant uterus	Secretes *human chorionic gonadotropin* (HCG), *human chorionic somatomammatropin* (HCS), estrogens, and progesterone

12.28 What are the endocrine functions of the pineal gland?

The pineal gland is the major source of plasma melatonin in humans. Melatonin is synthesized from serotonin (5-hydroxytryptamine). At the present time, the exact role of melatonin in humans is not known; however, clinical observations have revealed that precocious puberty may occur in males whose pineal gland has been destroyed by tumors. Therefore, it has been suggested that the pineal exerts an antigonadotropic effect. (In birds and rodents, melatonin has been implicated in the regulation of reproductive functions in relationship to diurnal light cycles.)

12.29 What are the functions of human chorionic gonadotropin (HCG)?

The trophoblastic tissue of the placenta begins to secrete HCG shortly after implantation of the fertilized ovum. Secretion increases up to about the seventh week of pregnancy and then declines to a comparatively low value at about the sixteenth week. The major function of HCG is to prevent the regression of the corpus luteum, with the result that menstruation is prevented during pregnancy. In the male fetus, HCG stimulates the production of testosterone, which is essential to male sexual differentiation and development.

Key Clinical Terms

Pituitary Disorders

Panhypopituitarism Multitropic pituitary hormone deficiency. *Symptoms* (depending on which of the tropic hormones are deficient; these are usually GH and one or more others): adult gonads may stop functioning, with amenorrhea (lack of menstruation) or aspermia (no sperm production); loss of pubic and axillary hair. *Treatment* (depending on which hormones are deficient): injections of thyroxine, cortisone, GH, or the gonadal hormones.

Dwarfism Decreased GH secretion before normal height is reached. *Symptoms*: small body, but the proportions are normal; mild obesity, with lack of appetite; skin tender and thin; normal intelligence. *Treatment*: GH injections.

Gigantism Excess GH before closure of the epiphyseal cartilage. *Symptoms*: pathological acceleration of growth; if a tumor is involved, the patient may have impaired vision. *Treatment*: surgical removal of pituitary (tumor).

Acromegaly Excess GH after closure of the epiphyseal cartilage. *Symptoms*: large jaw; nose thickened and puffy; large ears; large tongue; large head; increased BMR (basal metabolic rate); loss in visual fields. *Treatment*: surgical removal of pituitary (tumor); irradiation, radioisotope implantation.

Diabetes insipidus Lack of antidiuretic hormone (ADH). *Symptoms*: polyuria (5 to 25 liters of urine per day); polydypsia (increased drinking); severe electrolyte imbalance. *Treatment*: ADH injections.

Thyroid Disorders

Graves' disease, or **thyrotoxicosis** Hyperthyroid secretion (excessive secretion by thyroid). *Symptoms*: loss of weight; rapid pulse; warm, moist skin; increased appetite; increased BMR; tremor; goiter; exophthalmos (bulging of eyes); muscular weakness. *Treatment*: surgical removal of a portion of thyroid glands; radioiodine; antithyroid drugs.

Myxedema Hypothyroid secretion (insufficient secretion by thyroid) in adults. *Symptoms*: weight gain; slow pulse; dry, brittle hair; decreased BMR; lack of energy; sensation of coldness; diminished perspiration; weakness. *Treatment*: thyroid hormone (T_4 and T_3) administration.

Cretinism Hypothyroid secretion in infants and children. *Symptoms*: stunted growth; thickened facial features; large, protruding tongue; abnormal bone growth; mental retardation; decreased BMR; general lethargy. *Treatment*: thyroid hormone (T_4 and T_3) administration.

Adrenal Disorders

Cushing's disease (syndrome) Excess glucocorticoids (cortisol), with mineralocorticoid levels usually normal. *Symptoms*: thin arms, legs, and skin; red cheeks; poor wound healing; moon face; high blood pressure; decreased antibody formation; hyperglycemia (excessive blood sugar); muscle weakness. *Treatment*: surgical removal of portions of pituitary or adrenal glands; irradiation; hormone replacement therapy.

Addison's disease Insufficient glucocorticoids and mineralocorticoids. *Symptoms*: loss of electrolytes and body fluids; low blood pressure; hypoglycemia (insufficient blood sugar); weakness; loss of appetite; inability to withstand stress; increased pigmentation. *Treatment*: administration of glucocorticoids and mineralocorticoids.

Adrenogenital syndrome Excessive secretion of androgens from the adrenal cortex. *Symptoms*: (in young children) premature puberty and enlarged genitalia; (in mature women) development of masculine traits. *Treatment*: surgical removal, if tumor is causing hypersecretion.

Pheochromocytoma Tumor of the chromaffin cells of the adrenal medulla, with hypersecretion of epinephrine and norepinephrine. *Symptoms*: high blood pressure, increased BMR; hyperglycemia; nervousness; sweating. *Treatment*: surgical removal of tumor.

Review Questions

Multiple Choice

1. A hormone is best described as: (*a*) an internal secretion that is transported through ducts; (*b*) an internal secretion with many effects; (*c*) a chemical secreted by a gland; (*d*) a chemical produced in one part of the body, which is transported in the blood and acts in a regulatory capacity at another place in the body.

2. Which of the following is *not* a steroid hormone? (*a*) an estrogen, (*b*) cortisone, (*c*) adrenaline, (*d*) testosterone, (*e*) none of the above.

3. The portion of the pituitary that arises from the roof of the primitive oral cavity is the: (*a*) adenohypophysis, (*b*) pars nervosa, (*c*) neurohypophysis, (*d*) infundibulum, (*e*) hypothalamus.

4. The endocrine gland that is formed from two different germ layers is the: (*a*) ovary, (*b*) thyroid, (*c*) pituitary, (*d*) testis, (*e*) adrenal.

5. The alpha cells of the pancreas secrete: (*a*) insulin, (*b*) enzymes, (*c*) glucagon, (*d*) none of these.

6. The group of adrenocortical hormones concerned with electrolyte balance is: (*a*) the glucocorticoids, (*b*) the mineralocorticoids, (*c*) the androgens, (*d*) epinephrine and norepinephrine.

7. The adrenal medulla secretes: (*a*) cortisone, (*b*) cortisol, (*c*) epinephrine, (*d*) acetylcholine.

8. What hormone stimulates testosterone secretion? (*a*) LH, (*b*) progesterone, (*c*) FSH, (*d*) ACTH.

9. The release of ACTH from the pituitary stimulates the release of: (*a*) aldosterone from the adrenal medulla, (*b*) cortisol from the adrenal cortex, (*c*) epinephrine from the adrenal medulla, (*d*) renin from the kidney.

10. Oxytocin and ADH are stored in the: (*a*) adenohypophysis, (*b*) anterior pituitary, (*c*) posterior pituitary, (*d*) kidneys.

11. Hypersecretion of growth hormone after closure of the epiphyseal cartilage causes: (*a*) acromegaly, (*b*) myxedema, (*c*) Addison's disease, (*d*) gigantism, (*e*) none of the above.

12. When there is a marked deficiency of hormone secretion by the thyroid gland in a young child: (*a*) acromegaly soon results; (*b*) mental and physical growth are hindered; (*c*) the child's eyes gradually bulge; (*d*) there will be a high basal metabolic rate; (*e*) all the above.

13. Which of the following is *not* a pituitary hormone? (*a*) growth hormone, (*b*) luteinizing hormone, (*c*) prolactin, (*d*) testosterone, (*e*) oxytocin.

14. Calcium levels in the blood are increased by: (*a*) calcitonin, (*b*) heparin, (*c*) dicumarol, (*d*) parathyroid hormone, (*e*) vitamin E.

15. Milk ejection from the mammary glands is assisted by: (*a*) oxygen, (*b*) prolactin, (*c*) oxytocin, (*d*) prostate hormone, (*e*) ADH.

16. Releasing factors are synthesized in the: (*a*) hypothalamus, (*b*) hypophysis, (*c*) pancreas, (*d*) posterior pituitary, (*e*) ovary.

17. Which of the following hormones is secreted through the pars nervosa of the pituitary? (*a*) thyroid-stimulating hormone, (*b*) prolactin, (*c*) oxytocin, (*d*) luteinizing hormone, (*e*) growth hormone.

18. The largest endocrine gland is the: (*a*) thyroid, (*b*) pituitary, (*c*) adrenal, (*d*) pancreas, (*e*) gonad.

19. Which of the following are not influenced by parathyroid hormone? (*a*) kidneys, (*b*) bones, (*c*) intestines, (*d*) muscles, (*e*) none of the above.

20. The hormone whose action resembles stimulation through the sympathetic nervous system is: (*a*) epinephrine, (*b*) cortisol, (*c*) androgens, (*d*) aldosterone, (*e*) melatonin.

21. Secretion of which hormone would be *increased* in an iodine-deficiency goiter? (*a*) TSH, (*b*) thyroxine, (*c*) T_3, (*d*) all these.

22. The hormone that is released from the anterior pituitary and stimulates the development of the seminiferous tubules of the testes is called: (*a*) prolactin, (*b*) adrenocorticotropic hormone, (*c*) follicle-stimulating hormone, (*d*) luteinizing hormone.

23. Which of the following statements about glucocorticoids is/are true? (*a*) the major glucocorticoid in humans is cortisol; (*b*) they are secreted by the zona fasciculata of the adrenal cortex; (*c*) secretion of these hormones is decreased in Addison's disease; (*d*) all the above.

24. The basal metabolism rate can reflect dysfunction of the: (*a*) pituitary, (*b*) parathyroid, (*c*) adrenal, (*d*) thyroid, (*e*) pancreas.

25. What is the proper sequence of adrenal cortex zones, from the outside in? (*a*) zona glomerulosa, zona fasciculata, zona reticularis; (*b*) zona glomerulosa, zona reticularis, zona fasciculata; (*c*) zona reticularis, zona fasciculata, zona glomerulosa; (*d*) zona fasciculata, zona reticularis, zona glomerulosa.

26. Target cells for steroid hormones contain receptor proteins in their: (*a*) cell membrane, (*b*) nucleus, (*c*) nuclear membrane, (*d*) nucleoplasm.

27. A symptom of diabetes mellitus is: (*a*) glyconemia, (*b*) polydypsia, (*c*) gain in weight, (*d*) hypoglycemia, (*e*) all the above.

28. Which of the following is an endocrine-exocrine gland? (*a*) adrenal, (*b*) pituitary, (*c*) thyroid, (*d*) pancreas.

29. A hormone may, through negative feedback, shut off the secretion of an anterior pituitary hormone by: (*a*) stimulating the release of a (hypothalamic) releasing factor, (*b*) inhibiting the release of a (hypothalamic) inhibiting factor, (*c*) inhibiting the release of a (hypothalamic) releasing factor, (*d*) all the above.

30. Stimulation of the mother's nipples by the baby's nursing initiates sensory impulses which pass into the central nervous system and eventually reach the hypothalamus. These impulses result in the: (*a*) synthesis and release of prolactin from the posterior pituitary, (*b*) release of lactogenic hormone from the anterior pituitary, (*c*) release of oxytocin from the posterior pituitary, (*d*) release of prolactin-inhibiting factor.

31. Choose the true statement(s) about a person with type I (insulin-dependent) diabetes mellitus: (*a*) there is little or no insulin secretion; (*b*) dietary treatment may suffice; (*c*) there is hyperglycemia; (*d*) ketoacidosis and dehydration may develop; (*e*) all the above.

32. A person with untreated iodine-deficiency goiter has a high: (*a*) TSH secretion, (*b*) thyroxine secretion, (*c*) temperature, (*d*) metabolic rate, (*e*) all these.

33. Persistent headaches, visual disturbances, large ears, large tongue, apathy, and diabetic symptoms are characteristic of: (*a*) Cushing's syndrome, (*b*) acromegaly, (*c*) Addison's disease, (*d*) Rhees's syndrome.

34. The acidophils of the anterior pituitary secrete: (*a*) prolactin and GH, (*b*) LH and FSH, (*c*) TSH, (*d*) MSH.

True/False

1. Inhibition or stimulation of transport across the cell membrane is one of the major hormonal actions.

2. The major mode of action of steroid hormones is to increase protein synthesis in specific target-organ cells.

3. Two hormones never exist in the blood at the same time.

4. The adrenal medulla makes up 50% of the adrenal gland.

5. An enlarged thyroid gland is referred to as a goiter.

6. The cells of a parathyroid gland respond directly to the calcium concentration in the blood.

7. Aldosterone, secreted from the posterior pituitary, is involved in the regulation of sodium and potassium balance.

8. The posterior pituitary is not composed of true glandular tissue.

9. All hormones are steroids, amino-acid derivatives, peptides, or proteins.

Matching

1.	Dwarfism	(*a*)	Hyposecretion of thyroxin
2.	Graves' disease	(*b*)	Hypersecretion of thyroxin
3.	Precocious puberty in the male	(*c*)	Hyposecretion of somatotropin
4.	Cretinism	(*d*)	Hypersecretion of somatotropin
5.	Tetany	(*e*)	Hyposecretion of parathyroid hormone
6.	Diabetes mellitus	(*f*)	Hyposecretion of insulin
7.	Diabetes insipidus	(*g*)	Hyposecretion of insulin
8.	Acromegaly	(*h*)	Hypersecretion of testosterone

Clinical Cases

1. A 40-year-old man complained of polyuria, nocturia, and polydypsia. He was found to produce 7 to 10 liters of urine per day. Blood sugar level was 97 mg% (or 97 mg/dL serum) and PBI (protein-bound iodine) was 6 μg/dL serum. The patient enjoyed his Zoology 565 course, but didn't get his paper turned in until the last week of the semester.

    (a) What is the medical diagnosis?

    (b) What treatment should be used?

2. A 50-year-old man came to his physician complaining of dry skin and hair, constipation, intolerance to cold, and diminished vigor. In addition, he had gained weight and had a puffy look. His pulse rate was 55 beats/min and his blood pressure was 110/70 mmHg.

    (a) What is his condition?

    (b) What treatment should be used?

Chapter 13

Cardiovascular System: Blood Composition

OBJECTIVE A *To specify the composition of the blood.*

Survey

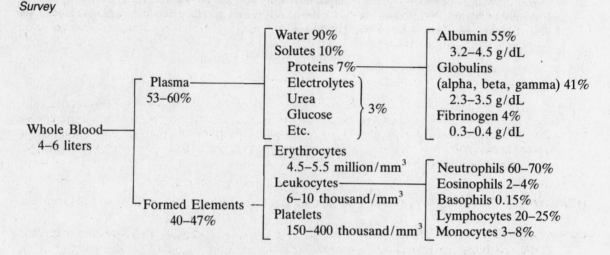

13.1 What is the blood volume of an average person?

The volume of whole blood is about 4–5 liters in women and 5–6 liters in men. To see that the average is indeed about 5 liters, recall that blood weight is about 7% of total body weight, and that 150 lb is a reasonable average body weight. The average person will thus have

$$(0.07)(150 \text{ lb}) = 10.5 \text{ lb}$$

of blood. Now, 1 lb of blood occupies approximately 1 pint or 500 mL; therefore, the average blood volume will be

$$(10.5 \text{ lb})(500 \text{ mL/lb}) = 5250 \text{ mL} = 5.25 \text{ L}$$

13.2 What is the *hematocrit*, and how is it measured?

The hematocrit is the percentage of total blood volume occupied by formed elements (for all practical purposes, red blood cells). It ranges from 40% to 54% in men, and from 38% to 47% in women. Measurement is by centrifuging a blood sample in a capillary tube: if the tube were 100 mm long and if the packed RBCs occupied the distal 45 mm, the hematocrit would be 45%.

13.3 What condition would cause a change in the hematocrit?

Anemia is one such.

13.4 Describe the genesis and function of platelets (cf. Problem 6.7).

Platelets, or thrombocytes, are small cellular fragments that originate in the bone marrow from a giant cell known as a *megakaryocyte*. The megakaryocytes form platelets by pinching off bits of cytoplasm and extruding them into the blood. Platelets contain several clotting factors, Ca^{2+}, ADP, serotonin, and various enzymes; they play an important role in *hemostasis* (arrest of bleeding).

OBJECTIVE B   *To learn the major components of the plasma.*

Survey   Plasma consists of (1) water; (2) proteins (albumins, globulins, fibrinogens); (3) electrolytes (Na^+, K^+, Ca^{2+}, Mg^{2+}, Cl^-, $HCO—_3^-$, HPO_4^{2-}, SO_4^{2-}); (4) nutrients (glucose, amino acids, lipids, cholesterol, vitamins, trace elements); (5) hormones; (6) dissolved gases (CO_2, O_2, N_2); (7) waste products (urea, uric acid, creatinine, bilirubin).

13.5   What are the characteristics and functions of the *albumins*?

Albumins ($MW = 69\,000$) are the smallest and most abundant proteins in the plasma. They are produced in the liver and play an important role in maintaining the osmotic pressure of the blood. They also increase the solubility of some substances and transport free fatty acids, ions, amino acids, hormones, drugs, and enzymes.

13.6   What are the major functions and types of *globulins*?

The globulin fractions of the blood protein contain numerous substances with a variety of functions, such as transport, enzymatic action, clotting, and immunity. They may be separated by electrophoresis into four types: *alpha 1* (e.g., fetoprotein, antitrypsin, lipoproteins); *alpha 2* (e.g., antithrombin, cholinesterase); *beta* (e.g., transferrin, plasminogen, prothrombin); *gamma* (e.g., IgG, IgA, IgM, IgD, IgE; Ig = *Immunoglobulin*).

OBJECTIVE C   *To detail the process of* erythropoiesis *and the formation and action of* hemoglobin.

Survey   *Erythrocytes* (RBCs) are flexible, biconcave, unnucleated cells [Fig. 4-4(*d*)] manufactured in the red bone marrow (Problem 6.7). The main constituent (about 1/3 by weight) of the RBCs is *hemoglobin*, and the essential function of these cells is to carry oxygen, reversibly trapped by the hemoglobin, to all parts of the body.

13.7   Show the sequence of cellular differentiation in erythropoiesis.

Hemocytoblast → *Proerythroblast* → *Erythroblast* → *Normoblast* → *Reticulocyte* → *Erythrocyte*
(loss of nucleus)

13.8   What substances are required for erythrocyte production?

Substance	Use
protein	cell membrane structure
lipid	cell membrane structure
amino acids	globin portion of hemoglobin
iron	incorporated into hemoglobin
vitamin B_{12}	DNA synthesis
folic acid	DNA synthesis
copper	catalyst for hemoglobin synthesis
cobalt	aids in hemoglobin synthesis

13.9   If about 2.5 million RBCs are formed each second in the bone marrow and if RBCs are destroyed at the same rate in the liver and spleen, calculate the average lifetime, *T*, of an RBC.

The concentration of RBCs is roughly 5 million per mm^3, and (Problem 13.1) the volume of blood is about $5\,L = 5$ million mm^3. Thus, the standing population of RBCs is approximately

$$(5 \times 10^6/mm^3)(5 \times 10^6\ mm^3) = 2.5 \times 10^{13}$$

Under homeostasis, this population will "turn over" once during the lifetime of the average RBC; that is,

$$(2.5 \times 10^6) T = 2.5 \times 10^{13}$$

or $T = 10^7$ seconds ≈ 120 days.

13.10 What factors cause fluctuations in erythrocyte number?

Any condition that *decreases oxygen* in the body tissues will, by a negative feedback mechanism, increase erythropoiesis; e.g., high altitudes (30% greater hematocrit at 14,000 ft than at sea level), muscle exercise, anemia, *ischemia* (Problem 14.15) or heart failure. *Temperature*: increased body temperature increases the number of RBCs. *Sex*: higher hematocrit in males after puberty (effect of testosterone). *Age*: higher hematocrit in infants. *Time of day*: highest RBC count in early evening.

13.11 Describe the feedback mechanism mentioned in Problem 13.10.

In response to low oxygen concentration, the kidneys release an *erythropoietic factor* that stimulates the conversion of a plasma protein (*liver protein*) to *erythropoietin*. Erythropoietin travels in the blood to the bone marrow, where it stimulates erythropoiesis. The increased number of erythrocytes transport more oxygen to the tissues.

13.12 What is the chemical makeup of hemoglobin?

Hemoglobin (Hb) consists of *globin* (four polypeptide chains; Fig. 13-1) and *heme* (four Fe^{2+}-porphyrin molecules; Fig. 13-2). Each heme molecule can reversibly bind an O_2 molecule.

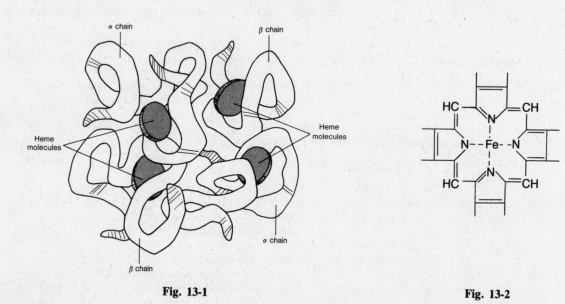

Fig. 13-1   **Fig. 13-2**

13.13 Can hemoglobin bind other gas molecules besides O_2?

Yes: carbon dioxide (CO_2) and carbon monoxide (CO) also bind to hemoglobin. Hemoglobin, when saturated with O_2, is called *oxyhemoglobin*; it is cherry red in color. When oxyhemoglobin loses its oxygen, the color reverts to purple-blue. Hemoglobin in combination with Co_2 is called *carbaminohemoglobin*. O_2 and CO_2 have distinct carrying sites on the Hb molecule.

Carbon monoxide combined with Hb is called *carboxyhemoglobin*. CO binds to a heme, and has 200 times the affinity for the heme that O_2 has. It is this exclusion of oxygen that makes carbon monoxide so dangerous a gas.

13.14 When disintegrating erythrocytes are phagocytosed in the spleen and liver (Problem 13.9), the hemoglobin molecule is broken down. Specify the catabolic events.

$$\text{Hemoglobin} \rightarrow \text{Heme} \; + \; \text{Globin}$$

$$Fe^{2+} \; + \; \text{Porphyrin} \qquad \text{Amino acids}$$

Porphyrin is changed from a ring structure (Fig. 13-2) to a straight-chain structure, called *biliverdin* ("green of bile"), which in turn is reduced to the straight-chain *bilirubin* ("red of bile"). Bilirubin, carried from the liver in the bile, may be excreted in the feces as *stercobilin* or in the urine as *urobilin*. Feces and urine owe their brown or yellowish color to these bilirubin products. When yellowish bilirubin accumulates in the blood to an abnormally high degree, it yellows the skin (*jaundice*). Causes of jaundice include liver disease, excess red cell destruction, and bile duct obstruction (feces will be gray).

OBJECTIVE D   *To describe the various types of* anemia.

Survey   See Table 13-1.

Table 13-1

Type	Cause	Symptoms	Treatment
Hemorrhagic	Blood loss	Shock	Transfusion
Aplastic	Bone marrow destruction by chemicals or radiation	If any, fatigue	Transfusion; removal of chemical or irradiator
Nutritional	Deficiency in folic acid, vitamin B_{12}, or iron	If any, fatigue	Folic acid, B_{12}, or iron administration
Hemolytic	Defective RBCs	If any, fatigue	*Various*

13.15 To which type does *pernicious anemia* belong?

Nutritional. The parietal cells of the stomach manufacture a substance (*intrinsic factor*) that is required for vitamin B_{12} to be absorbed later on in the small intestine. In the absence of intrinsic factor (owing to a stomach disorder), B_{12} is not absorbed and the result is pernicious anemia.

OBJECTIVE E   *To become familiar with the five types of* leukocytes (white blood cells).

Survey   Refer to Fig. 4-4(*d*) and Table 13-2.

Table 13-2

Type	Avg. No. per mm^3	Origin	Morphology	Functions
Neutrophils	5400	Bone marrow	Lobed nucleus, fine granules	Phagocytosis
Eosinophils	275	Bone marrow	Lobed nucleus, red or yellow granules	May phagocytize antigen-antibody complexes
Basophils	35	Bone marrow	Obscure nucleus, large purple granules	Release heparin, histamine, and serotonin
Lymphocytes	2750	Lymphoid tissues	Round nucleus, little cytoplasm	Phagocytosis
Monocytes	540	Lymphoid tissues	Kidney-shaped nucleus	Phagocytosis

13.16 Give examples of diseases that cause increases in the various leukocytes.

> *Neutrophils*: appendicitis, pneumonia, tonsillitis. *Eosinophils*: hay fever, asthma, parasitic infestations. *Basophils*: smallpox, nephrosis, myxedema. *Lymphocytes*: whooping cough, mumps, mononucleosis. *Monocytes*: tuberculosis, typhus.

13.17 Do leukocytes ever leave the circulatory system?

> Yes: leukocytes have the ability to "squeeze" through capillary walls (*diapedesis*) and move out into body tissues to fight infection.

13.18 How do leukocytes "know" where they are needed to combat infection?

> Infected tissue releases certain chemicals (e.g., *leukotaxine*) which locally increase the permeability of capillary walls. Circulating leukocytes are thereby attracted into the infected area—the process is termed *chemotaxis*.

OBJECTIVE F *To know the mechanisms of hemostasis (Problem 13.4).*

Survey The major mechanisms are: **constriction** of the blood vessels; **plugging** of the wound by aggregated platelets; **clotting** of the blood into a mass of *fibrin* which augments the plug in sealing the wound and providing a framework for repair.

13.19 List the chemicals or factors that are involved in the clotting process.

> The factors, nearly all of which are produced in the liver, are conventionally designated by roman numerals, as follows:

> I = fibrinogen
> II = prothrombin
> III = thromboplastin
> IV = calcium
> V = labile factor
> VII = SPCA (*serum prothrombin conversion accelerator*)
> VIII = AHF (*antihemophilic factor*)
> IX = PTC (*plasma thromboplastic component*)
> X = Stuart–Prower factor
> XI = PTA (*plasma thromboplastin antecedent*)
> XII = Hageman factor
> XIII = fibrin-stabilizing factor

> (Factor VI is no longer considered a separate entity.)

13.20 There are two systems that initiate clotting; describe them.

> Refer to Fig. 13-3. The *intrinsic system* is activated when blood is exposed to a foreign surface, such as collagen fibers (resulting from injury), or when blood is stored in a glass container. The *extrinsic system* is activated by tissue thromboplastin, which is released when vascular walls or other tissue is damaged.

13.21 Describe the actions of several *anticoagulants*.

> *Citrates* and *oxalates* tie up calcium, which is essential at several steps in the clotting process (see Fig. 13-3). *Heparin* prevents the activation of factor IX and interferes with thrombin action. *Dicoumarol* blocks the formation of prothrombin and factors VII, IX, and X, by interfering with vitamin K, which acts as a catalyst in the synthesis of these chemicals in the liver.

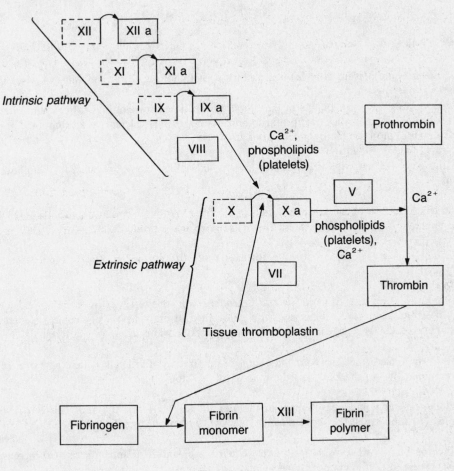

Fig. 13-3

13.22 Cite some disorders in which there is excessive bleeding.

Hemophilia is a hereditary lack, or altered biosynthesis, of a single clotting factor. Lack of VIII causes hemophilia A (*classical hemophilia*); lack of IX causes hemophilia B (*Christmas disease*). In *vitamin K deficiency*, clotting factors are not synthesized in the liver properly. In *thrombocytopenia*, the concentration of platelets is too low.

Review Questions

Multiple Choice

1. No granules are present in a: (*a*) neutrophil, (*b*) lymphocyte, (*c*) eosinophil, (*d*) basophil.

2. The following four components of the blood are necessary for clotting: (*a*) calcium, vitamin K, albumin, globulin; (*b*) calcium, heparin, prothrombin, fibrinogen; (*c*) calcium, prothrombin, fibrinogen, platelets; (*d*) calcium, prothrombin, platelets, vitamin A.

3. In the adult the majority of leukocytes are: (*a*) basophils, (*b*) eosinophils, (*c*) lymphocytes, (*d*) neutrophils.

4. The chief function of the serum albumin in the blood is to: (*a*) produce antibodies, (*b*) form fibrinogen, (*c*) maintain colloidal osmotic pressure, (*d*) remove waste products.

5. Calcium ions are necessary for the formation of: (*a*) fibrinogen, (*b*) thromboplastin, (*c*) thrombin, (*d*) prothrombin.

6. Among the blood tests ordered was a *differential count*; this test: (*a*) gives the number of red blood cells per cubic millimeter, (*b*) determines the percentage of erythrocytes per cubic millimeter, (*c*) gives the numbers of the various kinds of leukocytes per 100 white cells, (*d*) determines the platelet count.

7. The intrinsic factor necessary for the complete maturation of red cells is derived from: (*a*) bone marrow, (*b*) vitamin B_6, (*c*) the liver, (*d*) the mucosa of the stomach.

8. Bob has a hemoglobin measurement of 15 g per 100 mL (or 1 dL) of blood. This is: (*a*) within the normal limits; (*b*) subnormal; (*c*) above normal; (*d*) low, but satisfactory.

9. Which of the following would most likely confirm a diagnosis of appendicitis? (*a*) increase in monocytes, (*b*) increase in erythrocytes, (*c*) leukopenia, (*d*) increase in neutrophils.

10. Which concentration would be an indication of anemia? (*a*) thrombocytes—$300,000/\text{mm}^3$; (*b*) hematocrit—43%; (*c*) hemoglobin—17 g/dL; (*d*) erythrocytes—3.8 million/mm^3.

11. The leukocyte that is not involved in phagocytosis but does secrete the anticoagulant heparin is the: (*a*) basophil, (*b*) monocyte, (*c*) eosinophil, (*d*) lymphocyte.

12. Erythrocyte production: (*a*) is stimulated by high estrogen levels in the blood, (*b*) falls if the stomach loses the ability to produce gastric juice, (*c*) occurs in the spleen in normal adults at sea level, (*d*) is stimulated by a rise in the concentration of amino acids in the arterial blood.

13. Iron-deficiency anemia: (*a*) is more common in men than in women, (*b*) is characterized by increased numbers of leukocytes, (*c*) should generally be treated by intramuscular injections of iron, (*d*) is the form of anemia typically accompanying chronic blood loss from the body.

14. For blood clotting to occur normally: (*a*) heparin must be inactivated, (*b*) there must be a sufficient dietary intake of vitamin C, (*c*) tissue damage outside of the vessel must occur, (*d*) the liver must have adequate supply of vitamin K.

15. Which of the following is *not* a plasma protein? (*a*) albumin, (*b*) globulins, (*c*) fibrinogen, (*d*) platelets.

16. Which statement about erythrocytes is untrue? (*a*) the total number in the human body is approximately 25 trillion; (*b*) the shape of these cells is a biconcave disc; (*c*) their hemoglobin constitutes approximately one-sixth of the total blood weight; (*d*) an erythrocyte substance of great importance is citric acid.

17. Which statement about heparin is false? (*a*) it is found in various cells of the body; (*b*) it interferes with the ability of thrombin to split fibrinogen; (*c*) it is widely used in medicine as an anticoagulant; (*d*) it is a blood-thinner.

18. Which of the following does *not* stimulate erythropoietin production? (*a*) hemorrhage, (*b*) decreased heart pumping, (*c*) stress-induced release of epinephrine into the system, (*d*) decreased oxygen delivery to the tissues.

19. IgE immunoglobulins are involved in: (*a*) specific immunity against infectious bacteria, (*b*) certain allergic reactions, (*c*) protection of the urogenital area, (*d*) protection of the newborn because of placental transfer.

20. Insufficient vitamin B_{12} in the body may result in: (*a*) nutritional anemia, (*b*) pernicious anemia, (*c*) aplastic anemia, (*d*) an embolus.

21. Immunoglobulins of class IgA are produced by plasma cells and: (*a*) combat infectious microbes in the blood; (*b*) prevent certain allergic reactions; (*c*) protect the gastrointestinal, urogenital, and respiratory tracts; (*d*) none of these.

22. Leukocytes and platelets compose about what fraction of the blood volume? (*a*) 1%, (*b*) 45%, (*c*) 55%, (*d*) 11%.

23. The percent volume of whole blood occupied by packed red cells is referred to as the: (*a*) hematocrit, (*b*) formed elements, (*c*) erythrocytic fraction, (*d*) sedimentation index.

24. Production of red blood cells in a mature adult occurs in all the following areas except the: (*a*) sternum, (*b*) femora, (*c*) skull bones, (*d*) vertebrae.

25. The normal amount of hemoglobin per deciliter of blood is: (*a*) 60–63 g, (*b*) 13–16 g, (*c*) 1.3–1.6 g, (*d*) 4–6 g.

26. The structure of hemoglobin A is: (*a*) 2 α-polypeptide chains, 2 β-polypeptide chains, and 4 globin molecules; (*b*) 3 α-polypeptide chains, 1 β-polypeptide chain, and 4 heme molecules; (*c*) 4 α-polypeptide chains, 4 β-polypeptide chains, and 2 globin molecules; (*d*) 2 α-polypeptide chains, 2 β-polypeptide chains, and 4 heme molecules.

27. Plasma proteins constitute what percent of the plasma volume? (*a*) 17–19, (*b*) 7–9, (*c*) 25–27, (*d*) 52–55.

28. The blanket term for reactions that prevent or minimize loss of blood from the vessels if they are injured or ruptured is: (*a*) stabilization energy, (*b*) homeostasis, (*c*) syneresis, (*d*) hemostasis.

29. The most plentiful plasma protein is: (*a*) albumins, (*b*) globulins, (*c*) fibrinogen, (*d*) hemogen.

30. Hemostasis does not involve: (*a*) contraction of smooth muscles in blood vessel walls, (*b*) adherence of platelets to damaged tissue; (*c*) clot retraction, (*d*) increased renin-angiotensin activity.

31. Release of chemical substances from damaged tissues starts the: (*a*) intrinsic clotting mechanism, (*b*) phospholipid phase, (*c*) extrinsic clotting mechanism, (*d*) platelet plug.

32. The proper order of the events

      (1)   conversion of fibrinogen to fibrin

      (2)   clot shrinks and leaks serum

      (3)   thromboplastin production

      (4)   conversion of prothrombin to thrombin

is: (*a*) 3, 2, 1, 4; (*b*) 3, 4, 1, 2; (*c*) 3, 4, 2, 1; (*d*) 4, 1, 3, 2.

33. How many clotting factors are necessary for clot formation? (*a*) 8, (*b*) 10, (*c*) 12, (*d*) 17.

34. Which factor is not synthesized in hemophilia A? (*a*) VIII, (*b*) VII, (*c*) IX, (*d*) XIII.

35. Which blood constituent adheres to any rough surface? (*a*) thromboplastin, (*b*) fibrinogen, (*c*) leukocytes, (*d*) platelets.

Matching

1. Granules that show no dye preference	(*a*) Thrombus
2. Clot	(*b*) Phagocytosis
3. Granules that take up the red dye *eosin*	(*c*) Hematoma
4. Enzymatically decomposes fibrin	(*d*) Eosinophil
5. Cell with affinity for a basic dye	(*e*) Plasmin
6. Accumulation of blood	(*f*) Neutrophil
7. White blood cell	(*g*) Leukocyte
8. Selective defender against invaders	(*h*) Lymphocyte
9. Ingestion and digestion of particulate matter	(*i*) Basophil

Chapter 14

Cardiovascular System: The Heart

OBJECTIVE A *To describe the heart and locate it within the thorax.*

Survey The heart is a hollow, four-chambered, muscular organ that is specialized for pumping blood through the vessels of the body. It weighs about 255 grams in the female and 310 grams in the male; or approximately 5% of the body weight. As depicted in Fig. 14-1, the heart is located in the mediastinum (Problem 1.16), where it is enclosed in a loose-fitting serous sac called the *pericardial sac* or *parietal pericardium*.

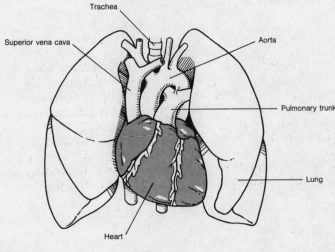

Fig. 14-1

14.1 Which portion of the heart is the *base*, and which the *apex*?

About two-thirds of the heart is to the left (the subject's left) of the midsagittal plane, with its *apex*, or cone-shaped end, pointing downward, in contact with the diaphragm. The *base* of the heart is the broad superior end, where the large vessels attach.

14.2 What is the *pluck*?

Ventilation of the lungs brings O_2 in contact with blood from the heart. The pumping action of the heart then circulates the oxygenated blood through the body and returns deoxygenated blood to the lungs for removal of CO_2. The vessels that connect the heart and lungs are called *pulmonary vessels*, and the *pluck* comprises heart, lungs, and pulmonary vessels.

14.3 *True or false*: The sole function of the pericardial sac is to sequester the heart from the other thoracic organs.

False: the inner, *serous layer* of the pericardial sac secretes *pericardial fluid* that lubricates the surface of the heart. The outer, *fibrous layer* of the pericardial sac has the protective and separative function.

14.4 Define *pericarditis*.

Pericarditis is an inflammation of the parietal pericardium that increases the secretion of pericardial fluid into the pericardial cavity. Because the fibrous layer of the pericardial sac is inelastic, an increase in fluid pressure impairs movement of blood through the chambers of the heart.

OBJECTIVE B *To trace the develoment of the embryonic heart from day 18 through day 25.*

Survey Development of the heart from undifferentiated mesoderm requires only seven to eight days. At 19 days after conception, specialized *heart cords* begin to migrate toward each other medially from the two longitudinal bands of splanchnic (visceral) mesoderm. By day 21, a hollow center has developed in each heart cord, and the structure is called a *heart tube* (Fig. 14-2). By day 23, the heart tubes have fused into a single, medial, *endocardial heart tube* (Fig. 14-3). By day 25, fusion is completed (Fig. 14-4), dilations are occurring, and blood is being pumped.

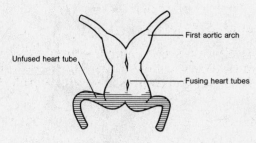

Fig. 14-2

Fig. 14-3

Fig. 14-4

14.5 When are congenital heart problems most likely to develop?

Congenital means "present at birth": partitioning of the heart chambers begins during the middle of the fourth week and is completed by the end of the fifth week. It is during this crucial period that conditions such as heart murmurs, septal defects, patent foramen ovale, and stenosis develop.

OBJECTIVE C   *To contrast the structures and functions of the three layers of the heart wall.*

Survey   Refer to Table 14-1.

Table 14-1

Layer	Structure	Function
Epicardium (visceral pericardium)	Serous membrane of connective tissue, covered with epithelium and including blood capillaries, lymph capillaries, and nerve fibers	Lubricative outer covering
Myocardium	Cardiac muscle tissue, separated by connective tissues and including blood capillaries, lymph capillaries, and nerve fibers	Contractile layer to eject blood from the heart chambers
Endocardium	Serous membrane of epithelium and connective tissues, including elastic and collagenous fibers, blood vessels, and specialized muscle fibers	Lubricative inner lining of the chambers and valves

14.6   Which of the three heart layers is the thickest?

The myocardium, especially in the ventricular walls, where forceful contraction is necessary to pump the blood throughout the body. The fibers of cardiac muscle are arranged in such a way that the intrinsic contraction results in an effective squeezing or wringing of the chambers of the heart.

14.7   What are the *trabeculae carneae*?

This latticelike arrangement of the endocardium (see Fig. 14-5) is composed primarily of dense, fibrous, connective tissue; it provides a strong flexible framework for the walls of the lower heart chambers.

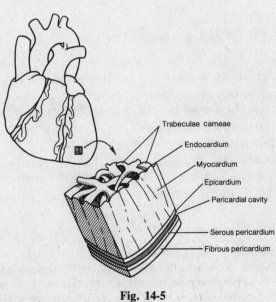

Trabeculae carneae
Endocardium
Myocardium
Epicardium
Pericardial cavity
Serous pericardium
Fibrous pericardium

Fig. 14-5

OBJECTIVE D   *To describe the chambers and valves of the heart.*

Survey   See Fig. 14-6. The heart is a four-chambered, double pump. It consists of upper-right and upper-left *atria*, that pulse together, and lower-right and lower-left *ventricles*, that also contract together. The atria are separated by the thin muscular *interatrial septum*, while the ventricles are separated by the thick, muscular, *interventricular septum*. Two *atrioventricular valves*, the *bicuspid* and *tricuspid* valves, are located between the chambers of the heart, and *semilunar valves* are present at the bases of the two large vessels (the pulmonary trunk and the aorta) that leave the heart.

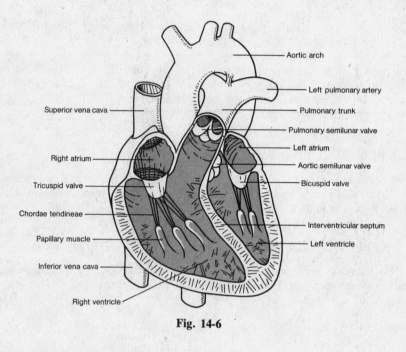

Fig. 14-6

14.8   Describe the working of each of the heart valves.

   Tricuspid valve.   Composed of three cusps that prevent a backflow of blood from the right ventricle into the right atrium during ventricular contraction.

   Pulmonary semilunar valve.   Composed of three half-moon-shaped flaps that prevent a backflow of blood from the pulmonary trunk into the right ventricle during ventricular relaxation.

   Bicuspid (mitral) valve.   Composed of two cusps that prevent a backflow of blood from the left ventricle to the left atrium during ventricular contraction.

   Aortic semilunar valve.   Composed of three half-moon-shaped flaps that prevent a backflow of blood from the aorta into the left ventricle during ventricular contraction.

14.9   *True or false: Papillary muscles and chordae tendineae are found only in the ventricles.*

   True: each cusp of the atrioventricular valves is held in position by strong tendinous cords, the *chordae tendineae*, which are secured to the ventricular wall by cone-shaped *papillary muscles*. As blood is ejected from the atria, the chordae tendineae are relaxed, with valvular opening. But as the ventricles, and with them the papillary muscles, contract, the chordae tendineae are pulled taut, preventing eversion of the valves and backflow of blood from the ventricles into the atria.

14.10 *True or false*: Of the heart chambers, the left ventricle has the largest capacity, because blood from there has to be pumped to the rest of the body.

   False: the capacity is the same for each of the four heart chambers (one and the same volume of blood moves from chamber to chamber).

OBJECTIVE E   *To distinguish between the* pulmonary *and* systemic circuits *of blood flow*.

Survey   The *pulmonary circuit* (through the lungs) involves the right side of the heart (right atrium and right ventricle), the pulmonary arteries, a capillary network in the lungs, and the pulmonary veins. The *systemic circuit* involves the left side of the heart (left atrium and left ventricle) and the remainder of the arteries, capillaries, and veins of the body.

14.11 Describe the flow of blood through the heart.

   Blood fills both atria and begins to flow into both ventricles [Fig. 14-7(*a*)]. Next, the atria contract, emptying the remaining blood into the ventricles [Fig. 14-7(*b*)]. Then the ventricles contract, forcing blood into the aorta and pulmonary trunk [Fig. 14-7(*c*)].

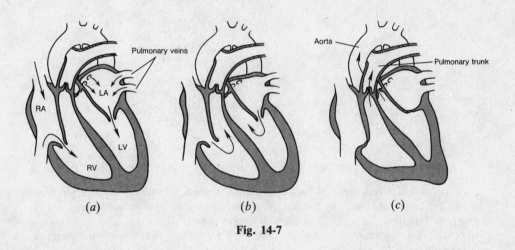

Fig. 14-7

14.12 Correlate the contractions of the heart chambers and the opening and closing of the heart valves.

   During atrial contraction, the atrioventricular valves are open and the semilunar valves are closed; during ventricular contraction, the reverse is true.

OBJECTIVE F   *To appreciate the differences between the circulatory systems of the fetus and of the newborn*.

Survey   Refer to Fig. 14-8. Fetal circulation involves an umbilical cord composed of an *umbilical vein* that transports oxygenated blood toward the heart and two *umbilical arteries* that return deoxygenated blood to the placenta. A *ductus venosus* allows blood to bypass the fetal liver; a *foramen ovale* permits blood to flow directly from the right atrium to the left; and a *ductus arteriosus* shunts blood from the pulmonary trunk to the aortic arch.

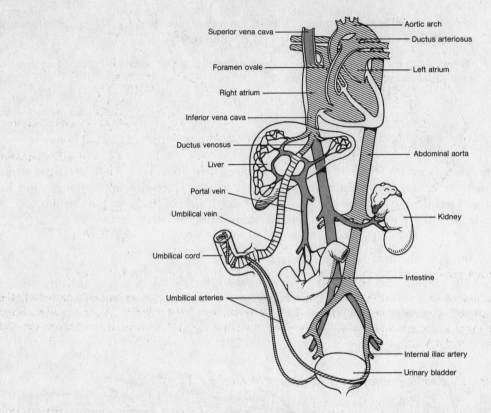

Fig. 14-8

14.13 What do the various cardiovascular structures of the fetus become in the neonatal infant?

The umbilical vein forms the *round ligament* of the liver; the umbilical arteries atrophy to become the *lateral umbilical ligaments*; the ductus venosus forms the *ligamentum venosum*, a fibrous cord in the liver; the foramen ovale closes at birth and becomes the *fossa ovalis*, a depression in the interatrial septum; and the ductus arteriosus closes shortly after birth, atrophies, and becomes the *ligamentum arteriosum*.

14.14 What is a "blue baby"?

Many newborn babies with congenital heart defects have insufficient oxygenated blood in the systemic circulation. The result is *cyanosis*, a bluish discoloration.

OBJECTIVE G   *To describe the myocardial portion of the circulatory system.*

Survey   Blood supply to the myocardium is provided by the right and left *coronary arteries*, which exit the ascending aorta just past the aortic semilunar valve (Fig. 14-9). Four arterioles branch from either coronary artery; these feed capillaries in the myocardium. The *great cardiac vein* and the *middle cardiac vein* return blood from the myocardial capillaries to the *coronary sinus*, and thence into the right atrium (Fig. 14-10).

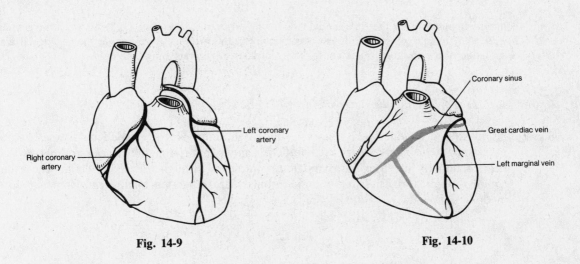

Fig. 14-9   **Fig. 14-10**

14.15 Distinguish among *ischemia*, *angina pectoris*, and *myocardial infarction* (or *infarct*).

   If a branch of a coronary artery becomes constricted or obstructed by an embolus (clot), the myocardial cells it supplies may experience a blood deficiency, called ischemia. Angina pectoris is the chest pain that accompanies ischemia. Death of a portion of the heart from ischemia is called myocardial infarction (*heart attack*).

OBJECTIVE H   *To describe the* conduction system *of the heart*.

Survey   The conduction system consists of *nodal tissue* (specialized cardiac muscle fibers) that initiates the conduction of depolarization waves through the myocardium. These cause the coordinated contractions that empty the heart chambers.

14.16 What is the "pacemaker" of the heart, where is it located, and what is its basic frequency?

   The normal pacemaker is the *sinoatrial node* (S.A. node), located in the posterior wall of the right atrium (Fig. 14-11); it initiates a basic depolarization rate of 70–80 per min that causes the atria to

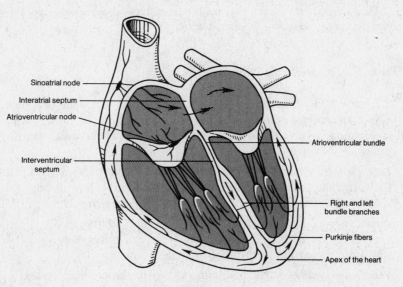

Fig. 14-11

contract. Impulses from the S.A. node pass to the *atrioventricular node* (A.V. node), *atrioventricular bundle* (A.V. bundle, or bundle of His), and finally to the *Purkinje fibers* within the ventricular walls. Stimulation of the Purkinje fibers causes the ventricles to contract simultaneously.

OBJECTIVE I   *To describe the innervation of the heart.*

Survey   As illustrated in Fig. 14-12, the S.A. and A.V. nodes are innervated by sympathetic and parasympathetic nerve fibers: sympathetic impulses through the *accelerator nerves* cause heart action to increase; parasympathetic impulses through the *vagus nerves* cause heart action to decrease. These autonomic impulses are regulated by the *cardiac center* in the medulla oblongata.

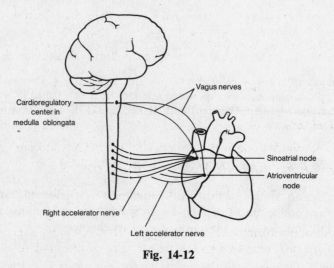

Fig. 14-12

14.17 *True or false*: Norepinephrine and acetylcholine have a synergistic (cooperative) rate-changing action on the heart.

   False: the effects of the two neurotransmitters—the one secreted by sympathetic, the other by parasympathetic, neurons—are antagonistic.

OBJECTIVE J   *To specify the* cardiac cycle.

Survey   The atria and ventricles go through a sequence of events that is repeated with each beat. This *cardiac cycle* consists of a phase of relaxation, called *diastole*, followed by a phase of contraction, referred to as *systole*. Major events of the cycle, starting in mid-diastole, are as follows.

   Late diastole. Atria and ventricles are relaxed, the tricuspid and mitral valves are open, and the aortic and pulmonary valves are closed. Blood is flowing passively from the atria into the ventricles, with 65–75% of ventricular filling occurring before the end of this phase.

   Atrial systole. Atria contract and pump the additional 25–35% of the blood into the ventricles. The orifices of the venae cavae and pulmonary veins narrow; however, there is still some regurgitation of blood into the veins.

   Ventricular systole. At the beginning of ventricular contraction the tricuspid and mitral valves close, causing the *first heart sound*, "*lub*". When pressure in the right ventricle exceeds the diastolic pressure in the pulmonary artery (10 mmHg) and the left ventricular pressure exceeds the diastolic pressure in the aorta (80 mmHg), then the pulmonary and aortic valves open and ventricular ejection begins. Under normal resting conditions, the pressure reaches 25 mmHg on the right side and 120 mmHg on the left side. The *stroke volume*, or volume of blood ejected from either ventricle, is 70 to 90 mL. About 50 mL of blood remains in either ventricle at the end of systole.

Early diastole. As the ventricles begin to relax, the pressure drops rapidly. The pulmonary and aortic valves close, preventing backflow into the ventricles from the arteries and causing the *second heart sound*, "*dub*". Also, the tricuspid and mitral valves open, and blood begins to flow from the atria into the ventricles.

14.18 What causes *heart murmurs*?

When blood flows smoothly through heart valves and blood vessels, it flows silently; turbulent flow generates sounds referred to as *murmurs*. Valves damaged by disease may fail to open completely (*stenosis*), or fail to close completely (*insufficiency*), thereby causing turbulence.

14.19 How is cardiac output calculated?

Cardiac output, which is the volume of blood pumped by the left ventricle in one minute, may be calculated from the formula

$$\text{Cardiac output (C.O.)} = \text{stroke volume (S.V.)} \times \text{heart rate (H.R.)}$$

For instance, if H.R. = 72 beats/min and S.V. = 80 mL/beat, then

$$\text{C.O.} = (72 \text{ beats/min})(80 \text{ mL/beat}) = 5760 \text{ mL/min} \approx 5.8 \text{ L/min}$$

14.20 The method of Problem 14.19 involves the stroke volume, the value of which is often unknown. Give an alternative procedure for determining cardiac output.

By the *Fick principle*, the amount (α) of a substance taken up by an organ (or the whole body) per unit time is equal to the arterial level (A.L.) of the substance minus the venous level (V.L.), times the blood flow. Since the blood flow equals the C.O., one obtains the formula

$$\text{C.O.} = \frac{\alpha}{\text{A.L.} - \text{V.L.}}$$

For instance, the body's O_2 consumption is typically given by $\alpha = 250$ mL/min; and typical blood levels of O_2 are A.L. = 190 mL/L blood, V.L. = 140 mL/L blood. Thus, a typical cardiac output is

$$\text{C.O.} = \frac{250 \text{ mL/min}}{50 \text{ mL/L blood}} = 5 \text{ L blood/min}$$

14.21 Which of the following factors influence(s) cardiac output? (i) increased activity of the sympathetic nervous system, (ii) increased end-diastolic volume, (iii) decreased venous return to the heart, (iv) various forms of anemia.

All the above affect cardiac output.

(i) Sympathetic stimulation increases the heart rate and the strength of heart contraction. It also initiates the release of epinephrine and norepinephrine from the adrenal medulla, which increase the cardiac output.

(ii) As the end-diastolic volume increases, the myocardium is increasingly stretched; as a result, the muscles contract with greater force, which leads to a greater stroke volume and cardiac output. This mechanism is known as *Starling's law of the heart* (or the *Frank–Starling law*).

(iii) If there is decreased venous return (hemorrhage, etc.), the heart does not fill properly, which causes a decreased stroke volume and a decreased cardiac output.

(iv) Under most conditions of anemia, there is a reduction in blood viscosity, as well as localized vasodilation, due to diminished oxygen transport to the tissues. Both conditions produce a decrease in the total peripheral resistance and therefore an increased cardiac output.

14.22 Where is the stethoscope best placed on the chest wall to hear the heart sounds?

See Fig. 14-13.

For sound of:	*Listen at*:
tricuspid valve	5th intercostal space at sternum
mitral valve	5th intercostal space inferior to left nipple
pulmonary valve	2nd intercostal space left of sternum
aortic valve	2nd intercostal space right of sternum

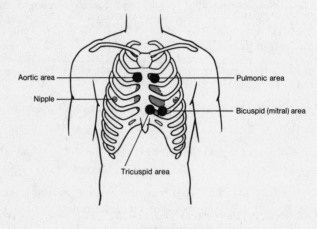

Fig. 14-13

OBJECTIVE K *To identify the normal features of an electrocardiogram.*

Survey Because the body is a good conductor of electricity, potential differences generated by the depolarization and repolarization of the myocardium can be detected on the surface of the body and recorded as an electrocardiogram (ECG or EKG). In Fig. 14-14, *P* marks the

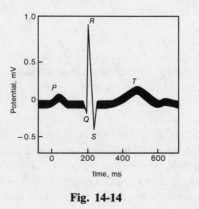

Fig. 14-14

peak of the *P wave*, the signature of atrial depolarization (conduction through the atria). The *QRS complex* is the record of ventricular depolarization; the *T wave*, of ventricular repolarization. The short flat segment between *S* and *T* represents the refractory state of the ventricular myocardium; that between *P* and *Q*, a nonconductive phase of the A.V. node, during which atrial systole can be completed.

14.23 Describe the three conventional systems for placement of electrocardiographic leads.

　　Standard limb leads (Fig. 14-15).　Each lead joins two electrodes of opposite polarities.

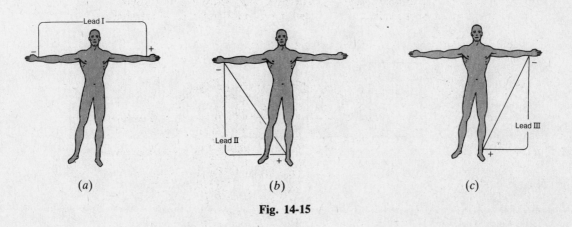

(a)　　　　　　　*(b)*　　　　　　*(c)*

Fig. 14-15

　　Augmented unipolar limb leads (Fig. 14-16).　Each lead has one positive and two negative electrodes. Signal aVR is inverted relative to the other two.

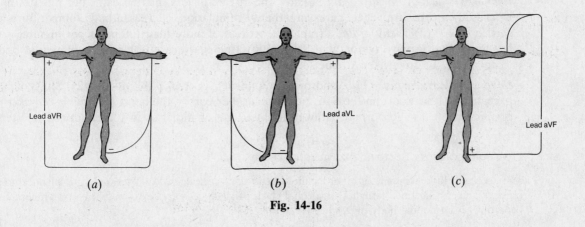

(a)　　　　　　*(b)*　　　　　*(c)*

Fig. 14-16

　　Chest (precordial) leads.　Each lead joins a positive electrode, attached at one of the six sites shown in Fig. 14-17, with three negative electrodes (arms and leg). Because the heart surfaces are close to the chest wall, each chest lead records the electrical potential of that part of the heart which lies immediately beneath the positive electrode. Typical signals are shown in Fig. 14-18.

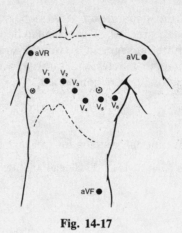

Fig. 14-17

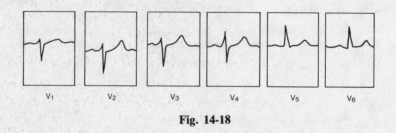

Fig. 14-18

OBJECTIVE L *To become familiar with some common abnormalities in heart rhythm.*

Survey Deviations from normal heart rate or from normal electrical activity of the conduction system are referred to as *cardiac arrhythmias*.

Rate arrythmias. (1) *Bradycardia*, a (slow) heart rate of less than 60 beats per minute, may be caused by excessive vagal (parasympathetic) stimulation, decreased body temperature, or certain drugs. (2) *Tachycardia*, a (rapid) heart rate of more than 100 beats per minute, may be caused by excessive sympathetic stimulation (increased catecholamines), increased body temperature, or drugs like caffeine.

Conduction arrythmias. (1) Abnormal rhythmicity of the S.A. node. (2) Shift of the pacemaking function from the S.A. node to another part of the heart (*ectopic pacemaker* or *ectopic focus*). (3) Abnormal pathway or blockage of impulses in the conduction system.

14.24 Give some causes of ectopic pacemaker activity.

Causes include: ischemia or other localized heart damage; calcified or arteriosclerotic plaques; toxic irritants such as nicotine, caffeine, or alcohol; lack of sleep; anxiety; extremes in body temperature; departures from normal body pH.

14.25 What may cause *heart block*?

Impulses through the heart are sometimes blocked at critical points in the conduction system, owing to: (i) infection in the conduction system, (ii) excessive stimulation of the vagus nerves, (iii) pressure on the conduction system by arteriosclerotic plaques, (iv) localized destruction of the conduction system as a result of infarct (Problem 14.15).

14.26 What are the characteristics of EKGs from patients who experience *premature beats*?

Premature beats are caused when an ectopic pacemaker fires, creating waves that appear earlier in the cycle than is usual.

Atrial premature beat (Fig. 14-19) is due to premature depolarization of an ectopic pacemaker in the atria. It may precede an interval of atrial flutter or fibrillation (Problem 14.27). This kind of premature beat is usually considered innocuous.

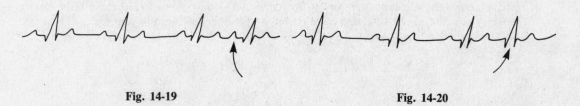

Fig. 14-19   Fig. 14-20

A.V.-nodal premature beat (Fig. 14-20) originates from an ectopic discharge in the A.V. node. The EKG shows a normal QRS complex which generally is not preceded by a P wave.

Premature ventricular depolarization (PVD) (Fig. 14-21) originates from an ectopic pacemaker in the ventricles. The P wave is lacking in the EKG. The QRS complex is wide (because conduction is mainly through the muscle cells of the ventricle rather than through the Purkinje system) and tall (one ventricle depolarizes slightly before the other). Often the T wave is inverted (*altered repolarization*). One PVD may be coupled with one or several normal beats.

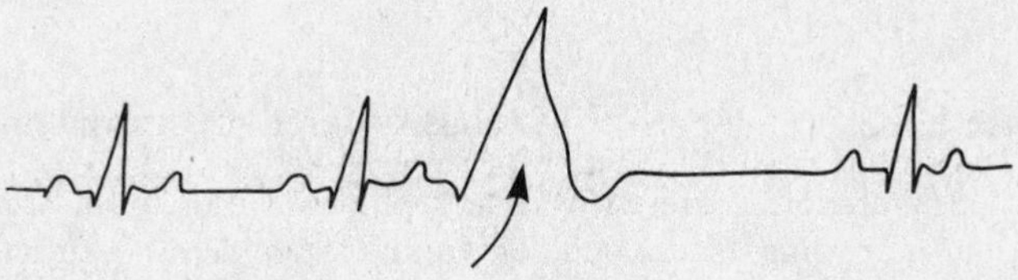

Fig. 14-21

14.27 What are the characteristics of EKGs from patients who experience *flutter* or *fibrillation*?

The atria or ventricles may begin to contract extremely rapidly: coordinated contractions are called *flutter*, and incoordinated contractions *fibrillation*.

Atrial flutter (Fig. 14-22) originates from an ectopic pacemaker in the atria; it is marked by a *2 : 1* or *3 : 1 rhythm* (the atria contract 2 or 3 times for each ventricular contraction). Packed, regular P waves.

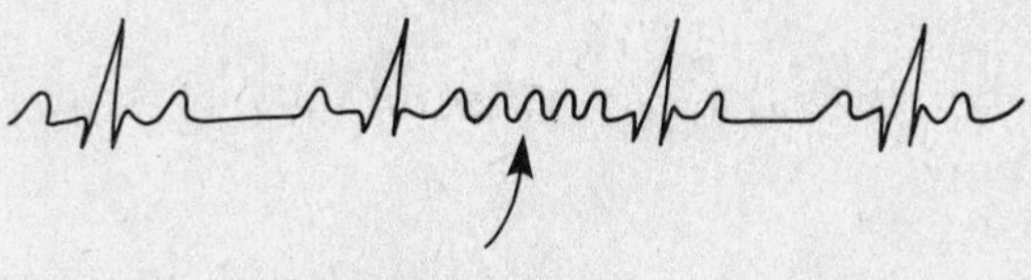

Fig. 14-22

Atrial fibrillation (Fig. 14-23) is caused by many ectopic pacemakers in the atria. The P waves are packed and irregular; the QRS complexes and T waves usually look normal.

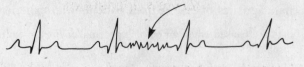

Fig. 14-23

Ventricular flutter (Fig. 14-24) is due to a single ectopic pacemaker in the ventricles. The EKG usually resembles a smooth sine-wave. This condition is dangerous, because the heart does not fill properly and cardiac output is decreased. Moreover, it may develop into ventricular fibrillation.

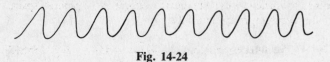

Fig. 14-24

Ventricular fibrillation (Fig. 14-25) caused by a plurality of ectopic pacemakers in the ventricles. The EKG is chaotic—like "random noise." In this the most grave among the spasmodic conditions, the blood pressure drops.

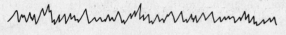

Fig. 14-25

14.28 Characterize an EKG from a patient who has suffered myocardial infarction (heart attack).

A myocardial infarction is caused by a lack of blood flow to an area of the heart as a result of coronary vascular narrowing (spasmodic or from *atherosclerosis*) or vascular blockage (*embolism*; Problem 14.15). The EKG reflects in its features the *classical triad* of heart attack. ***Ischemia*** is evidenced by a symmetrical inversion of the T wave [Fig. 14-26(*a*)]. ***Injury*** is manifested by an elevation of the ST segment of the curve [Fig. 14-26(*b*)]. ***Infarction*** is indicated by a pronounced Q wave [Fig. 14-26(*c*)], which may have 1/3 the height of the QRS complex. (This feature may be observable for many years after the heart attack.)

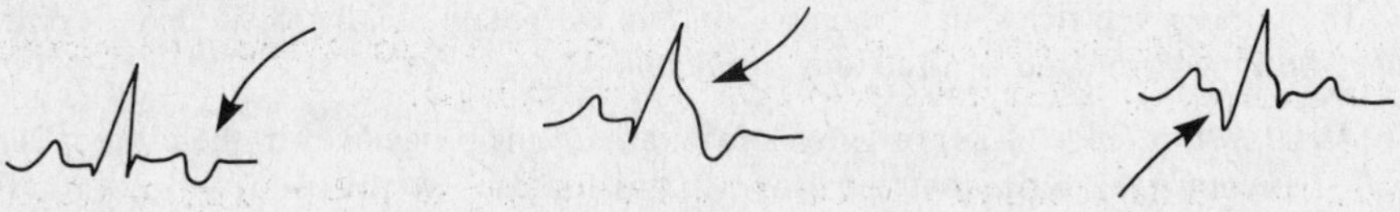

          (*a*)   (*b*)   (*c*)

Fig. 14-26

Review Questions

Multiple Choice I

1. The prenatal heart begins to pump blood in the: (*a*) fourth week, (*b*) fifth week, (*c*) sixth week, (*d*) seventh week.

2. Which of the following organs or structures is *not* located within the mediastinum? (*a*) heart, (*b*) esophagus, (*c*) portion of trachea, (*d*) lungs, (*e*) vagus nerves.

3. Which is the correct nesting? (*a*) heart, mediastinum, visceral sac, thoracic cavity; (*b*) heart, pericardial sac, pleural space, thoracic cavity; (*c*) heart, visceral sac, pleural space, thoracic cavity; (*d*) heart, pericardial sac, mediastinum, thoracic cavity.

4. Which pairing is valid for fetal circulation? (*a*) foramen ovale—right ventricle to left ventricle, (*b*) ductus venosus—umbilical vein to inferior vena cava, (*c*) foramen ovale—right atrium to pulmonary trunk, (*d*) ductus arteriosus—pulmonary artery to pulmonary vein.

5. The valve that is located on the same side of the heart as the pulmonary semilunar valve is the: (*a*) tricuspid, (*b*) mitral, (*c*) bicuspid, (*d*) aortic semilunar.

6. A stenosis of the bicuspid heart valve might cause blood to back up into the: (*a*) coronary circulation, (*b*) venae cavae, (*c*) pulmonary circuit, (*d*) left ventricle.

7. In the fetus, fully oxygenated blood is carried by the: (*a*) ductus arteriosus, (*b*) umbilical artery, (*c*) placental vein, (*d*) umbilical vein.

8. After birth, the ductus arteriosus develops into the: (*a*) fossa ovalis, (*b*) ligamentum arteriosum, (*c*) lateral umbilical ligament, (*d*) ligamentum venosum, (*e*) round ligament of the liver.

9. The outermost of the three layers of the heart is the: (*a*) epicardium, (*b*) supracardium, (*c*) pericardium, (*d*) endocardium.

10. The correct sequence for blood entering the heart through the venae cavae and leaving through the aorta is: (*a*) right atrium, left atrium, left ventricle, right ventricle; (*b*) left ventricle, left atrium, right ventricle, right atrium; (*c*) right atrium, right ventricle, left atrium, left ventricle; (*d*) left atrium, left ventricle, right atrium, right ventricle.

11. The sinoatrial node is situated in the wall of the: (*a*) right atrium, (*b*) interventricular septum, (*c*) pulmonary trunk, (*d*) superior vena cava, (*e*) left ventricle.

12. Impulses through the conduction system of the heart follow the ordered path: (*a*) A.V. node, S.A. node, Purkinje fibers, atrioventricular bundle; (*b*) S.A. node, Purkinje fibers, atrioventricular bundle, A.V. node; (*c*) S.A. node, A.V. node, atrioventricular bundle, Purkinje fibers; (*d*) A.V. node, atrioventricular bundle, S.A. node, Purkinje fibers.

13. Which pairing is incorrect? (*a*) chordae tendineae—semilunar valves, (*b*) right ventricle—papillary muscle, (*c*) left ventricle—trabeculae carneae, (*d*) right atrium—coronary sinus, (*e*) left atrium—pulmonary veins.

14. The heart is covered by: (*a*) pericardium, (*b*) epicardium, (*c*) supracardium, (*d*) endocardium.

15. An increase in cardiac output is required by all the following, except: (*a*) physical exercise, (*b*) fever, (*c*) digestion, (*d*) stimulation of the vagus nerves.

16. The "lub" sound of the heart is caused by the: (a) closing of the A.V. valves, (b) closing of the aortic and pulmonary valves, (c) blood rushing out of the ventricles, (d) filling of the ventricles.

17. Which occurs during systole? (a) ventricular filling, (b) atrial filling, (c) ventricular contraction, (d) atrial relaxation.

18. When the bundle of His is completely interrupted, the: (a) atria beat at an irregular rate; (b) ventricles typically contract at about 30–40 beats/min; (c) PR intervals in the EKG are longer than normal, but remain constant from beat to beat; (d) QRS complex varies in shape from beat to beat.

19. Which of the following is *not* a condition of late diastole? (a) the atria and ventricles are relaxed, (b) the A.V. valves are open, (c) the aortic valve is open, (d) the ventricles receive blood from the atria.

20. During ventricular contraction: (a) all the blood is forced out of the ventricles, (b) some blood is left in the ventricles, (c) no blood is forced out of the ventricles, (d) some blood is forced back into the atria.

21. The following are events in the cardiac cycle. Number them in order, starting at mid-diastole.

   (a) The pressure in the left ventricle reaches the diastolic pressure in the aorta, and the right ventricular pressure reaches the diastolic pressure in the pulmonary artery.

   (b) The aortic and pulmonary valves close as ventricular ejection ends.

   (c) The mitral and tricuspid valves are open, the aortic and pulmonary valves are closed, and blood is flowing into the atria and ventricles.

   (d) The mitral and tricuspid valves close as the ventricles start to contract.

   (e) Ventricular pressure falls below atrial pressure, and the mitral and tricuspid valves open, permitting the ventricles to fill again.

   (f) The aortic and pulmonary valves open and the phase of ventricular ejection begins.

   (g) The atria contract, propelling additional blood into the ventricles.

True/False

1. Initiation of the heartbeat, at 8 weeks after conception, marks the transition from embryo to fetus.

2. The pericardial sac secretes fluids that lubricate the surface of the heart.

3. Cutting the vagus nerves where they innervate the heart would increase the heart rate.

4. A patent (open) ductus arteriosus in an adult permits a blood flow from the pulmonary trunk to the aorta.

5. The right atrium of the fetal heart receives relatively well-oxygenated blood.

6. Epinephrine increases the rate, but not the strength, of heart contraction.

7. The pluck includes all the thoracic organs.

8. The heart is totally derived from embryonic mesoderm.

9. Chordae tendineae, papillary muscles, and trabeculae carneae are structural features of only the ventricles of the heart.

10. *Angina pectoris* is the comprehensive term for heart attack.

11. The walls of arteries are not as thick and strong as the walls of veins.

12. Blood entering the left ventricle is poorer in oxygen than blood in the vena cava.

Matching

1. Atrial depolarization
2. Cardiac output
3. Ventricular depolarization
4. Ventricular repolarization
5. Closure of the A.V. valves at the onset of systole
6. Closure of the pulmonary and aortic valves at the onset of diastole

(a) P wave
(b) First heart sound
(c) Second heart sound
(d) QRS complex
(e) S.V. × H.R.
(f) T wave
(g) A.L. − V.L.

Labeling

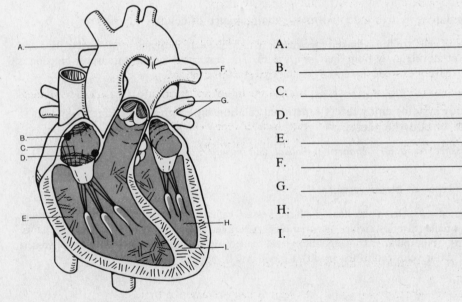

A. _____

B. _____

C. _____

D. _____

E. _____

F. _____

G. _____

H. _____

I. _____

Fig. 14-27

Multiple Choice II

Pick letters from Fig. 14-27 to answer the following descriptions:

1. The tissue with the fastest inherent rhythm.
2. The blood returns to the heart from the lungs via this structure.
3. Propagation of a wave of excitation through this structure is delayed about 100 ms.
4. This structure pumps blood to the systemic circuit.
5. This structure pumps blood to the pulmonary circuit.
6. This structure receives venous blood from all regions of the body except the lungs.
7. In the fetus, most blood passes directly from this structure to the left atrium.

Chapter 15

Cardiovascular System: Vessels and Blood Circulation

OBJECTIVE A   *To characterize the functions of the* circulatory system (*cardiovascular-plus lymphatic system*).

Survey   **Transport**: of nutrients and oxygen to all body cells; of waste products and carbon dioxide from the cells of the body to the organs of excretion; of hormones from the endocrine glands to body tissues. **Thermoregulation**: the amount of heat lost from the body is regulated by the degree of blood flow through the skin. **Acid-base balance**: in cooperation with the respiratory and urinary systems, the circulatory system regulates (through buffer substances in the blood) the body pH. **Protection against disease**: the leukocytes are adapted to defend against foreign microbes and toxins.

15.1   Why is the circulatory system essential to the maintenance of cellular function in the body?

The majority of cells are firmly linked together in tissues; they are incapable of procuring food and oxygen, or even of moving away from their own wastes. The circulatory system's transport function is thus crucial to their very existence.

15.2   Which other body systems enjoy direct functional relationships with the circulatory system in the maintenance of body homeostasis?

Urinary, *digestive*, *respiratory*, *endocrine*, *skeletal*, and *integumentary* systems.

15.3   List the various types of blood vessels.

Blood is carried away from the heart in large vessels called *arteries*. These divide into smaller arteries, and the smaller arteries divide into *arterioles*. Arterioles divide into tiny capillaries (exchange area of the system). The capillaries converge into vessels called *venules*, which join to form larger vessels called *veins*. The major veins return the blood to the heart.

OBJECTIVE B   *To compare arteries, capillaries, and veins as to structure and function.*

Survey   Table 15-1 makes reference to the layers, or *tunics*, out of one or more of which the walls of blood vessels are composed. Specifically, these are: the *tunica intima*, an inner layer of squamous epithelium, called *endothelium*, resting on a layer of connective tissue; the *tunica media*, a middle layer of smooth muscle fibers mixed with elastic fibers; and the *tunica adventitia*, an outer layer of connective tissue containing elastic and collagenous fibers.

15.4   Do capillaries exchange substances between the blood and the interstitial fluid in the same way throughout the body?

No. Among other things (see Problem 15.6), the size and number of *fenestrations* (openings or pores) in capillaries vary with the function of the organ or tissue. Large fenestrations, and therefore increased exchange, are characteristic of endothelial cells of capillaries in the gastrointestinal tract, renal glomeruli, and some glands. In the brain, capillaries have small or no fenestrations, and the exchange of many substances is retarded (the blood–brain barrier, cf. Problem 9.30).

Table 15-1

Vessel	Structure	Function
Artery	Strong, elastic vessel, consisting of all three layers; lumen diameter large relative to wall thickness	Distributive channel to body tissues; blood carried under high pressure (muscular wall and large lumen minimize pressure drop)
Arteriole	Thick layer of smooth muscle in tunica media, relatively narrow lumen	Alters diameter to control blood flow, dampens pulsatile flow to a steady flow
Capillary	Wall composed of a single layer of endothelium (tunica intima); has a cuff of smooth muscle (*precapillary sphincter*) at its origin, which regulates flow	Allows exchange of fluids, nutrients, and gases between the blood and the interstitial fluids
Vein	Thin walls composed of all three layers; contains small amount of elastic tissue and smooth muscle; has valves	Carries blood from tissues to heart; serves as fluid reservoir (veins hold 60–75% of circulating blood volume); constricts in response to sympathetic stimulation; valves ensure unidirectional blood flow (toward the heart)

15.5 Compare arterial blood pressure and venous blood pressure.

The three most important variables affecting blood pressure are cardiac rate, blood volume, and total peripheral resistance. Arterial blood pressure is much greater than venous blood pressure due to the immediate effect of ventricular contraction within the heart pulsating the blood into arteries and the recoiling of the arterial walls. Blood pressure decreases rapidly within the capillaries and is near zero as venous blood returns to the heart.

15.6 List factors that influence exchange between the blood and the interstitial fluid.

(1) Large surface area (about 700 m^2) for exchange, because of large number of capillaries in the body. (2) Fenestration. (3) Diffusion—the principal mechanism of exchange. (4) Capillary hydrostatic pressure, the force that pushes fluids out into the interstitial space; it ranges from 10–45 mmHg in most tissues. (5) Interstitial pressure (varies depending on physiological conditions). (6) Capillary osmotic pressure, mainly due to plasma proteins (albumin). Normal osmotic pressure (23–28 mmHg) tends to cause reabsorption of fluid into the capillaries. (7) Interstitial osmotic pressure (movement of some protein out of capillaries induces outward filtration of fluid into the interstitial space).

15.7 What are the causes of *varicose veins*?

The term is applied to superficial veins that are overdistended, irregular, and tortuous. Principal causes are *malfunctioning valves* (because of increased back-pressure from gravity or pregnancy) and *vessel blockage* (owing to thrombophlebitis).

OBJECTIVE C *To identify the principal* systemic arteries.

Survey See Fig. 15-1.

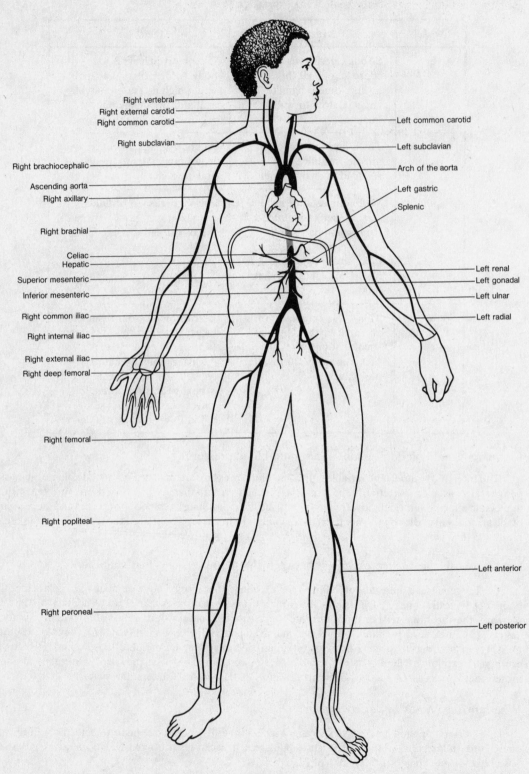

Right vertebral
Right external carotid
Right common carotid

Left common carotid

Right subclavian

Left subclavian

Right brachiocephalic

Arch of the aorta

Ascending aorta
Right axillary

Left gastric

Splenic

Right brachial

Celiac
Hepatic
Superior mesenteric

Left renal
Left gonadal
Left ulnar

Inferior mesenteric

Left radial

Right common iliac

Right internal iliac

Right external iliac
Right deep femoral

Right femoral

Right popliteal

Left anterior

Right peroneal

Left posterior

Fig. 15-1

15.8   Specify the arteries that branch off from the arch of the aorta.

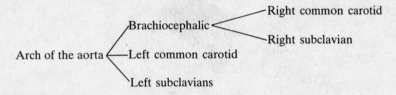

15.9   Supply the missing labels in Fig. 15-2.

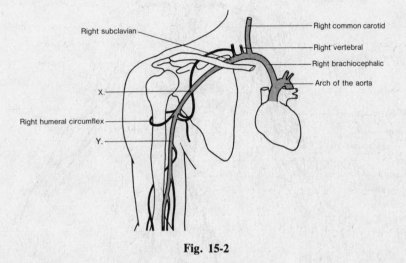

Fig. 15-2

X. = Right axillary, Y. = Right brachial

15.10   What are the four vessels that supply blood to the brain?

The paired *internal carotid arteries* and the paired *vertebral arteries*. See Fig. 15-3.

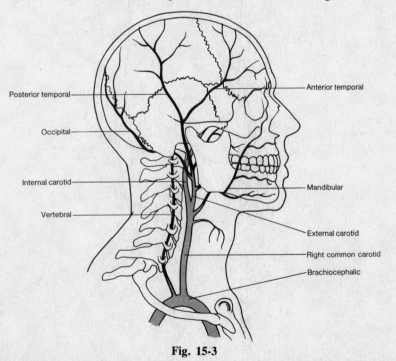

Fig. 15-3

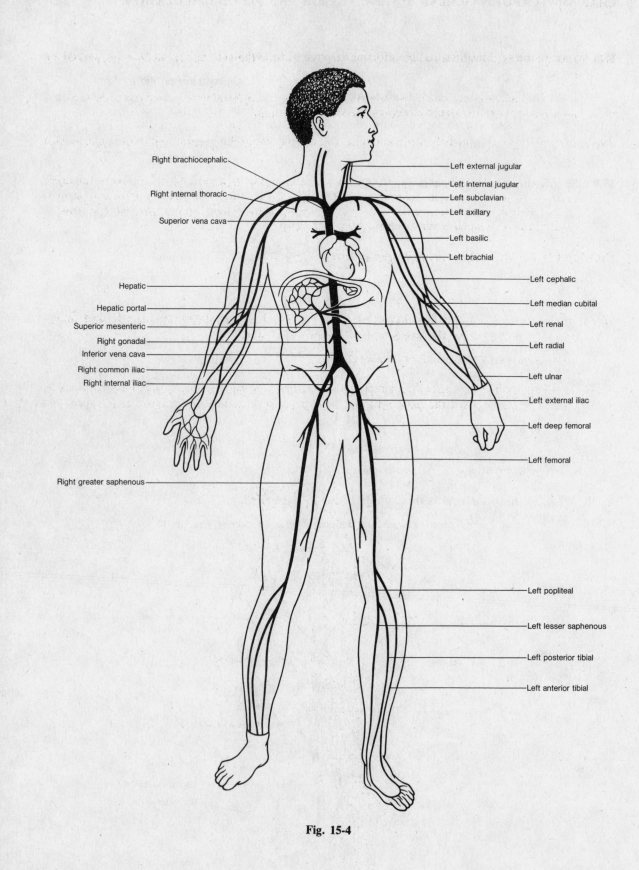

Right brachiocephalic

Right internal thoracic

Superior vena cava

Hepatic

Hepatic portal

Superior mesenteric

Right gonadal

Inferior vena cava

Right common iliac

Right internal iliac

Right greater saphenous

Left external jugular

Left internal jugular

Left subclavian

Left axillary

Left basilic

Left brachial

Left cephalic

Left median cubital

Left renal

Left radial

Left ulnar

Left external iliac

Left deep femoral

Left femoral

Left popliteal

Left lesser saphenous

Left posterior tibial

Left anterior tibial

Fig. 15-4

226

5.11 List the arterial branches of the thoracic aorta and name the general region or organ served by each.

> *Pericardials*—pericardium; *intercostals*—thoracic wall; *superior phrenics*—diaphragm; *bronchials*—bronchi; *esophageals*—esophagus; *inferior phrenics*—diaphragm.

15.12 List the arterial branches of the abdominal aorta and name the general region or organ served by each.

> *Celiac (common hepatic)*—upper digestive system; *celiac (splenic)*—spleen, pancreas, stomach; *celiac (left gastric)*—stomach, esophagus; *renals*—kidneys; *suprarenals*—adrenals; *superior mesenteric*—large intestine; *inferior mesenteric*—large intestine; *gonadals (testiculars)*—testes; *gonadals (ovarians)*—ovaries; *common iliacs*—genital organs, lower extremities.

OBJECTIVE D *To identify the principal* systemic veins.

Survey See Fig. 15-4.

15.13 Name the two major veins that return blood to the heart from the head, neck, and upper extremities, and from the abdomen and lower extremities.

> *Superior vena cava* and *inferior vena cava*.

15.14 Name the paired vein that drains blood from the brain, meninges, and cranial nervous sinuses, and that passes down the neck adjacent to the common carotid artery and vagus nerve.

> *Internal jugular*.

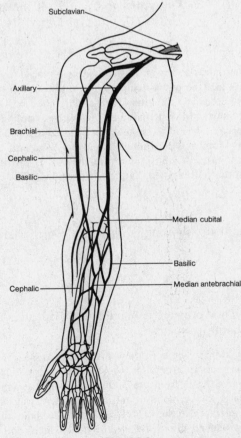

Fig. 15-5

15.15 Classify the veins that drain the upper extremity (Fig. 15-5) as *deep* or *superficial*.

Deep: brachial, axillary, subclavian. *Superficial*: basilic, cephalic, median cubital, median ante-brachial.

15.16 Which vein in the arm is punctured to extract a blood sample?

Median cubital.

15.17 Give the region(s) drained by the (*a*) renal veins, (*b*) lumbar veins, (*c*) inferior phrenic veins, (*d*) internal iliac veins, (*e*) suprarenal veins.

(*a*) kidneys; (*b*) posterior abdominal wall, spinal cord; (*c*) diaphragm; (*d*) urinary bladder, rectum, prostate gland; (*c*) adrenal glands.

OBJECTIVE E   *To define* blood pressure *and to explain how it is measured and controlled.*

Survey   Blood pressure is the force per unit area exerted by the blood against the inner walls of blood vessels; it is due primarily to the force of the heart. The body adjusts blood pressure by altering the heart rate (increased heart rate increases pressure), blood volume (increased volume increases pressure), and peripheral resistance (decreased vessel diameter increases resistance and, with it, pressure). Normal blood pressure is about 120/80; i.e.,

Systolic pressure	120 mmHg
Diastolic pressure	80 mmHg
Pulse pressure	40 mmHg

15.18 Describe the use of the *sphygmomanometer*.

A cuff is wrapped around the arm and a stethoscope is placed over the brachial artery near the elbow. The cuff is inflated with air (the pump being a hand bulb) until the pressure is greater than the *systolic pressure*; this occludes (closes up) the artery, preventing blood flow to the lower arm. Now the pressure in the cuff is lowered slowly; when it falls just below the systolic pressure, a spurt of blood will flow through the artery, causing a sound to be heard in the stethoscope. A tapping sound is heard with each successive heartbeat as blood passes through the artery. The cuff pressure at which the first sound is heard is the *systolic pressure*. As the pressure in the cuff is further lowered, the tapping sounds get louder, then become softer and muffled, and finally disappear. The cuff pressure at the last sound is the *diastolic pressure*.

15.19 Make a sketch of the body, showing the *pressure points* where arterial pulsations can best be detected.

See Fig. 15-6.

15.20 Give mechanisms of blood pressure regulation via (*a*) changes in blood volume, (*b*) changes in heart rate and peripheral resistance.

(*a*)   **Neural mechanism.** *Baroreceptors* in the walls of large vessels and the chambers of the heart detect a decrease in blood pressure. Impulses from these receptors reach the hypothalamus, which orders increased secretion of ADH from the pituitary. Under the action of ADH (see Table 12-2), the kidneys restore water to the bloodstream, thereby increasing blood volume. **Renal mechanism.** A decrease in blood pressure in the kidneys activates the renin-angiotensin system (Problem 12.22). The aldosterone produced alters the electrolyte balance, and along with it the water balance, between kidneys and bloodstream. The net effect is the same as in the neural mechanism.

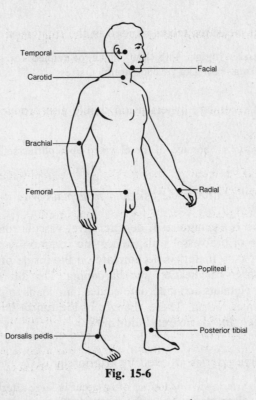

Fig. 15-6

(b)   A change in blood pressure is reported by baroreceptors to the vasomotor center (Problem 9.23). The vasomotor center sends sympathetic impulses to the heart, which change the heart rate, and to the smooth muscles of the vessels, which change vessel diameter and thus peripheral resistance. In addition, the vasometer center can effect release of epinephrine and norepinephrine from the adrenal medulla. These two hormones likewise alter peripheral resistance and heart rate.

OBJECTIVE F   *To define* hypertension *and to give its causes.*

Survey   *Hypertension* is a sustained elevation of the systemic arterial pressure. It is generally characterized by a systolic pressure that exceeds 160 mmHg and a diastolic pressure that exceeds 95 mmHg. Hypertension is classified as being of two types. *Primary* or *essential hypertension* accounts for 85 to 90 percent of all cases and occurs without any known cause. It is found more often in females than in males, more often in black people than in white people, and runs in some families. Excessive salt intake, a too great volume of body fluids, psychoemotional stress, faulty responses of baroreceptors, and increased sensitivity of vessels to catecholamines are *perhaps* factors in primary hypertension. *Secondary hypertension*, which accounts for the remaining 10 to 15 percent of cases, has definite structural and physiological causes:

> *renal diseases*, including *renal ischemic disease* (narrowing of the renal artery), *glomerulonephritis*, and *pyelonephritis*;
>
> *adrenal diseases*, including Cushing's syndrome (see Key Clinical Terms, Chapter 12), primary *aldosteronism* (excess aldosterone), and pheochromocytoma (see Key Clinical Terms, Chapter 12);
>
> *narrowing of the aorta*;
>
> *hypercalcemia* (too much blood calcium);
>
> *oral contraceptives*;
>
> severe *polycythemia* (too many red blood cells).

15.21 List the general measures recommended in the treatment of hypertension.

Regular exercise, weight loss, low intake of refined carbohydrates, restriction of salt intake, cessation of smoking, easing of psychoemotional stress.

15.22 Which cause of secondary hypertension should also produce headache, vertigo, and dimmed vision?

Pheochromocytoma, because of the elevated epinephrine secretion and excretion.

OBJECTIVE G *To define* arteriosclerosis *and to explain why it creates a serious health problem.*

Survey *Arteriosclerosis* is a generalized, degenerative, vascular disorder that results in a thickening and hardening of the vessel wall (hence the common name, "hardening of the arteries"). Soft masses of fatty materials accumulate on the inside of the arterial wall (*atherosclerosis*) and later undergo calcification and hardening. The altered wall presents a rough surface which attracts platelets and macromolecules, and leads to proliferation of the smooth muscle cells of the tunica media. These changes in the tunica intima and tunica media result in a narrowed lumen and decreased blood flow.

15.23 Are only the large arteries affected by arteriosclerosis?

No. Although arteriosclerotic lesions often occur in large arteries, such as the aorta, they also occur in medium and small arteries, such as the coronary, renal, mesenteric, and iliac arteries.

15.24 Although the cause of arteriosclerosis is not yet well understood, the disease does appear to be positively correlated with six, and negatively correlated with one, of the following conditions: (*a*) high intake of saturated fats, (*b*) high intake of refined carbohydrates, (*c*) elevated blood pressure, (*d*) regular sustained exercise, (*e*) cigarette smoking, (*f*) obesity, (*g*) family history of heart disease. Which is the odd condition?

(*d*)

Review Questions

Multiple Choice

1. Veins, as compared to arteries: (*a*) contain more muscle, (*b*) appear more round, (*c*) stretch more, (*d*) are under a greater pressure.

2. Return of blood to the heart is not facilitated by: (*a*) venous valves, (*b*) the skeletal-muscle pump, (*c*) skeletal-muscle groups, (*d*) venous pressure.

3. Resistive vessels of the circulatory system are: (*a*) large arteries, (*b*) large veins, (*c*) small arteries and arterioles, (*d*) small veins and venules.

4. Discontinuous or fenestrated capillaries are found in: (*a*) muscles, (*b*) adipose tissue, (*c*) the CNS, (*d*) the intestines.

5. Compared to veins, arteries contain a thicker: (*a*) endothelium, (*b*) tunica intima, (*c*) tunica media, (*d*) tunica adventitia.

6. The blood vessels that are under the greatest pressure are: (*a*) large arteries, (*b*) small arteries, (*c*) veins, (*d*) capillaries.

7. Capillaries provide a total surface area of: (*a*) 50 ft^2, (*b*) 200 ft^2, (*c*) 7500 ft^2, (*d*) 1 square mile.

8. Interstitial fluid enters capillaries at the venular end through the action of: (*a*) negative pressure, (*b*) colloid osmotic pressure, (*c*) active transport, (*d*) capillary pores.

9. A hormone that is significantly involved in regulation of blood volume is: (*a*) ACTH, (*b*) osmoretic hormone, (*c*) ADH, (*d*) LH.

10. Edema is *not* caused by: (*a*) high blood pressure, (*b*) increased plasma protein concentration, (*c*) leakage of plasma proteins into tissue fluid, (*d*) obstruction of lymphatic drainage.

11. A person with a blood pressure of 135/75 has a pulse pressure of: (*a*) 60, (*b*) 80, (*c*) 105, (*d*) 210.

12. Arteries are: (*a*) strong, rigid vessels that are adapted for carrying blood under high pressure; (*b*) thin, elastic vessels that are adapted for transporting blood through areas of low pressure; (*c*) elastic blood vessels that form the connection between arterioles and venules; (*d*) strong, elastic vessels that are adapted for carrying blood under high pressure.

13. The innermost layer of an artery is composed of: (*a*) stratified squamous epithelium, (*b*) simple cuboidal epithelium, (*c*) simple columnar epithelium, (*d*) endothelium.

14. The tunica adventitia is relatively thin and consists chiefly of: (*a*) collagenous fibers, (*b*) elastic fibers, (*c*) connective tissue, (*d*) epithelium.

15. The *vasa vasorum* are minute vessels within the: (*a*) tunica adventitia, (*b*) tunica intima, (*c*) tunica media, (*d*) metarterioles.

16. Sympathetic impulses to the smooth muscles in the walls of arteries and arterioles produce: (*a*) vasodilation, (*b*) vasoconstriction, (*c*) vasomotor inhibition, (*d*) arteriosclerosis.

17. *Arteriovenous shunts* are: (*a*) metabolic pathways used by the cells in order to maximize the availability of oxygen; (*b*) composed of *metarterioles* which directly join to venules, thus bypassing the capillaries; (*c*) capillary networks which are able to distribute the blood to meet the cells' need at a specific time; (*d*) used only in major heart surgery.

18. Identify the true statement: (*a*) oxygen and nutrients leave the blood of capillaries at their venular ends, while wastes enter the blood at the arteriolar ends; (*b*) oxygen and nutrients leave the blood of capillaries at their arteriolar ends, while wastes enter the blood at the venular ends; (*c*) the density of capillaries within a tissue is inversely proportional to the tissue's rate of metabolism; (*d*) the pattern of capillary arrangement is basically uniform throughout the body.

19. Substances exchanged at the capillary level move through the capillary walls primarily by: (*a*) diffusion, (*b*) filtration, (*c*) osmosis, (*d*) active transport.

20. In the brain, the endothelial cells of the capillary walls are more tightly fused than those in other body regions. This fact underlies the operation of the: (*a*) precapillary sphincter network, (*b*) astrocyte network, (*c*) blood–brain barrier, (*d*) impermeable membrane region.

21. The substance in the blood that helps to maintain the osmotic pressure is: (a) lipids, (b) plasma proteins, (c) lipid-soluble vitamins, (d) histamine.

22. If hydrostatic pressure in the arteriole end of a capillary is 36.6 mmHg and if osmotic pressure of the blood is 28 mmHg, then the net movement of water and dissolved substances is: (a) inward at 64.6 mmHg, (b) inward at 8.6 mmHg, (c) outward at 64.6 mmHg, (d) outward at 8.6 mmHg.

23. Which venous layer is poorly developed? (a) tunica adventitia, (b) tunica intima, (c) tunica media.

24. The accumulation of soft masses of fatty materials, particularly cholesterol, on the inside of the arterial wall is known as: (a) ischemia, (b) atherosclerosis, (c) arteriosclerosis, (d) phlebitis.

25. Identify the false statement(s): (a) arterial blood pressure rises and falls in a pattern corresponding to the phases of the cardiac cycle; (b) during ventricular systole, the pressures in the arteries rise sharply; (c) the maximum pressure achieved during ventricular contraction is called *diastolic* pressure; (d) "blood pressure" most commonly means "arterial pressure"; (e) the radial pulse rate is usually equal to the rate at which the right ventricle is contracting.

26. In the measurement of blood pressure, the cuff of the sphygmomanometer usually surrounds the: (a) radial artery, (b) dorsalis pedis artery, (c) brachiocephalic artery, (d) subclavian artery, (e) brachial artery.

27. If the blood pressure of an individual is measured at 125 over 81, the approximate mean arterial pressure is: (a) 206, (b) 44, (c) 103, (d) 96.

28. Identify the true statement(s): (a) blood pressure varies inversely with cardiac output; (b) if either the stroke volume or the heart rate increases, so does the cardiac output and, with it, the blood pressure; (c) if the stroke volume or the heart rate decreases, so should the cardiac output, but the blood pressure should remain the same; (d) blood pressure decreases as the distance from the left ventricle increases.

29. Arterial blood pressure is independent of: (a) blood volume, (b) heart rate, (c) peripheral resistance, (d) blood viscosity, (e) an influx of calcium ions, (f) a deficiency of sodium ions.

30. Identify the true statement(s): (a) an increased cardiac output is reflected in an elevated diastolic pressure; (b) an increased cardiac output is reflected in a decreased diastolic pressure; (c) an increase in the force of ventricular contraction produces an elevated systolic pressure; (d) an increase in the force of ventricular contraction produces a decreased systolic pressure.

Chapter 16

Lymphatic System
and Body Immunity

OBJECTIVE A *To describe the functional relationship between the lymphatic system and the cardiovascular system.*

Survey The lymphatic system cooperates with the cardiovascular system in three ways: (1) it transports interstitial fluid, renamed *lymph*, from the tissues into the bloodstream, where it becomes *plasma*; (2) it assists in fat absorption in the small intestine; and (3) it plays a key role in protecting the body from bacterial invasion (via the bloodstream).

16.1 What is *edema*?

Much of the fluid of the body (approximately 11%) surrounds the cells in body tissues as interstitial fluid. Excessive accumulation of interstitial fluid is known as *edema*.

16.2 Could edema result from: (*a*) obstruction of lymphatic drainage? (*b*) hypertension? (*c*) leakage of plasma proteins into interstitial fluid (or other decrease in plasma protein concentration)? (*d*) allergy?

All the above are potential causes of edema. (*a*) The condition *elephantiasis* is caused by a tropical nematode parasite that blocks lymphatic drainage. (*b*) Hypertension increases capillary pressure and causes excessive outflow of plasma into the interstitial fluid. (*c*) Lowered concentration of plasma proteins—perhaps because of liver or kidney disease—provokes osmosis of plasma into the interstitial fluid. (*d*) Allergic reaction or inflammation may reduce the osmotic flow of water from the afflicted tissue into the capillaries.

OBJECTIVE B *To specify the routes of fluid transport in the lymphatic system.*

Survey Interstitial fluid enters the lymphatic system through the walls of *lymph capillaries*, composed of simple squamous epithelium. From merging lymph capillaries, the lymph is carried

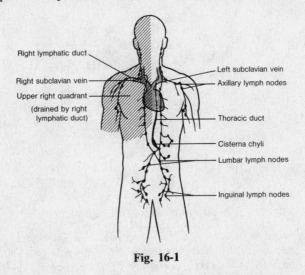

Fig. 16-1

233

into larger *lymph ducts*. Interconnecting lymph ducts eventually empty into one of two principal vessels: the *thoracic duct* and the *right lymphatic duct*. These drain into the *left subclavian vein* and *right subclavian vein*, respectively. See Fig. 16-1.

16.3 Compare lymph ducts and veins with regard to structure.

Although thinner, the walls of lymph ducts are similar to those of veins in that they have the same three layers and contain valves to prevent backflow.

16.4 What is the *cisterna chyli*, and how does it relate to *lacteals*?

The *cisterna chyli* is a saclike enlargement of the thoracic duct in the abdominal area. *Lacteals* are specialized lymph capillaries within the villi of the small intestine; they transport certain products of fat absorption out of the digestive tract and into the cisterna chyli.

16.5 Which two body regions are drained by the two principal lymphatic vessels?

The right lymphatic duct drains lymph from the upper-right quadrant of the body (shaded area in Fig. 16-1). The larger thoracic duct drains lymph from the remainder of the body.

16.6 What constitutes the impetus for lymph flow?

Contraction of skeletal muscles, intestinal peristalsis, and body movement that acts to massage the lymph vessels. Gravity also aids movement of lymph.

OBJECTIVE C   *To elucidate the structure of* lymph nodes.

Survey   *Lymph nodes* are small, oval bodies enclosed in fibrous capsules; they contain phagocytic, reticular, *cortical tissue* adapted to filter lymph. Specialized bands of connective tissue, called *trabeculae*, divide a node. *Afferent lymphatic vessels* carry lymph into the node, where it is circulated through the *cortical sinuses*. The filtered lymph leaves the node through the *efferent lymphatic vessels*, which emerge through the concave *hilus*. See Fig. 16-2.

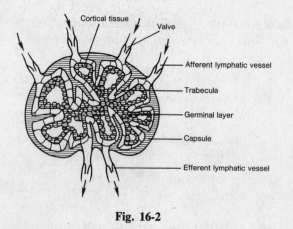

Cortical tissue
Valve
Afferent lymphatic vessel
Trabecula
Germinal layer
Capsule
Efferent lymphatic vessel

Fig. 16-2

16.7 What is the function of the *germinal layer* of a lymph node?

This layer serves as a harbor for *lymphocytes*, the leukocytes (white blood cells) which, in circulation, are responsible for body immunity. They have large nuclei, long life spans (years; cf. Problem 13.9), and account for about one-fourth of all leukocytes.

16.8   Describe *lymphoid leukemia*.

Lymphoid leukemia is a form of cancer characterized by an uncontrolled production of lymphocytes that remain immature. These leukemic cells eventually appear in such great numbers that they crowd out the normal, functioning cells. Chemotherapeutic drugs are fairly effective in treating lymphoid leukemia.

16.9   What are *macrophages*?

Macrophages are large phagocytic cells, found in lymphatic cortical tissue, that engulf and destroy foreign substances, damaged cells, and cellular debris before these materials can enter the bloodstream. Thus, harboring of lymphocytes and macrophagic cleansing are the two major functions of lymph nodes.

OBJECTIVE D   *To chart the distribution of lymph nodes.*

Survey   Lymph nodes usually occur in clusters or chains (see Fig. 16-3). Some of the principal groupings are the *popliteal* and *inguinal nodes* of the lower extremity, the *lumbar nodes* of the pelvic region, the *cubital* and *axillary nodes* of the upper extremity, and the *cervical nodes* of the neck. A cluster of mesenteric lymph nodes located in the abdominal cavity and associated with the small intestine is referred to as *Peyer's patches*.

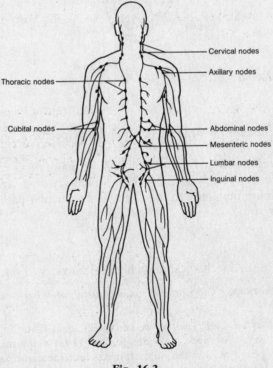

Fig. 16-3

16.10   Are the *tonsils* to be included among the lymph nodes?

The three pairs of tonsils—*pharyngeal* (adenoids), *palatine*, and *lingual*—are not specifically lymph nodes, but are **lymphatic organs** of the pharyngeal region. The function of the tonsils is to combat infections of the ear, nose, and throat regions. Because of the persistent infections that some children suffer, however, the tonsils may become so overrun as to be prime sources of infection themselves. When this happens, removal of certain tonsils may be necessary.

16.11 Characterize the *spleen* and the *thymus* as lymphoid organs.

The spleen (Fig. 16-4) is found in the upper-left part of the abdominal cavity, beneath the diaphragm and hanging from the stomach. Though not too important an organ (in the adult), it participates in the lymphoid tasks of lymphocyte production, blood filtration, and destruction of old erythrocytes.

The thymus (Table 12-4) adapts lymphocytes for the destruction of antigens (see Fig. 16-6 and Problem 16.17). It is thus crucial to the body's immunity system (particularly in the preadult).

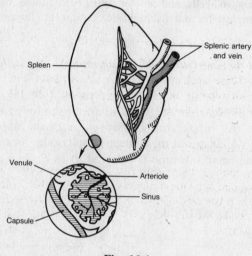

Fig. 16-4

OBJECTIVE E *To distinguish between* specific *and* nonspecific *defenses against infection.*

Survey **Nonspecific mechanisms** give protection against many types of pathogens. Among such are *mechanical barriers, enzymes, interferon, phagocytosis,* and *species resistance.* **Specific mechanisms** furnish immunity to the effect of a particular pathogen (e.g., the disease caused by a particular virus).

16.12 What are some of the mechanical and chemical barriers to infection?

Mechanical barriers. The *skin* and *mucous membranes* bar infectious organisms so long as they remain unbroken.

Chemical barriers. *Lysozyme,* an antimicrobial agent found in tears, saliva, and blood plasma; *pepsin,* an enzyme in the stomach that lyses (disintegrates) many microorganisms; *hydrochloric acid,* which creates a low pH (1 to 2) in the stomach that is lethal to many pathogens; *complement,* a series of enzymatic proteins which are activated by antigen-antibody interactions; *interferon,* a group of proteins that are produced by virus-infected cells and that inhibit viral growth, thus aiding in immunity to that virus.

16.13 What kind of cells provide a second line of defense if the mechanical and chemical barriers have been breached?

The phagocytes: neutrophilis, monocytes, and macrophages.

OBJECTIVE F   *To define* immunity *and to explain how it may be acquired.*

Survey   *Immunity* is the resistance to specific foreign agents (*antigens*), such as microorganisms, viruses, and their toxins, as well as to foreign tissue and other substances. Immunity may be gained in two ways (or, as one says, the body has two modes of *immune response*):   In **antibody-mediated immunity**, an antigen stimulates the body to produce special proteins, called *antibodies*, which can destroy that antigen through an *antigen-antibody reaction*. In **cell-mediated immunity**, lymphocytes become sensitized to an antigen, attach themselves to that antigen, and destroy it.

16.14 What are the chemical characteristics of antigens?

Antigens are usually large ($MW > 10^4$), complex molecules, such as proteins, polysaccharides, and mucopolysaccharides. Some small molecules can act as incomplete antigens, or *haptens*. Alone, these small molecules are not capable of stimulating immune responses, but upon combining with *carrier proteins* they create a complex that is antigenic.

16.15 What are the chemical characteristics of antibodies?

Antibodies are gamma globulins (Problem 13.6) composed of four interlinked polypeptide chains, two short (*light chains*) and two long (*heavy chains*); see Fig. 16-5. All antibodies are structurally similar, except for variations in a few amino acids at the end of the chains, near the *binding sites*. These small differences, however, make the antibodies very specific, and they will combine only with the antigen that stimulated their production.

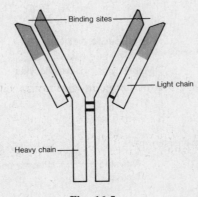

Fig. 16-5

As immunoglobulins, antibodies are subdivided into five classes: *IgG* are the most abundant; they cross the placenta. *IgM* are the largest molecules; they are very active against Gram-negative organisms. *IgA* inhibit the entrance of antigens into the body; they are found in nasal, salivary, lacrimal, bronchial, intestinal, and vaginal secretions. *IgD* are of unknown function. *IgE* mediate allergic responses and cause degranulation of mast cells (Problem 4.15), with release of heparin, histamine, and vasoactive substances.

16.16 Does vaccination against a disease confer *active* or *passive* immunity?

Active immunity is conferred when the body manufactures antibodies in response to direct contact with an antigen; e.g., when a person has had a disease or has been exposed to an antigen or has been *vaccinated* with dead or weakened pathogens or altered toxins.

Passive immunity is conferred by the transfer of antibodies from one person to another; the recipient does not produce his or her own antibodies. For example, a gamma globulin shot (someone else's antibodies) can give passive immunity against hepatitis B.

OBJECTIVE G   *To identify the components of the* immune system.

Survey   The immune system is composed of lymphocytes (*B-lymphocytes* and *T-lymphocytes*), substances released from lymphocytes (antibodies and *lymphokines*), and macrophages (Problem 16.9). Figure 16-6 shows the development of the two kinds of lymphocytes and Fig. 16-7 is a scheme of the immune system as a whole.

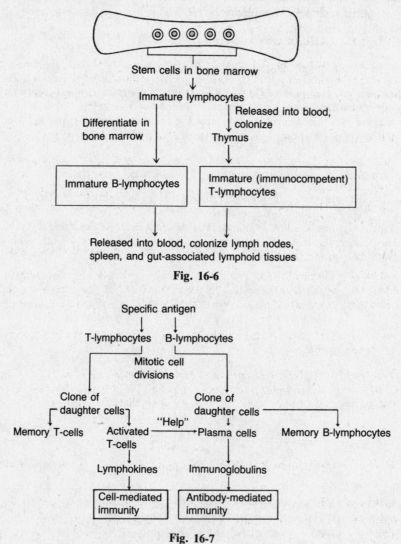

Fig. 16-6

Fig. 16-7

16.17   Describe the functions of the T- and B-lymphocytes.

   T-lymphocytes (thymus-derived lymphocytes) produce cell-mediated immunity. They comprise 70–80% of circulating lymphocytes and become associated with lymph nodes, spleen, and other lymphoid tissues. Upon interaction with a specific antigen, they become sensitized and differentiate into several types of daughter cells, such as: *memory T-cells*, which remain inactive until future exposure to the same antigen; *killer T-cells*, which combine with the antigen on the surface of the foreign cells, causing lysis of the foreign cells and the release of lymphokines; *helper T-cells*, which stimulate B-lymphocytes to differentiate into plasma cells that produce antibodies; *suppressor T-cells*, which suppress antibody formation and cytotoxic action of other T-lymphocytes; and *delayed-hypersensitivity*

T-cells, which initiate a type of cell-mediated immunity called "delayed hypersensitivity" and also release lymphokines.

B-lymphocytes (bone-marrow-derived lymphocytes) produce antibody-mediated immunity. They comprise 20–30% of circulating lymphocytes and (like T-lymphocytes) become associated with lymph nodes, spleen, and other lymphoid tissues. As B-lymphocytes become sensitized to an antigen they proliferate and differentiate to form clones of daughter cells that produce antibodies specifically against that antigen. Some of the B-lymphocytes are "helped" by T-cells to differentiate into plasma cells which also produce large amounts of specific antibodies.

16.18 Give some examples of *lymphokines*.

Interferon (Problem 16.12); *chemotactic factors*, which attract phagocytes; *macrophage-activating factors*; *migration-inhibiting factors*, which inhibit the movement of macrophages and so keep them at the site of immune response; and *transfer factors*, which cause lymphocytes to become sensitive to, and to attack, the invading organism.

16.19 What role does complement (Problem 16.12) play in the immune response?

The activated *complement system* (nine enzyme precursors, designated C-1, . . . , C-9) helps to provide protection against an invading organism by: (i) causing *lysis* of bacteria or other invading cells; (ii) enhancing the *inflammation* response; (iii) attracting phagocytes to the area (*chemotaxis*); (iv) enhancing *phagocytosis* by coating microbes so that phagocytes can better hold them; *neutralizing* viruses (rendering them nonvirulent).

16.20 What part does cell-mediated immunity play in the prevention of cancer?

Potential cancer cells are marked by certain antigens on their surface. These can sensitize T-lymphocytes, which then interact with the antigens and destroy the abnormal cells. Clinical cancer may, therefore, result when the cell-mediated immune system does not function properly.

16.21 Do both the cell-mediated and antibody-mediated immune systems play a role in the rejection of tissue transplants?

Yes: tissue transplants (except when obtained from an identical twin) contain protein molecules, called *histocompatibility antigens*, that are foreign to the recipient and therefore trigger the recipient's immune mechanisms. Both the cell- and antibody-mediated responses may act to destroy the donated tissue.

OBJECTIVE H   *To understand* transfusion reactions *as special tissue rejections.*

Survey   Red blood cells have large numbers of antigens (*agglutinogens*) present on their plasma membrane; these can initiate antibody (*agglutinin*) production and, therefore, antigen-antibody reactions. One of the groups of antigens most likely to cause blood transfusion reactions is the *ABO system* (Table 16-1).

Table 16-1

Genotype	Blood Group	Antigens (agglutinogens)	Antibodies (agglutinins)
OA or AA	A	A	Anti-B
OB or BB	B	B	Anti-A
AB	AB	A and B	*none*
OO	O	*none*	Anti-A and Anti-B

16.22 Antigens of the ABO system are inherited factors and are present on the RBC membranes at the time of birth. Can the same be said of the corresponding antibodies?

No: the antibodies begin to appear about 3 to 8 months after birth and reach maximum concentrations at about 10 years of age. This phenomenon is unexplained.

16.23 What happens if the recipient of a blood transfusion and the donor(s) are improperly matched in blood type?

An antigen-antibody reaction (*transfusion reaction*) occurs in the recipient, causing the red cells to clump together (*agglutination*). The clumps may block or occlude small vessels in the circulatory system (*thromboembolism*); the red cells may also rupture (*hemolyze*) and release hemoglobin into the plasma. A severe transfusion reaction will raise plasma bilirubin levels, leading to jaundice. In extreme cases, renal tubular damage, anuria, and death are possible.

16.24 What are the preferred and permissible blood types for transfusions?

See Table 16-2 (AB = universal recipient, O = universal donor).

Table 16-2

Recipient's Type	Preferred Donor's Type	Permissible Donor's Type
A	A	O
B	B	O
AB	AB	A, B, O
O	O	*none* (only O)

OBJECTIVE I   *To describe the events leading to* erythroblastosis fetalis.

Survey   *Erythroblastosis fetalis* is a hemolytic disease of the newborn resulting from an antigen-antibody reaction associated with the *Rh system* of the blood. Rh antigens (first found in the *Rh*esus monkey) are present on the RBC membranes of about 85% of the population; these people are classed as Rh^+, while the 15% or so remaining are classed as Rh^-. Rh^- individuals do not develop antibodies against Rh antigens until they are exposed to Rh^+ blood.

16.25 Give the steps in the development of erythroblastosis fetalis.

(1) An Rh^- mother and an Rh^+ father have an Rh^+ baby. (2) At birth, some of the Rh^+ red cells from the baby enter the mother's circulation via damaged placental tissues. (3) As the Rh antigens are foreign to the mother, she begins to produce anti-Rh antibodies. (4) The mother becomes pregnant again, and the fetus is Rh^+. (5) The anti-Rh antibodies cross the placenta and enter the fetal blood. (6) The anti-Rh antibodies from the mother react with the Rh antigens on the fetal red cells, causing agglutination and hemolysis.

16.26 How can erythroblastosis fetalis be prevented?

Give the (Rh^-) mother an injection of anti-Rh antibodies (*Rhogam injection*) within 72 hours after delivery or abortion. These anti-Rh antibodies will tie up and destroy any absorbed Rh^+ cells and thus prevent the mother from being sensitized and producing her own anti-Rh antibodies. The antibodies the mother received by passive immunity only last for several months; therefore the fetus of the next pregnancy is protected.

Review Questions

Multiple Choice

1. The immune system is involved with: (*a*) destruction of abnormal or mutant cell types which arise within the body, (*b*) the process of aging, (*c*) obstructing organ transplants, (*d*) all the above.

2. Active immunity is: (*a*) borrowed from an active disease case, (*b*) developed in direct response to a disease agent, (*c*) the product of borrowed antibodies, (*d*) passive immunity that is activated.

3. In the cell-mediated immune response, T-lymphocytes divide and secrete: (*a*) antigens, (*b*) plasmogens, (*c*) collagens, (*d*) lymphokines.

4. B-lymphocytes are primarily involved in: (*a*) humoral immunity, (*b*) autoimmune disorders, (*c*) graft rejection, (*d*) cell-mediated immunity.

5. Plasma cells are: (*a*) responsible for specific immunity, (*b*) derived from lymphocytes, (*c*) involved in the production of antibodies, (*d*) all the above.

6. Transfusing a person with plasma proteins from a person or an animal that has been actively immunized against a specific antigen provides: (*a*) active immunity, (*b*) passive immunity, (*c*) autoimmunity, (*d*) anti-immunity.

7. A person with type AB blood has: (*a*) both anti-A and anti-B antibodies, (*b*) only anti-O antibodies, (*c*) neither anti-A nor anti-B antibodies, (*d*) no antigens.

8. When an Rh-negative mother and an Rh-positive father produce an Rh-negative baby: (*a*) the mother may develop Rh antibodies unless she is treated with Rhogam within 72 hours after birth of the baby, (*b*) the baby will be born with a yellowish color, (*c*) the mother will not develop any Rh antibodies, (*d*) the baby has a high risk of congenital defects.

9. Which of the following is *not* a body mechanism of defense? (*a*) mucous membranes, (*b*) microglia, (*c*) white blood cells, (*d*) mRNA.

10. Small molecules that combine with larger ones to create an antigenic compound are called: (*a*) trypsin, (*b*) adipose, (*c*) ionic, (*d*) haptens.

11. Lymphocytes from the bone marrow that do not reach the thymus gland mature into: (*a*) T-lymphocytes, (*b*) macrophages, (*c*) B-lymphocytes, (*d*) neutrophils.

12. Nonself substances against which the body launches an immune response are called: (*a*) antibodies, (*b*) antigens, (*c*) anticlines, (*d*) agglutinins.

13. The antibodies produced and secreted by B-lymphocytes are soluble proteins called: (*a*) immunoglobulins, (*b*) immunosuppressants, (*c*) lymphokines, (*d*) histones.

14. Which of the following is *not* a major organ of the lymphatic system? (*a*) lymph nodes, (*b*) thymus, (*c*) kidney, (*d*) spleen.

15. Which is the proper order of events in cell-mediated immunity? (*a*) antigen enters tissue, macrophages engulf antigen, antigen passed to members of a clone of lymphocytes, T-lymphocytes attack antigen-bearing agents after sensitization; (*b*) antigen enters tissues, antigen passed to members of a clone of lymphocytes, sensitization of lymphocytes, macrophages engulf antigen, T-lymphocytes attack antigen-

bearing agents; (c) antigen enters tissues, macrophages engulf antigen, antigen passed to members of a clone of lymphocytes, sensitization of lymphocytes, B-lymphocytes secrete antibodies that react with antigen-bearing agents; (d) antigen enters tissues, sensitization of lymphocytes, antigen passed to members of a clone of lymphocytes, macrophages engulf antigen, T-lymphocytes attack antigen-bearing agents.

16. A dilation of the lymphatic duct in the lumbar region that marks the beginning of the thoracic duct is the: (a) cisterna chyli, (b) right lymphatic duct, (c) Peyer's patches, (d) mesenteric lymph nodes.

17. The spleen does not: (a) house lymphocytes; (b) filter from the blood foreign particles, damaged red blood cells, and cellular debris; (c) contain phagocytes; (d) change undifferentiated lymphocytes into T-lymphocytes.

18. The small depression on one side of a lymph node where the artery and vein enter and leave is called: (a) lymph sinus, (b) hilus, (c) glenoid cavity, (d) calyx.

19. An Rh⁻ mother and an Rh⁺ father are preparing for the birth of their first child. (a) They should be prepared for the mother to receive a Rhogam injection. (b) There should be no problem with this pregnancy. (c) There may be no problems with future pregnancies. (d) All the above.

True/False

1. Valves are present in lymphatic vessels.

2. A person with type B blood has B antibodies.

3. Immunoglobulins are composed of two heavy and two light chains. It is the light chains exclusively that are responsible for binding with the antigen.

4. When antigenically stimulated, B-lymphocytes proliferate and form plasma cells.

5. A negative immune response leads to tolerance of the antigen.

6. A person who encounters a pathogen and has a primary immune response develops passive immunity.

7. There are five major types of immunoglobulins: IgG, IgA, IgD, IgL, and IgE.

8. The interaction of antigen with antibody is highly specific.

9. An antigen is a substance produced in the body to prevent a disease.

10. Animals cannot produce antibodies against laboratory-synthesized antigens.

11. Passive immunity is the transfer of antibodies developed in one individual into the body of another.

12. T- and B-lymphocytes may cooperate in response to a particular antigen.

Completion   (Insert + for *clumping* or − for *no clumping*.)

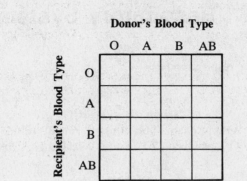

Chapter 17

Respiratory System

OBJECTIVE A *To define* respiration.

Survey Any cell requires a continuous supply of oxygen (O_2) and must continuously eliminate a metabolic waste product, carbon dioxide (CO_2). The term *respiration*, which on the macroscopic level simply means "breathing," on the cellular level denotes all processes by which cells utilize O_2, produce CO_2, and convert energy into biologically useful forms such as ATP.

17.1 Distinguish between *external*, *internal*, and *cellular* respiration.

    External respiration is the process by which gases are exchanged between the blood and the air. *Internal* respiration is the process by which gases are exchanged between the blood and the cells. *Cellular* respiration is the process by which cells use O_2 for metabolism and give off CO_2 as a waste.

OBJECTIVE B *To identify the basic components of the respiratory system.*

Survey Refer to Fig. 17-1. The major passages of the respiratory system are the *nasal cavity*, *pharynx*, *larynx*, and *trachea*. Within the *lungs*, the *trachea* branches into *bronchi*, *bronchioles*, and finally, *alveoli* (Fig. 17-5).

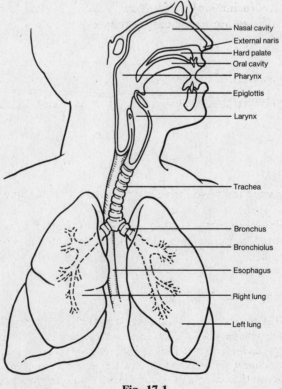

Nasal cavity
External naris
Hard palate
Oral cavity
Pharynx
Epiglottis
Larynx
Trachea
Bronchus
Bronchiolus
Esophagus
Right lung
Left lung

Fig. 17-1

17.2   Distinguish between the *conducting division* and the *respiratory division* of the respiratory system.

The conducting division includes all cavities and structures that transport gases to and from the microscopic air pockets (alveoli) in the lungs. The alveoli constitute the respiratory division.

17.3   Which of the following help(s) to ventilate the lungs? (*a*) rib cage, (*b*) thoracic muscles, (*c*) oral cavity, (*d*) pleural membranes, (*e*) specialized neural tissue in vessels and brain, (*f*) all the above.

(*f*).

17.4   What physical requirements must be satisfied for the respiratory system to function effectively?

(1) The membranes through which gases are exchanged with the circulatory system must be thin and differentially permeable so that diffusion can occur easily. (2) These membranes must be kept moist so that O_2 and CO_2 can be dissolved. (3) A rich blood supply must be present. (4) The surfaces for gas exchange must be located deep in the body so that the incoming air can be sufficiently warmed, moistened, and filtered. (5) There must be an effective pumping mechanism to replenish the air continually.

OBJECTIVE C   *To identify the functions of the respiratory system.*

Survey   (1) *Gaseous exchange* for the cellular respiratory process; (2) *sound production* (vocalization) as expired air passes over the vocal cords; (3) assistance in *abdominal compression* during micturition (urination), defecation (passing of feces), and parturition (childbirth); (4) *coughing* and *sneezing* (self-cleaning reflexes).

17.5   What are the two phases of breathing?

Breathing, or pulmonary ventilation, consists of an *inspiration* (inhalation) phase, and an *expiration* (exhalation) phase.

OBJECTIVE D   *To examine the* nose, nasal cavity, *and* paranasal sinuses *as respiratory structures.*

Survey   The nose includes an external portion that juts out from the face, and an internal cavity for the passage of air. The paranasal sinuses (Problem 6.19) help, in a small way, to warm and moisten inspired air.

17.6   Specify the anatomy of the nasal cavity.

See Fig. 17-2. The nasal cavity is divided into two lateral halves, each referred to as a *nasal fossa*, by the *nasal septum*. The *vestibule* is the anterior expanded portion of a nasal fossa. In the lateral walls of either fossa are three shell-like concavities called *conchae* (*superior*, *middle*, and *inferior*); air passageways, or *meatuses*, connect the conchae. The external *nares* (nostrils) open anteriorly into the nasal cavity, and the *choanae* (posterior nares) communicate posteriorly with the *nasopharynx*.

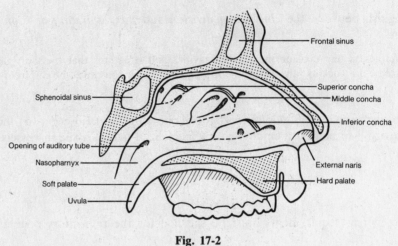

Frontal sinus

Superior concha
Middle concha

Sphenoidal sinus

Inferior concha

Opening of auditory tube

Nasopharynx

External naris

Soft palate

Hard palate

Uvula

Fig. 17-2

17.7   What sorts of tissue line the nasal cavity?

The lining of the vestibules is nonkeratinized stratified squamous epithelium (Table 4-2); this epithelium rapidly divides and supports protective nasal hairs, or *vibrissae*. The conchae of the nasal fossae are lined with pseudostratified ciliated columnar epithelium (Table 4-1) that secretes mucus to trap inspired airborne particles, such as dust, pollen, or smoke. Specialized columnar epithelium, called *olfactory epithelium*, lines the upper medial portion of the nasal cavity, where it responds to odors.

17.8   Why are nosebleeds common?

The nasal epithelia are highly vascular, with the capillaries close to the surface, and are extensive. This serves to warm inspired air, but it also makes us susceptible to *epistaxes* (nosebleeds).

OBJECTIVE E   *To distinguish three regions of the* pharynx.

Survey   The pharynx is divided on the basis of location and function into a *nasopharynx*, an *oropharynx*, and a *laryngopharynx* (see Fig. 17-3). The auditory (Eustachian) canals, uvula, and pharyngeal tonsils (Problem 16.10) are in the nasopharynx; the palatine and lingual tonsils are in the oropharynx. The oropharynx and laryngopharynx have respiratory and digestive functions, while the nasopharynx serves only the respiratory system.

17.9   How does the uvula serve both the respiratory system and the digestive system?

The pendulous uvula hangs from the middle of the lower border of the soft palate. During swallowing, the soft palate and uvula are elevated, closing off the nasal cavity so that food cannot enter.

OBJECTIVE F   *To identify the anatomical structures of the* larynx *associated with sound production and breathing*.

Survey   The larnyx (*voice box*) forms the entrance into the trachea. A primary function of the larynx is to prevent food or fluid from entering the trachea and lungs during swallowing, and to permit the passage of air into the trachea at other times. A secondary function is to produce sound vibrations.

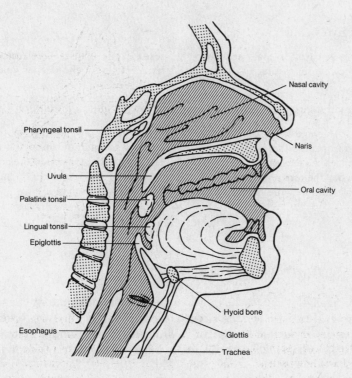

Fig. 17-3

17.10 Which cartilages of the larynx are paired? which are unpaired? which is the largest and most prominent?

Refer to Fig. 17-4. The larynx is a roughly triangular box composed of nine hyaline cartilages; three are large, single structures and six are smaller, paired structures. The *anterior thyroid cartilage* ("Adam's apple") is the largest. The spoon-shaped *epiglottis* has a cartilaginous framework. The lower portion of the larynx is formed by the ring-shaped *cricoid cartilage*. The three paired cartilages are the *arytenoid cartilages*, which support the vocal cords, and the *cuneiform* and *corniculate cartilages*, which aid the arytenoid cartilages.

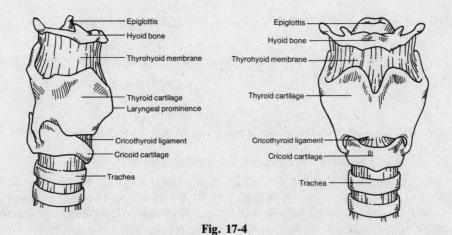

Fig. 17-4

17.11 Why not "Eve's apple"?

During puberty, the male sex hormone, testosterone, causes accelerated growth of the larynx, especially the thyroid cartilage. The larger larynx accounts for the deeper voice of males.

17.12 Explain how the laryngeal muscles aid in swallowing and in *phonation* (speech).

Extrinsic laryngeal muscles elevate the larynx during swallowing, closing the glottis over the epiglottis and opening the esophagus for food or fluid to enter. Contraction of the *intrinsic* laryngeal muscles changes the tension of the vocal cords. The greater the tension of the cords, the more rapid their vibration under the airstream and the higher the pitch of the sound (Problem 11.16). The greater the amplitude of vibration, the louder the sound. *Whispering* is phonation in which the vocal cords do not vibrate.

OBJECTIVE G   *To describe the* bronchial tree.

Survey   Refer to Fig. 17-5. The trachea divides inferiorly to form the *right* and *left primary bronchi*. These branch into *secondary* (*lobar*) *bronchi*, which in turn branch into many *tertiary* (*segmental*) *bronchi* that terminate in *bronchioles*. The entire system of branches is called the *bronchial tree*. Hyaline cartilaginous rings or partial rings support the trachea and the tree.

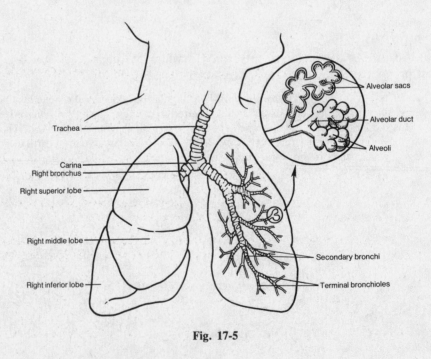

Fig. 17-5

17.13 Describe three protective features of the trachea and bronchial tree.

The cartilaginous framework maintains patent (open) lumina. Mucus-secreting pseudostratified ciliated columnar epithelium (cf. Problem 17.7) lines the lumina, which traps airborne particles and moves this debris toward the pharynx where it may be swallowed. Irritation to the epithelial lining of the trachea or bronchial tree elicits a violent coughing reflex which has a cleansing action on the alveoli.

OBJECTIVE H   *To continue the bronchial tree to the terminal* alveoli.

Survey   Branchioles end in *terminal bronchioles*, which branch into many *alveolar ducts* that lead directly into *alveolar sacs*, themselves clusters of many microscopic *alveoli*. Alveolar ducts are lined with simple cuboidal epithelium, whereas alveoli are lined with simple squamous epithelium. Gas exchange with the blood of the circulatory system occurs through the thin-walled, moistened alveoli (see Problem 17.2).

17.14   What are the functions of *septal cells* and *alveolar macrophages*?

   Small septal cells, dispersed among the cells of the simple squamous epithelium lining an alveolus, secrete a phospholipid *surfactant* that lowers the surface tension. Also found in the alveolar wall are phagocytic alveolar macrophages (*dust cells*) that remove dust particles or other debris from the alveolus.

17.15   Diagram the process of external respiration (Problem 17.1) on the alveolar level.

   See Fig. 17-6.

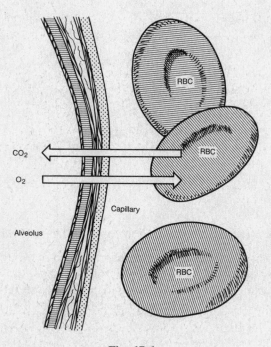

Fig. 17-6

OBJECTIVE I   *To describe the* lungs.

Survey   The lungs are large, paired organs within the thoracic cavity, separated from each other by the mediastinum. Each lung is composed of *lobes*, and these, in turn, of *lobules* that contain the alveoli. The left lung has a *cardiac notch*, has one *fissure* and two lobes, and contains eight *bronchial segments*. The right lung lacks a notch, has two fissures and three lobes, and contains ten bronchial segments.

17.16   In reference to the lung, define the terms *apex*, *base*, *hilum*, *costal surface*, *mediastinal surface*, and *diaphragmatic surface*.

See Fig. 17-7. Each lung presents four borders that match the contour of the thoracic cavity. The *mediastinal surface* is slightly concave and contains a vertical slit, the *hilum*, through which pulmonary vessels, primary bronchus, and branches of the vagus nerve pass. The inferior *base* of the lung has a *diaphragmatic surface* in contact with the diaphragm. The top of the lung is the *apex*, and the broad, rounded surface in contact with membranes covering the ribs is the *costal surface*.

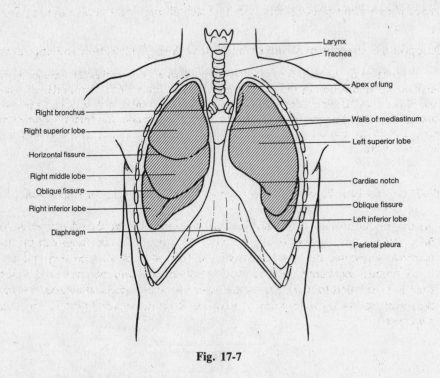

Fig. 17-7

OBJECTIVE J   *To describe the* pleurae *and to explain their respiratory significance.*

Survey   Pleurae are two-layer, serous membranes associated with the lungs. The inner layer, or *visceral pleura*, is attached to the surface of the lungs; the outer layer, or *parietal pleura*, lines the thoracic cavity. Pleurae serve to lubricate the lungs, and they assist in creating respiratory pressure. (For an important nonrespiratory function, see Problem 1.17.)

17.17   How do the pleurae perform their respiratory functions?

A damp *pleural cavity* exists between the visceral and parietal pleurae of each lung; this moist environment serves to lubricate the lungs in their constant motion. Air pressure in each pleural cavity (the *intrathoracic pressure*) is slightly below atmospheric (−2.5 mmHg, approximately) in the resting lungs, and becomes even lower during inspiration, causing air to inflate the lungs.

17.18   Define *pleurisy*.

Pleurisy, or inflammation of the pleurae, is usually secondary to some other respiratory disease. Inspiration may become painful, and fluid may collect in the pleural cavity.

OBJECTIVE K *To examine the mechanics of breathing.*

Survey Inspiration (Problem 17.5) occurs when contraction of the respiratory muscles (diaphragm, internal intercostals; Table 7-6) causes an increase in thoracic volume, with expansion of the lungs and a decrease in intrathoracic and *intrapulmonic* (alveolar) pressures. Air enters the lungs when intrapulmonic pressure falls below atmospheric pressure (760 mmHg at sea level). Expiration follows passively as thoracic volume decreases and intrapulmonic pressure rises above atmospheric, with recoil of the rib cage and contraction of the lungs.

17.19 Describe the changes in shape of the thorax during inspiration and expiration.

Contraction of the dome-shaped diaphragm downward increases the thoracic vertical dimension. A simultaneous contraction of the external intercostals (Table 7-6) increases the side-to-side and front-to-back dimensions. During deep inspiration or forced breathing, the scalenus (Fig. 7-13) and pectoralis minor (Table 7-9) become involved. During forced expiration, the internal intercostals are contracted, depressing the rib cage. Contracting the abdominal muscles also forces air from the lungs by elevating the diaphragm.

OBJECTIVE L *To identify the various volumes of air exchanged in respiration.*

Survey As diagrammed in Fig. 17-8, *total lung capacity* may be expressed as the sum of four volumes: *tidal volume*, the volume of air moved into and out of the lungs during normal breathing; *inspiratory reserve*, the maximum volume beyond tidal volume that can be inspired in one deep breath; *expiratory reserve*, the maximum volume beyond tidal volume that can be forcefully exhaled following a normal expiration; and *residual volume*, the air that remains in the lungs following a forceful expiration. Respiratory air volumes are measured with the *spirometer*.

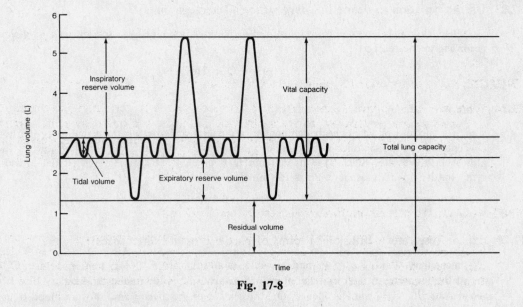

Fig. 17-8

17.20 Account for the variability of respiratory air volumes.

Clinically speaking, it is important to know the amount of air that is breathed in at a given time, as well as the difficulty in breathing. The amount of air exchanged during pulmonary ventilation varies from person to person according to age, sex, activity, and general health.

17.21 Calculate the *minute respiratory volume* of an individual who has a tidal volume of 500 mL and a respiratory rate of 12 breaths per minute.

Minute respiratory volume is the volume of air moved in normal ventilation in one minute. Therefore,

$$\text{Minute respiratory volume} = (\text{tidal volume}) \times (\text{respiratory rate})$$
$$= (0.500 \text{ L}) \times (12 \text{ min}^{-1}) = 6 \text{ L/min}$$

a typical value.

17.22 Define *alveolar ventilation*.

Alveolar ventilation is the volume of air exchanged in one minute in the alveoli (for transport to the cells). Thus,

$$\text{Alveolar ventilation} = [(\text{tidal volume}) - (\text{dead-space volume})] \times (\text{respiratory rate})$$

OBJECTIVE M   *To describe the transport of gases between the lungs and the cells.*

Survey   Of the O_2 transported in the blood, only a small amount is dissolved in the plasma; up to 99% is carried on hemoglobin molecules in the erythrocytes (Problem 13.12). CO_2 carried in the blood is mostly converted to bicarbonate ions in the erythrocytes and released into the plasma; unconverted CO_2 is also carried dissolved in the plasma and on hemoglobin molecules and certain plasma proteins.

17.23 Use an equation to define the *oxygenation* of hemoglobin.

Hemoglobin is converted from bluish-red *deoxyhemoglobin* (Hb) to scarlet *oxyhemoglobin* (HbO_2) according to the reaction

$$Hb + O_2 \rightleftharpoons HbO_2$$

17.24 What is meant by *partial pressure*?

In a mixture of gases, each component gas exerts a *partial pressure* that is proportional to its concentration in the mixture. For example, since air is 21% O_2, this gas is responsible for 21% of the atmospheric pressure. Since 21% of 760 mmHg is equal to 160 mmHg, the partial pressure of O_2, symbolized P_{O_2}, in atmospheric air is 160 mmHg. Similarly,

$$P_{CO_2} = 0.3 \text{ mmHg}$$

17.25 Explain respiratory diffusion in terms of partial pressure differences.

In capillary blood, $P_{CO_2} = 46$ mmHg; while, in alveolar air, $P_{CO_2} = 39$ mmHg. Hence, CO_2 diffuses out of the bloodstream and into the alveoli. Similarly, $P_{O_2} = 100$ mmHg in capillary blood; while, in alveolar air, $P_{O_2} = 101$ mmHg. Hence, O_2 diffuses out of the alveoli and into the bloodstream.

17.26 What factors precipitate release of O_2 from hemoglobin to body tissues?

(1) A decreased concentration of CO_2 in the plasma; (2) a decreased blood pH (i.e., an increased H^+ concentration); (3) an increased body temperature.

OBJECTIVE N   *To know the role of the respiratory system in the acid-base balance of the body.*

Survey   The presence of the enzyme *carbonic anhydrase* in the erythrocytes causes about 67% of the CO_2 in blood quickly to combine with water to form carbonic acid, most of which dissociates into bicarbonate and hydrogen ions:

$$CO_2 + H_2O \rightleftharpoons H_2CO_3 \rightleftharpoons HCO_3^- + H^+$$

17.27   Define *alkali reserve* and *chloride shift*.

Bicarbonate ions (HCO_3^-), which make up a large part of the blood buffer system, constitute the *alkali reserve*. As these ions leave erythrocytes, they cause an excess of negative charge, which is relieved by the diffusion of chloride ions (Cl^-) from the blood into the cells. This movement of chloride ions is the *chloride shift*.

17.28   Define *respiratory acidosis* and *respiratory alkalosis*.

Respiratory acidosis, or a blood pH below 7.35, occurs when CO_2 is not eliminated at a normal rate, so that vascular P_{CO_2} increases. Lung disease or an obstructed respiratory passageway may cause respiratory acidosis.

Respiratory alkalosis, or a blood pH above 7.45, occurs when CO_2 is eliminated too rapidly, so that vascular P_{CO_2} decreases. Either hyperventilation or the action of certain drugs (such as excessive aspirin) on the respiratory control center of the brain may produce respiratory alkalosis. The effects of hyperventilation are rapidly reversed if the person breathes into a paper bag, inhaling expired air and so causing the vascular P_{CO_2} to rise.

OBJECTIVE O   *To become familiar with the neural and chemical regulation of respiration.*

Survey   Controlling centers for breathing in the CNS were identified in Fig. 9-7; the rhythmicity area of the medulla is actually composed of separate *expiratory* and *inspiratory centers*. The medulla also contains chemoreceptors concerned with respiration (Fig. 17-9), as do the *carotid* and *aortic bodies* in the thorax (Fig. 17-10).

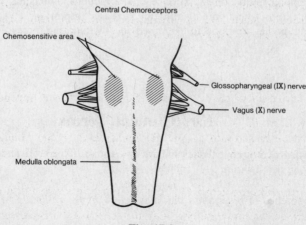

Fig. 17-9

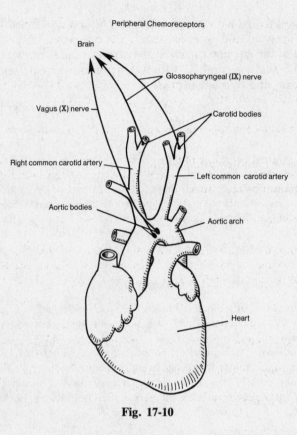

Fig. 17-10

17.29 How do the neural respiratory centers operate?

Refer to Fig. 9-7. The rhythmicity area of the medulla oblongata consists of two intermingled groups of neurons. When the *inspiratory group* is excited (via the apneustic center), the respiratory muscles are signaled to accomplish inbreathing; at the same time, the other, *expiratory group* is inhibited. After about two seconds, the reciprocal process occurs: the pneumotaxic center stimulates the expiratory group to signal for outbreathing, with simultaneous inhibition of the inspiratory group.

17.30 How do the respiratory chemoreceptors operate?

Central chemoreceptors, on the surface of the medulla oblongata, respond to increased P_{CO_2} or decreased pH of cerebrospinal fluid by initiating increased inspiration. Peripheral chemoreceptors, in the carotid and aortic bodies, respond to decreased P_{O_2} by initiating increased inspiration.

Key Clinical Terms

Anoxia　A severe shortage of oxygen in tissues and organs. Anoxia of the brain results in cell destruction within 30 seconds, and, generally, in death within 5–10 minutes.

Apnea　A temporary absence of respiration that may follow *hyperventilation* (q.v.).

Asphyxia　Suffocation.

Asthma   A disease characterized by recurrent attacks of *dyspnea* (q.v.). It is usually an allergic response to disagreeable plants, animals, or food products, resulting in contraction of the bronchial muscles; sometimes, however, psychological factors are responsible.

Bronchitis   Inflammation of the mucous membrane lining the bronchial tubes. Cigarette smoking, air pollution, and allergies may cause the disease or be contributing factors.

Cleft lip (*harelip*)   A genetic developmental disorder in which the two sides of the upper lip fail to fuse.

Cleft palate   A developmental deformity of the hard palate, resulting in a persistent opening between the oral and nasal cavities. The condition may be hereditary or a complication of some disease (e.g., German measles) contracted by the mother during pregnancy.

Dyspnea   Difficult breathing.

Emphysema   A breakdown of the alveolar walls that decreases alveolar surface area and increases the size of air spaces distal to the terminal bronchioles. It is a frequent cause of death among heavy cigarette smokers, and may also result from severe air pollution.

Epistaxis   Nosebleed.

Hiccup   Spasmodic contraction of the diaphragm, causing a short, abrupt inspiration. Also spelled *hiccough*.

Hyperventilation   Excessive inhalation and exhalation.

Laryngitis   Inflammation of the larynx.

Pleurisy   Inflammation of the pleurae.

Pneumonia   Acute infection and inflammation of the lungs, with exudation of fluids into, and consolidation (collapse) of, lung tissue.

Rhinitis   Inflammation of the nasal mucosa.

Sinusitis   Inflammation of the mucous membrane of one or another of the paranasal sinuses—usually secondary to a nasal infection.

Tuberculosis   An inflammatory disease of the lungs, caused by the tubercle bacillus, in which the tissue caseates (becomes cheesy) and ulcerates. The disease is usually contracted by inhaling air sneezed or coughed by someone with an active case.

Review Questions

Multiple Choice

1. Which is *not* a structure of the respiratory system? (*a*) pharynx, (*b*) bronchus, (*c*) larynx, (*d*) hyoid, (*e*) trachea.

2. The roof of the nasal cavity is formed primarily by the: (*a*) hard palate, (*b*) cribriform plate, (*c*) superior concha, (*d*) vomer, (*e*) sphenoid.

3. The cartilages upon which the vocal cords are attached are the: (a) thyroid and arytenoid, (b) thyroid and cricoid, (c) cuneiform and cricoid, (d) thyroid and corniculate.

4. Pulmonary vessels, nerves, and a bronchus enter or leave the lung at the: (a) cardiac notch, (b) apex, (c) capsule, (d) hilus, (e) base.

5. Neither the trachea nor the bronchi contain: (a) hyaline cartilage, (b) ciliated columnar epithelium, (c) goblet cells, (d) simple squamous epithelium.

6. Pharyngeal tonsils are located in the: (a) nasopharynx, (b) oral cavity, (c) nasal cavity, (d) oropharynx, (e) lingualopharynx.

7. It is false that: (a) slacker vocal cords produce lower sounds; (b) during swallowing, the epiglottis is depressed to cover the glottis; (c) in whispering, the vocal cords do not vibrate; (d) testosterone secretion influences laryngeal development during puberty.

8. The serous membrane in contact with the lung is the: (a) parietal pleura, (b) pulmonary mesentery, (c) pulmonary peritoneum, (d) visceral pleura.

9. In the blood, most of the CO_2 is transported as: (a) carboxyhemoglobin, (b) HCO_3^-, (c) carbaminohemoglobin, (d) dissolved CO_2.

10. Peripheral chemoreceptors are located in the: (a) lung tissue, (b) pons and medulla, (c) aortic and carotid bodies, (d) myocardium.

11. As CO_2 produced in the tissues combines with H_2 in the blood: (a) carbonic acid is formed, (b) Cl^- enters the blood, (c) most of the HCO_3^- from the carbonic acid leaves the RBCs for the plasma, (d) all the above.

12. When blood CO_2 levels rise: (a) only the rate of breathing decreases, (b) respiratory acidosis may occur, (c) peripheral pressure receptors respond, (d) both rate and depth of breathing decrease.

13. The amount of air that is moved in and out of the lungs during normal quiet breathing is called the: (a) vital capacity, (b) tidal volume, (c) residual volume, (d) vital volume, (e) inspiratory volume.

14. Which of the following is not a structural feature of the left lung? (a) superior lobe, (b) cardiac notch, (c) inferior lobe, (d) middle lobe.

15. Which combination of muscles permits inspiration? (a) internal intercostals–diaphragm, (b) diaphragm–abdominal complex, (c) external intercostals–diaphragm, (d) external–internal intercostals.

16. The maximum amount of air that can be expired after a maximum inspiration is called the: (a) forced expiratory volume, (b) maximum expiratory flow, (c) tidal volume, (d) vital capacity.

17. Surfactant: (a) reduces the surface tension in alveoli, (b) increases the P_{CO_2} levels in blood, (c) is a mucus secreted by goblet cells, (d) reduces friction in the pleural cavity.

18. The basic inspiratory and expiratory centers are located in the: (a) lungs, (b) medulla oblongata, (c) carotid and aortic bodies, (d) pons.

19. Factors determining the extent to which O_2 will combine with hemoglobin are: (a) P_{O_2} in blood, (b) temperature, (c) blood H^+ concentration, (d) all the above.

True/False

1. The nasal septum divides the nose into right and left *nasal cavities*.

2. The pleural cavities are closed, separate cavities within the thorax.

3. In normally functioning lungs, the intrathoracic pressure is always greater than the intrapulmonic pressure.

4. Expiration is usually passive and occurs with the cessation of inspiratory contractions.

5. Active transport mechanisms effect exchange of gases in the lungs and tissues.

6. An increase or decrease in blood P_{CO_2} is always accompanied by a change in plasma H^+ concentration (plasma pH).

7. In the respiratory system, simple squamous epithelium is restricted to the alveoli.

8. The vomer and sphenoid form the bony framework of the nasal septum.

9. Hyperventilation may cause body fluids to become more acidic.

10. The auditory canals open into the nasopharynx.

11. An elevation in pH causes peripheral chemoreceptors to increase the rate and depth of breathing.

12. The partial pressure of a gas is directly proportional to its molecular weight.

13. An increased body temperature increases the ability of hemoglobin to supply O_2 to tissues.

14. The release of O_2 from hemoglobin is termed oxygenation.

15. The fraction of the tidal volume that is ventilated in one minute is known as the minute respiratory volume.

16. An increased concentration of O_2 in the plasma contributes to the release of O_2 from hemoglobin to body tissues.

17. As bicarbonate ions attach to erythrocytes, the balance between positively and negatively charged ions is disturbed, causing the chloride shift.

18. Apnea frequently follows hyperventilation.

Chapter 18

Digestive System

OBJECTIVE A   *To define* digestion *as a mechanical, chemical, and absorptive process.*

Survey   The food we eat is utilized at the cellular level in chemical reactions involving synthesis of enzymes; cellular division, growth, and repair; and production of heat. To become usable by the cells, most food must be mechanically and chemically reduced to forms that can be absorbed through the intestinal wall and transported to the cells by the blood.

18.1   Define as a mechanical and/or a chemical process: *ingestion, mastication, deglutition, peristalsis, absorption, defecation.*

    Ingestion—taking of food into the digestive system by way of the mouth; *mechanical.* **Mastication**—chewing movements to pulverize food and mix it with saliva; *mechanical* (chewing) and *chemical* (salivary action). **Deglutition**—swallowing of food; *mechanical.* **Peristalsis**—rhythmic, wavelike contractions that move food through the digestive tract; *mechanical.* **Absorption**—passage of food molecules through the mucous membrane of the small intestine and into the circulatory or lymphatic systems, for distribution to cells; *mechanical* and *chemical.* **Defecation**—discharge of indigestible wastes, called *feces,* from the digestive tract; *mechanical.*

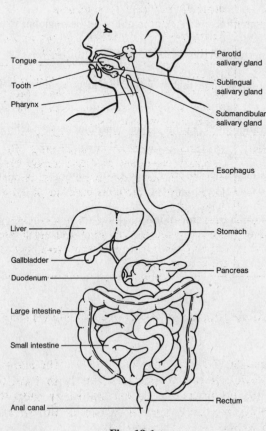

Fig. 18-1

OBJECTIVE B   *To specify the structural and functional organization of the digestive system.*

Survey   See Fig. 18-1. The digestive system can be divided into a tubular *alimentary canal*, or *gastrointestinal (GI) tract*, and *accessory organs*. The GI tract is about 9 m (30 ft) long and extends from the mouth to the anus. The organs of the GI tract include the oral (buccal) cavity, pharynx, esophagus, stomach, small intestine, and large intestine. The accessory digestive organs include the teeth, tongue, salivary glands, liver, gallbladder, and pancreas.

18.2   Distinguish between *viscera* and *gut.*

Although *viscera* is frequently used in reference to the abdominal organs of digestion, the term actually applies to all thoracic and abdominal organs. *Gut* generally refers to the developing stomach and intestines in the embryo.

18.3   What are the basic functions of the different regions of the GI tract?

See Table 18-1.

Table 18-1

Region	Functions
Oral cavity	Ingests food; receives saliva; mastication; initiates digestion of carbohydrates; forms *bolus* (food mass); deglutition
Pharynx	Receives bolus from oral cavity; autonomically continues deglutition of bolus to esophagus
Esophagus	Transports bolus to stomach by peristalsis; esophageal sphincter restricts backflow of food
Stomach	Receives bolus from esophagus; churns bolus with gastric juice to form *chyme*; initiates digestion of proteins; limited absorption; moves chyme into duodenum and prohibits backflow of chyme; regurgitates when necessary
Small intestine	Receives chyme from stomach, along with secretions from liver and pancreas; chemically and mechanically breaks down chyme; absorbs nutrients; transports wastes through peristalsis to large intestine; prohibits backflow of intestinal wastes from large intestine
Large intestine	Receives undigested wastes from small intestine; absorbs water and electrolytes; forms and stores feces, and expels them through defecation reflex

OBJECTIVE C   *To detail the structure of the* serous membranes *associated with the digestive organs of the abdominal cavity.*

Survey   Review Problems 1.20–1.23 and inspect Fig. 18-2. The *parietal peritoneum* lines the wall of the abdominal cavity; it comes together dorsally to form the double-layered *mesentery*, which supports the GI tract. (The *dorsal mesentery* supports the small intestine and the *mesocolon* supports the large intestine.) The peritoneal covering continues around the intestinal viscera as the *visceral peritoneum.*

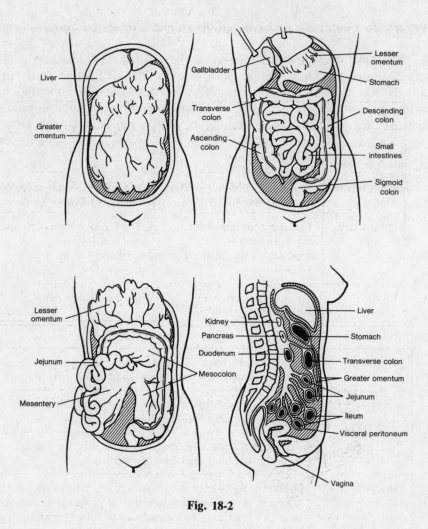

Fig. 18-2

Extensions of the parietal peritoneum serve specific functions. The *falciform ligament* attaches the liver to the diaphragm and anterior abdominal wall. The *lesser omentum* runs between the liver and the lesser curvature of the stomach. The *greater omentum* extends from the greater curvature of the stomach to the transverse colon, forming an apronlike structure over the small intestine. The omentum stores fat, cushions visceral organs, supports lymph nodes, and protects against the spread of infections.

18.4 Distinguish between the *abdominal* and the *peritoneal* cavities.

The abdominal cavity is the space within the confines of the abdominal wall. The peritoneal cavity is the space between the parietal and visceral portions of the peritoneum. Most of the abdominal visceral organs are inside both cavities; a few, such as the retroperitoneal organs (Problem 1.22), are within the abdominal cavity only.

18.5 What is *peritonitis*?

Peritonitis is a bacterial inflammation of the peritoneum caused by trauma, rupture of a visceral organ, ectopic pregnancy, or postoperative complications.

OBJECTIVE D *To identify the four basic layers* (tunicae) *of the GI tract.*

Survey See Table 18-2 and Fig. 18-3.

Table 18-2

Tunica	Structure	Function
Mucosa	Simple columnar epithelium with goblet cells	Secretion and absorption
Submucosa	Highly vascular; contains autonomic nerve fibers	Absorption of nutrients and fluids into capillaries
Muscularis	Circular and longitudinal layers of smooth muscle; modified for sphincters or valves	Segmental contractions and peristalsis
Serosa	Areolar connective tissue and visceral peritoneum	Binding and protection

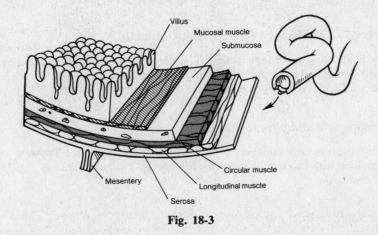

Fig. 18-3

18.6 Specify the autonomic innervation of the tunicae in the abdominal region.

The *submucosal plexus*, or *plexus of Meissner*, provides autonomic innervation to the *muscularis mucosae* (a thin layer of smooth muscle of the tunica mucosa). The *myenteric plexus*, or *plexus of Auerbach*, located between the longitudinal and circular muscle layers of the tunica muscularis, provides the major nerve supply to the GI tract and includes fibers and ganglia from both autonomic divisions.

OBJECTIVE E *To detail the anatomy of the* oral cavity, pharynx, *and* esophagus, *and to state the digestive events of these regions of the GI tract.*

Survey Refer to Fig. 18-4. The *oral cavity* (mouth), or *buccal cavity*, formed by the lips, cheeks, hard and soft palates, and tongue, is a receptacle for food; initiates digestion through mastication with the *teeth*; participates in swallowing; and forms words in speech. The *pharynx*, located posterior to the oral cavity, is a common passageway for the respiratory and digestive systems. The *esophagus* transports food and fluid from the pharynx to the stomach.

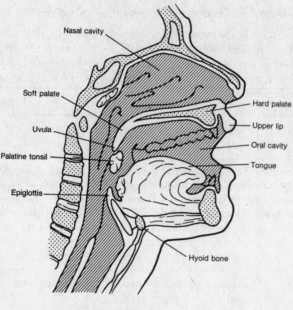

Fig. 18-4

18.7 Give the special adaptations of the four kinds of teeth.

Each jaw holds two pairs of (anteriormost) *incisors*, adapted for cutting and shearing food; one pair of cone-shaped *canines*, adapted for holding and tearing; two pairs of *premolars* (*bicuspids*) and three pairs of *molars*, all with cusps (like spearpoints) for crushing and grinding.

18.8 Define *heterodontia* and *diphyodontia*.

Heterodontia means differentiation of teeth for different tasks (see Problem 18.7). *Diphyodontia* means the provision with two sets of teeth in a lifetime; in humans, there are 20 *deciduous* (*milk*) *teeth*, and 32 *permanent teeth*.

18.9 What are "wisdom teeth"?

If the *third molars*, or *wisdom teeth*, erupt, it is generally between the ages of 17 and 25. Impacted wisdom teeth are common because the jaws are formed and the other teeth are in place at this time.

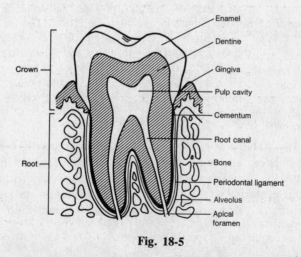

Fig. 18-5

18.10 Diagram a typical tooth and give the functions of the principal structures.

See Fig. 18-5. The exposed crown of a tooth is covered with protective *enamel*. The *dentine* provides structural support to the tooth and surrounds the *pulp cavity* containing nerves and blood and lymph vessels. A thin layer of *cementum* fastens the tooth to the *peridontal ligament* within a socket called an *alveolus*. A *root canal*, in each root, for passage of vessels and nerves communicates with the pulp cavity through an *apical foramen*.

18.11 Where are the *salivary glands* located, and what are the functions of *saliva*?

See Fig. 18-6 and Table 18-3. The 1000 to 1500 mL of saliva secreted daily, in response to parasympathetic stimulation, cleanses the teeth, initiates carbohydrate digestion through the action of *amylase* (Problem 18.24), helps form the bolus, lubricates the oral cavity and pharynx, and dissolves food chemicals so that they can be tasted.

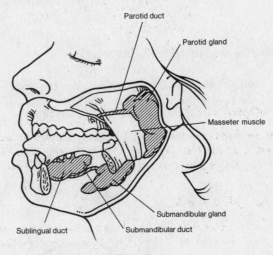

Fig. 18-6

Table 18-3

Gland	Location	Duct	Entry into Oral Cavity	Type of Secretion
Parotid	Anterior and inferior to auricle; subcutaneous over *masseter muscle*	Parotid (*Stensen's duct*)	Lateral to upper second molar	Watery serous fluid, salts, and enzyme
Submandibular	Inferior to the base of the tongue	Submandibular (*Wharton's duct*)	At papilla lateral to lingual frenulum	Watery serous fluid, with some mucus
Sublingual	Anterior to submandibular; under tongue	Several small ducts (*Rivinus's ducts*)	With submandibular duct	Mostly thick, stringy mucus, salts, and enzyme

18.12 What are the digestive functions of the tongue?

It moves food around in the mouth during mastication and assists in swallowing. Taste buds (Chapter 11) in the tongue sense food flavors, provoking the flow of digestive juices.

18.13 Describe the location and structure of the palate.

The *palate* is the roof of the oral cavity, consisting anteriorly of the bony *hard palate* and posteriorly of the *soft palate*. Transverse ridges, called *palatal rugae*, are located along the mucous membranes of the hard palate, where they serve as friction bands against which the tongue is placed during swallowing. The *uvula* is suspended from the soft palate. During swallowing, the soft palate and uvula are drawn upward, closing the nasopharynx and preventing food and fluid from entering the nasal cavity.

18.14 Is deglutition voluntary or involuntary?

Only the first stage of deglutition (swallowing) is voluntary: the chewing of the food, forming a bolus that is forced against the soft palate with the elevated tongue. The second stage is the (involuntary) *deglutition reflex*, which begins when pharangeal sensory receptors are stimulated. During this stage, the uvula is elevated, sealing off the nasal cavity; the hyoid bone and larynx are elevated, so that food or fluid is less likely to enter the trachea; and the esophagus is opened. During the third stage of deglutition, the bolus or fluid enters the esophagus and is transported to the stomach by peristalsis.

OBJECTIVE F   *To describe the* stomach *and its workings*.

Survey   The distensible stomach (Fig. 18-7) is positioned in the upper-left quadrant of the peritoneal cavity, immediately below the diaphragm. Its functioning is as summarized in Table 18-1.

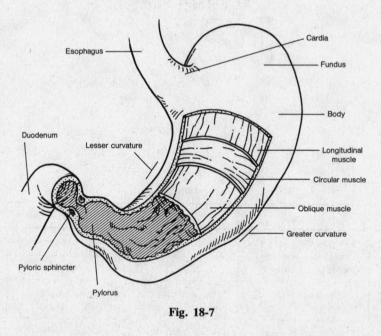

Fig. 18-7

18.15 Name the four regions of the stomach.

The **cardia** is the upper, narrow region immediately below the *gastroesophageal* (lower esophageal) *sphincter*; the **fundus** is the dome-shaped portion to the left; the **body** is the large central portion; and the **pylorus** is the funnel-shaped terminal portion that contains the *pyloric sphincter*.

In addition, the stomach presents to the right a concave *lesser curvature*, and to the left a convex *greater curvature*.

18.16 How are the inner tunicae of the stomach specialized for their tasks?

 The longitudinal, circular, and oblique sublayers of the tunica muscularis enable churning movements in the formation of chyme. *Gastric rugae*, which are longitudinal folds of the tunica mucosa, permit distension. Also, the tunica mucosa has *gastric glands* (Fig. 18-8) with three kinds of secretory cells: *chief* (*zygogenic*), *parietal*, and *mucous*; furthermore, the mucosal wall is permeable enough to allow some absorption.

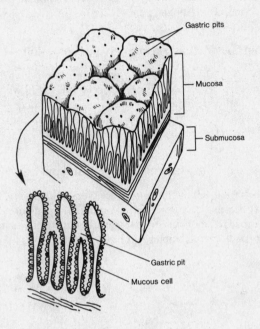

Fig. 18-8

18.17 What is *gastric juice*?

 The joint secretory product of the mucous, chief, and parietal cells; see Table 18-4.

Table 18-4

Component	Source	Function
Pepsinogen	Chief cells	Inactive form of pepsin
Pepsin	Formed from pepsinogen in presence of HCl	Protein-splitting enzyme
Hydrochloric acid (HCl)	Parietal cells	Strong acid for killing pathogens; conversion of pepsinogen to pepsin
Mucus	Goblet and mucous cells	Viscous, alkaline, protective lining of lumen
Intrinsic factor	Parietal cells	Aids absorption of vitamin B_{12}

18.18 Describe the phases of gastric secretion.

>   ***Cephalic phase.*** In response to sight, taste, smell, etc., parasympathetic impulses in the vagus nerves initiate secretion of 50–150 mL of gastric juice.
>
>   ***Gastric phase.*** Food-induced distension of the tunica mucosa plus chemical breakdown of protein stimulates release of gastrin (Table 12-4), which causes 600–750 mL of gastric juice to be produced.
>
>   ***Intestinal phase.*** Chyme entering the duodenum stimulates the release of intestinal gastrin, which leads to the production of additional small quantities of gastric juice.

18.19 What do *ulcers* and *pernicious anemia* have in common?

>   Not much, outside of being associated with the same region (the stomach). *Peptic ulcers* may be caused by an increase in HCl and pepsinogen secretion, along with insufficient protective mucus being secreted during sympathetic stimulation (stress). A *gastric ulcer* refers to the stomach location rather than to the cause, which could be peptic or other, such as alcohol. *Pernicious anemia* is a vitamin-B_{12}-deficiency disease brought about by lack of intrinsic factor.

18.20 Compare the activities of the gastroesophageal and pyloric sphincters.

>   The gastroesophageal sphincter constricts after food passes into the stomach; it is not a true sphincter, however, because during reflexive *vomiting* it opens to permit flow of the regurgitated matter into the esophagus toward the mouth. The pyloric sphincter regulates the movement of chyme into the small intestine and, as a true sphincter, prohibits backflow.

OBJECTIVE G   *To identify the regions of the* small intestine *and to discuss the process of food absorption.*

Survey   The small intestine is divided on the basis of function and histological structure into the *duodenum*, *jejunum*, and *ileum* (Fig. 18-2). The functions of the small intestine are as given in Table 18-1.

18.21 How may the three regions of the small intestine be distinguished?

>   The C-shaped *duodenum* measures 25 cm (10 in) from the pyloric sphincter of the stomach to the duodeno-jejunal flexure. It receives bile secretions through the *common bile duct* from the liver and gallbladder, and pancreatic secretions through the pancreatic duct (see Fig. 18-9). Mucus-secreting *duodenal* (*Brunner's*) *glands* are numerous in the submucosa of the duodenum.

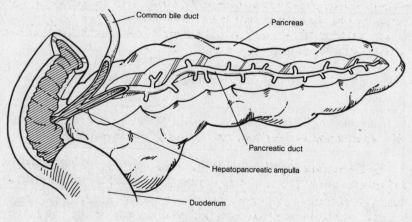

Fig. 18-9

The 1-m (3-ft) *jejunum* extends from the duodenum to the ileum. It is characterized by deep folds in the mucosa and submucosa, called *plicae circulares*.

The 2-m (6.5-ft) *ileum* joins the *cecum* of the large intestine (Fig. 18-12) at the *ileocecal valve*. Aggregates of lymph nodes (Peyer's patches; Fig. 16-3) mark the ileum.

18.22 What structural modifications of the small intestine facilitate absorption?

The plicae circulares (Problem 18.21) obviously increase the absorptive area. Plicae, in turn, are covered by numerous fingerlike projections of the mucosa, called *villi*. Each villus (Fig. 18-10) contains a capillary network, smooth muscle, and a specialized lymph vessel called a *lacteal*. Absorption is accomplished as food molecules enter the minute vessels of the villi through *microvilli* (Problem 3.8). At the bases of the villi are *intestinal glands* (or *crypts of Lieberkuhn*) that secrete digestive enzymes.

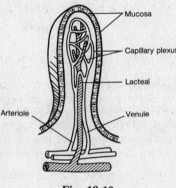

Fig. 18-10

18.23 What kinds of motions are the small intestine capable of?

Contractions of the longitudinal and circular muscles of the tunica muscularis produce three distinct types of movements. *Rhythmic segmentations*, of 12 to 16 per minute, in regions containing chyme churn the chyme with digestive juices and bring it in contact with the villi. Irregular *pendular movements*—constriction waves first in one direction, then back again—further mix the chyme. *Peristalsis*, of 15 to 18 pulses per minute, propels the chyme through the small intestine.

18.24 List the digestive enzymes secreted by the intestinal glands, along with their actions.

Peptidase converts proteins into amino acids; *sucrase* (*maltase* and *lactase*) converts disaccharides into monosaccharides; *lipase* converts fats into fatty acids and glycerol; *amylase* converts starch and glycogen into disaccharides; *nuclease* converts nucleic acids into nucleotides; *enterokinase* activates trypsin secreted from the pancreas.

OBJECTIVE H   *To describe the location, structure, and functions of the* liver.

Survey   The 1.3-kg (3-lb) liver is positioned beneath the diaphragm, in the right hypochondrium of the abdominal cavity. The four *lobes* of the liver (see Fig. 18-11) are composed of *lobules* that carry out numerous functions, including synthesis, storage, and release of vitamins and of glycogen; synthesis of blood proteins; phagocytosis of old red blood cells and certain bacteria; removal of toxic substances; and production of bile.

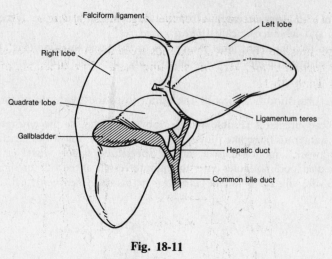

Fig. 18-11

18.25 Is the gallbladder a part of the liver?

No: the gallbladder is a separate organ that stores and concentrates bile secreted from the liver. Bile is ejected from the gallbladder into the common bile duct (Problem 18.21), under the influence of cholecystokinin (Table 12-4). The contribution of bile to digestion is the emulsification of neutral fats and the absorption of fatty acids, cholesterol, and certain vitamins.

18.26 What is meant by the *double blood supply* to the liver?

The hepatic artery brings oxygenated blood to the liver, while the hepatic portal vein brings digested food substances. Arterial and venous blood is mixed in the *sinusoids* (tiny endothelium-lined passages in the lobules), where O_2, most of the nutrients, and certain toxic substances are extracted by the hepatic cells. When needed by other cells of the body, the nutrients are returned to the venous blood drainage.

OBJECTIVE I   *To consider the digestive role of the* pancreas.

Survey   The pancreas (Fig. 12-8) lies horizontally along the posterior abdominal wall, adjacent to the greater curvature of the stomach. Its exocrine function is summarized in Problem 12.2.

18.27 Which digestive enzymes are secreted in the pancreatic juice?

Besides amylase, lipase, and nuclease (Problem 18.24), pancreatic juice contains three *peptidases*—trypsin, chymotrypsin, and carboxypeptidase—that act to convert proteins into amino acids.

18.28 Describe the regulation of pancreatic exocrine secretion.

Secretin (Table 12-4) from the duodenum stimulates the release of pancreatic juice that contains few digestive enzymes, but has a high bicarbonate-ion concentration. It is cholecystokinin, from the intestinal wall, which stimulates the release of pancreatic juice having a high concentration of digestive enzymes.

OBJECTIVE J   *To identify the regions of the* large intestine *and to discuss its functions.*

Survey   The large intestine (Fig. 18-12) extends from the ileocecal valve to the anus. It is subdivided into the *cecum, colon, rectum,* and *anal canal.* The functions of the large intestine are as given in Table 18-1.

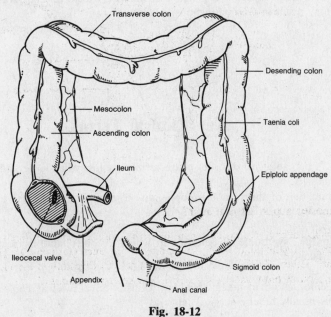

Fig. 18-12

18.29 Has the vermiform appendix a digestive function?

No; it does have an abundance of lymphatic tissue, however, that may serve to resist infection.

18.30 What are the four regions of the colon?

The colon has *ascending, transverse, descending,* and *sigmoid* portions. The *hepatic flexure* divides the ascending and transverse portions, and the *splenic flexure* divides the transverse and descending portions.

18.31 Contrast the tunicae of the large intestine to those of the small intestine.

The mucosa and submucosa of the large intestine lack plicae, but do have sacculations, or *haustra,* along their length. The tunica muscularis consists of longitudinal bands of smooth muscle, called *taeniae coli.* In the tunica serosa, attached superficially to the taeniae coli, are fat-filled pouches called *epiploic appendages.*

18.32 Are the movements of the large intestine the same as those of the small intestine?

Only the peristaltic movements of the colon are similar to those of the small intestine, though they are usually more sluggish (3 to 12/min). *Haustral churning* is the contraction of a haustrum stimulated by distended tunicae. *Mass movement* (two or three times a day) is gross motion of fecal material, brought on by contraction of the taeniae coli.

18.33 Cite a function of the large intestine omitted from Table 18-1.

The large intestine absorbs vitamins; in particular, vitamin K.

18.34 How does the large intestine (and the rest of the digestive system) respond to parasympathetic stimulation and to sympathetic stimulation?

Parasympathetic stimulation of the GI tract and digestive organs generally *increases* digestive activity; specifically, glandular secretion and autonomic muscular movement. Sympathetic stimulation *inhibits* digestive activity. It is for this reason that excessive and prolonged stress (sympathetic stimulation) may result in GI dysfunctions.

Key Clinical Terms

Anorexia nervosa   A psychological disorder characterized by loss of appetite and extreme weight loss.

Appendicitis   Inflammation of the vermiform appendix. The appendix is prone to infection, and so appendicitis is one of the most common surgical emergencies.

Bulimia   A "binge-purge" syndrome of behavior in which uncontrollable overeating is followed by forced vomiting or excessive laxatives.

Cholelithiasis   Stone formation in the gallbladder. Gallstones, as well as *cholecystitis*, are treated with a low-fat diet; gallstones usually require surgical removal.

Cirrhosis   A condition in which normal liver epithelium is replaced by connective tissue and, sometimes, fat, causing blockage of the sinusoids. Alcohol and malnutrition may cause cirrhosis.

Colostomy   Formation of an abdominal exit of the digestive tract by bringing a loop of the colon to the surface of the abdomen. If the rectum is removed because of cancer, the *stoma* (opening created by the colostomy) provides a permanent outlet for feces.

Constipation   A condition in which the feces are retained for a longer-than-normal period of time; infrequent or difficult defecation.

Cystic fibrosis   An inherited disease of the exocrine glands, particularly the pancreas. Pancreatic secretions are too thick to drain easily, causing the ducts to become inflamed and stimulating connective tissue formation, which occludes the drainage passageway.

Dental caries   Tooth decay, involving a gradual disintegration of the enamel and dentine. Caries are caused by certain bacteria; improper diet; improper hygiene; or crowded, uneven teeth.

Diarrhea   Frequent defecation of loose and unformed feces. There are many causes, including physical and mental stress, disagreeable foods, a number of diseases, and intestinal parasites or bacteria.

Diverticulitis   Inflammation of a *diverticulum*, an abnormal side-pouch of the colon.

Enteritis   An inflammation of the intestinal mucosal lining, commonly called "intestinal flu." The symptoms of enteritis are cramps, nausea, and diarrhea. It may be caused by viruses or certain foods.

Gastric ulcer   An open sore or lesion of the mucous membrane of the stomach wall. Among the things that are believed to cause ulcers are certain foods and medications, alcohol, coffee, aspirin, and overstimulation of the vagus nerve due to stress.

Gingivitis Inflammation of the gums. It may result from improper hygiene, poorly fitted dentures, improper diet, or certain body infections.

Halitosis Offensive odor of the breath. It may be due to dental caries, certain diseases, or particular foods eaten.

Heartburn A burning sensation in the region of the esophagus and stomach. Frequently associated with overeating, it may result from the regurgitation of gastric juices into the lower portion of the esophagus.

Hemorrhoids Varicose veins of the rectum and anus.

Hepatitis Inflammation of the liver. It may be caused by organisms (e.g., viruses, protozoa, and bacteria) or by the absorption of toxic materials. Hepatitis is of two major types: *serum hepatitis*, which is transmitted through blood transfusion, and *infectious hepatitis*, which is transmitted through contaminated foods.

Mumps (*parotitis*) A contagious viral disease resulting in the inflammation of the parotid and other salivary glands. Mumps is particularly dangerous in adults because the disease causes swelling of the gonads.

Nausea Gastric discomfort and sensations of illness, with a tendency to vomit. The feeling is due to nervous and physical factors, and is symptomatic of many conditions (e.g., motion sickness, foul odors or sights, pregnancy, diseases).

Periodontal disease A collective term for conditions characterized by degeneration of cementum, periodontal ligament, alveolar bone, and gingivae.

Peritonitis An inflammation of the peritoneum.

Peptic ulcers Craterlike lesions that develop in the mucosa of the GI tract in areas exposed to gastric juice.

Pyorrhea Discharge of pus at the base of the teeth at the gum line.

Regurgitation (*vomiting*) Forceful expulsion of gastric contents into the mouth. Along with nausea, a symptom of almost any dysfunction of the digestive system.

Trench mouth (*Vincent's angina*) Contagious bacterial infection that causes inflammation, ulceration, and swelling of the floor of the mouth. Generally, it is contracted by kissing an infected person. Trench mouth can be treated by cleaning the teeth and gums, and applying medication.

Vagotomy Surgical removal of a section of the vagus nerve where it enters the stomach, in order to eliminate nerve impulses that stimulate gastric acid secretion. This procedure helps to cure ulcers.

Review Questions

Multiple Choice

1. If an incision had to be made in the small intestine to remove an obstruction, which tunica would be cut first? (*a*) muscularis, (*b*) mucosa, (*c*) submucosa, (*d*) serosa, (*e*) endothelium.

2. The hepatic flexure of the large intestine occurs between the: (*a*) transverse colon and descending colon, (*b*) cecum and ascending colon, (*c*) ascending colon and transverse colon, (*d*) descending colon and rectum, (*e*) descending colon and sigmoid colon.

3. Obstruction of the common bile duct by gallstones would most likely affect the digestion of: (*a*) carbohydrates, (*b*) fats, (*c*) proteins, (*d*) nucleic acids, (*e*) none of the above.

4. Formation of gallstones is referred to as: (*a*) jaundice, (*b*) cirrhosis, (*c*) hepatitis, (*d*) cholelithiasis.

5. Which of the following is *not* a function of saliva? (*a*) to initiate protein digestion, (*b*) to aid in cleansing the teeth, (*c*) to lubricate the pharynx, (*d*) to assist in formation of the bolus.

6. Peristalsis moves food material: (*a*) in the stomach and small intestine only, (*b*) in the small and large intestines only, (*c*) in the stomach and small and large intestines only, (*d*) from the pharynx to the anal canal.

7. A gastrointestinal tumor involving villi and plicae circulares might interfere with: (*a*) deglutition, (*b*) absorption, (*c*) peristalsis, (*d*) defecation, (*e*) emulsification.

8. The greater omentum does not participate in: (*a*) secretion of enzymes, (*b*) support and cushioning of the viscera, (*c*) storage of lipids, (*d*) protection against the spread of infection.

9. All enzymes involved in protein digestion are: (*a*) secreted by the pancreas, (*b*) activated by HCl, (*c*) present in the stomach, (*d*) secreted in an inactive form, (*e*) stimulated by enterokinase.

10. The terminal portion of the small intestine is the: (*a*) ileum, (*b*) cecum, (*c*) duodenum, (*d*) jejunum, (*e*) colon.

11. The large intestine lacks: (*a*) goblet cells, (*b*) epiploic appendages, (*c*) plicae circulares, (*d*) haustra, (*e*) taeniae coli.

12. Which of the following is not a major gastrointestinal hormone? (*a*) epinephrine, (*b*) secretin, (*c*) gastrin, (*d*) cholecystokinin.

13. A set of permanent teeth contains: (*a*) 20 teeth, (*b*) 30 teeth, (*c*) 32 teeth, (*d*) 24 teeth.

14. Teeth adapted to shear food are: (*a*) premolars, (*b*) canines, (*c*) incisors, (*d*) molars.

15. A patient who has undergone *gastrectomy* (removal of the stomach) may suffer from: (*a*) cirrhosis, (*b*) pernicious anemia, (*c*) gastric ulcer, (*d*) inability to digest fats, (*e*) inability to digest proteins.

16. Amylase in saliva initiates digestion of: (*a*) lipids, (*b*) proteins, (*c*) fats, (*d*) carbohydrates.

17. Pancreatic juice contains a protein-splitting enzyme called: (*a*) trypsin, (*b*) zymogen, (*c*) pepsin, (*d*) amylase, (*e*) nuclease.

18. Secretin is a hormone that: (*a*) stimulates the release of pancreatic fluid, (*b*) converts trypsinogen into trypsin, (*c*) activates chymotrypsin, (*d*) inhibits the action of pancreatic lipase.

19. Secretion of *enterogastrone* would be stimulated by the presence of: (*a*) protein in the small intestine, (*b*) protein in the stomach, (*c*) fat in the small intestine, (*d*) fat in the stomach.

20. A dysfunction of the parietal cells of gastric glands would result in a decreased production of which two of the following? (*a*) mucus, (*b*) pepsinogen, (*c*) hydrochloric acid, (*d*) pepsin.

21. The part of the stomach that meets the esophagus at the gastroesophageal sphincter is the: (*a*) fundus, (*b*) cardia, (*c*) pylorus, (*d*) body.

22. Which organ is retroperitoneal? (*a*) liver, (*b*) ileum, (*c*) stomach, (*d*) pancreas, (*e*) esophagus.

23. The small intestine is held to the posterior abdominal wall by the: (*a*) mesentery, (*b*) falciform ligament, (*c*) greater omentum, (*d*) lesser omentum, (*e*) visceral peritoneum.

True/False

1. The principal function of the digestive system is to prepare food for cellular utilization.

2. Parasympathetic impulses to the gastrointestinal tract decrease peristaltic activity and muscle tone.

3. The GI tract is innervated by both sympathetic and parasympathetic fibers.

4. The tongue is a mass of smooth muscles covered by mucous epithelial membrane.

5. Jaundice is a disease of the liver.

6. Pancreatic juice is secreted from acinar cells of the pancreas.

7. The falciform ligament attaches the liver to the diaphragm.

8. Intrinsic factor is necessary for the normal absorption of amino acids from the small intestine.

9. The spleen is not an accessory organ of the digestive system.

10. The process by which bile causes the breakdown of fat globules into smaller droplets is deglutition.

11. The primary tissue of serous membranes is simple squamous epithelium.

12. Intestinal rugae are folds of the mucosa within the small intestine that greatly increase the surface area for absorption.

13. Elevation of the uvula prevents food or fluid from entering the trachea.

14. The serosa is the intestinal layer lining the lumen of the GI tract.

15. Cirrhosis is a chronic disease of the liver in which fibrous tissue replaces functional hepatic cells.

Chapter 19

Metabolism, Nutrition, and
Temperature Regulation

OBJECTIVE A   *To define* metabolism.

Survey   Foods are first digested, then absorbed, and, finally, metabolized. *Metabolism* refers to all the chemical reactions of the body. There are two aspects of metabolism: *catabolism*, the breaking-down side, and *anabolism*, the building-up side. (Glycolysis, in which glucose is broken down to yield products and energy, is an example of catabolism; the biosynthesis of proteins, in which amino acids are joined together, is an example of anabolism.) Energetically, metabolism may be thought of as a balancing of catabolism, which provides energy (stored in ATP), against anabolism, which requires energy.

19.1   Define *nutrients*.

Chemical substances in food that enter into metabolism are referred to as nutrients; they are classified into *carbohydrates*, *lipids*, *proteins*, *vitamins*, *minerals*, and *water*.

OBJECTIVE B   *To describe the major events in* carbohydrate metabolism.

Survey   The average human diet consists largely of carbohydrates (polysaccharides and disaccharides; review Chapter 2). These are changed by digestion into the monosaccharides glucose, fructose, and galactose. In the liver, fructose and galactose are converted to glucose. Thus, carbohydrate metabolism is essentially glucose metabolism, for which the overall equation is

$$C_6H_{12}O_6 + 6O_2 \rightarrow 6CO_2 + 6H_2O + energy \text{ (in 36 ATP)}$$

glucose   oxygen   carbon   water   adenosine
                        dioxide   triphosphate

The way in which the energy-laden ATP is formed is diagramed in Fig. 19-1.

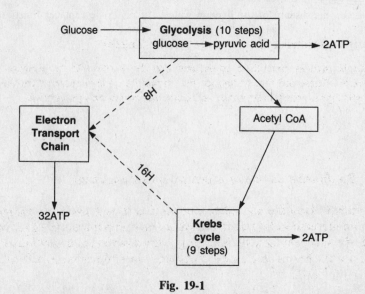

Fig. 19-1

19.2   Does glucose readily move into and out of the body cells?

No: insulin is required for the rapid movement of glucose into most cells of the body. Upon entry into a cell, glucose combines with a phosphate group, and this *phosphorylation* serves to capture the glucose in the cell. Several cell types (liver, kidney tubule, intestinal epithelial) have *phosphatase*, which removes the phosphate group and thereby allows the glucose to pass out of the cell.

19.3   Is oxygen required for the chemical reactions associated with glycolysis?

No: glycolysis occurs under *anaerobic* conditions (as during strenuous exercise). The resultant pyruvic acid is reduced to lactic acid, which remains in the cell until *aerobic* conditions are restored; it is then converted back to pyruvic acid. Aerobic conditions must persist if the catabolism of glucose is to be completed via the *Krebs cycle* and the *electron transport chain.*

19.4   In what parts of the cell does catabolism of glucose take place?

The 10 steps of glycolysis, each step mediated by a different enzyme, take place in the cytoplasm (Table 3-1). The Krebs cycle (9 steps, 8 enzymes; also called the *citric acid cycle*) and the electron transport chain of oxidation-reduction reactions both take place in mitochondria (Table 3-1), the latter specifically on cristae.

19.5   Is all the energy released during glucose metabolism utilized to form ATP?

No: less than half of the energy released is captured by ATP; the balance is given off as heat.

19.6   Is all of the glucose that enters body cells immediately catabolized to form energy and heat?

No: depending on energy demands, a portion of the glucose entering many cell types is converted into glycogen (*glycogenesis*). When the body needs energy, glycogen stored in liver and muscle cells is broken down and glucose is released into the blood. This inverse process, *glycogenolysis*, is spurred by the pancreatic hormone glucagon (cf. Chapter 12, Objective I) and the adrenal hormones epinephrine and norepinephrine (cf. Table 12-3).

19.7   Can glucose be formed from noncarbohydrate sources?

Yes: both protein and lipid molecules can be converted to glucose; the process is called *gluconeogenesis*. There are five hormones that stimulate gluconeogenesis (review Chapter 12): cortisol, thyroxine, glucagon, growth hormone, and epinephrine (or norepinephrine).

OBJECTIVE C   *To describe the major events in* lipid metabolism.

Survey   Lipids (mainly, fats) are second to carbohydrates as a source of energy for ATP synthesis. There is an increase in fat utilization when the carbohydrate level is low. Fats enter into the building of many cellular structures (e.g., membranes), and cholesterol is a precursor in the synthesis of sex hormones, adrenocorticoids, and bile salts. A scheme of lipid metabolism is given in Fig. 19-2.

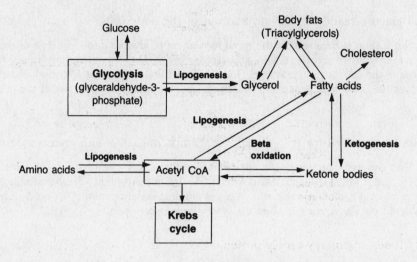

Fig. 19-2

19.8   What happens when total food intake—whether carbohydrate, protein, or fat—exceeds the body's needs?

The excess (or that portion which is not already fat) is converted into fat and stored in *adipose cells* (cf. Table 4-5). These are located in subcutaneous connective tissue and in deeper supporting tissue of the viscera, mesenteries, and omenta. Before entering adipose cells, fats are hydrolyzed to glycerol and fatty acids by *lipoprotein lipase*, which is present in capillary endothelium. After entering the adipose cells, the glycerol and fatty acids are recombined into fats [technically, into *triacylglycerols*; the terminology is made clear in Fig. 2-7].

19.9   Discuss *beta oxidation* and *lipogenesis* as inverse processes.

Refer to Fig. 19-2. When stored triacylglycerols are catabolized—for energy or for synthesis of new nutrients—the glycerol components may be converted to *phosphoglyceraldehyde* (PGAL), which enters the glycolytic pathway and is utilized for energy production or synthesis of glucose. The fatty acid components undergo stepwise breakdown of the fatty acid chain (two carbon fragments are serially split off) into acetyl molecules, which combine with *coenzyme A* to form *acetyl CoA*. This breaking down of fatty acids to form acetyl molecules constitutes *beta oxidation*. The oppositely directed, anabolic process, leading from glucose or amino acids to lipids, is known as *lipogenesis*.

19.10   How are *ketone bodies* formed, and what are the clinical consequences of an excess of them?

As part of fatty acid catabolism, the liver condenses two molecules of acetyl CoA to form *acetoacetic acid*, which is converted to *β-hydroxybutyric acid* and *acetone*. These three substances are collectively known as the *ketone bodies*. Ketone bodies are normally decomposed into acetyl CoA and utilized in the Krebs cycle. If they are not—in consequence of a meal rich in fats, a period of starvation or fasting, or a case of diabetes mellitus—the condition *ketosis* results. Extreme or prolonged ketosis can lead to ketoacidosis (Problem 12.25).

OBJECTIVE D   *To describe the major events in* protein metabolism.

Survey   The body uses very little protein for energy, so long as the intake or stores of carbohydrates and fats are sufficient. Proteins play an essential role in cellular structure and function (see Problem 2.15). Figure 19-3 schematizes protein metabolism.

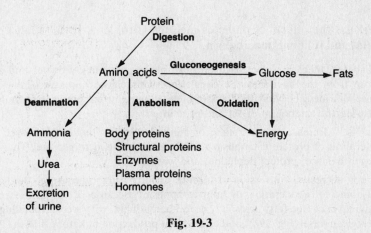

Fig. 19-3

19.11 What foods commonly supply proteins?

Eggs, milk, meats, fish, poultry, cereals, cheese, nuts, and various legumes (beans, peas, etc.).

19.12 According to Fig. 19-3, protein (amino acids) may be oxidized to provide energy when other sources prove inadequate. What is the mechanism for this oxidation?

The Krebs cycle. Depending on the amino acid, it may enter the cycle via acetyl CoA, as in Fig. 19-2, or may directly enter at another stage of the cycle.

19.13 What is meant by a negative or a positive *nitrogen balance*?

Negative nitrogen balance: protein catabolism exceeds protein anabolism; occurs with a protein-poor diet or during starvation. *Positive nitrogen balance*: protein anabolism exceeds protein catabolism; occurs during growth, during recovery from illness, or after administration of anabolic steroids.

OBJECTIVE E   *To review the hormonal regulation of metabolism.*

Survey   Pertinent information from Chapter 12 is abstracted in Table 19-1.

Table 19-1

Hormone	Metabolic Effects
Insulin	Increases glucose uptake into cells; increases glycogenesis; increases lipogenesis and decreases lipolysis; increases amino-acid uptake into cells; increases protein synthesis
Glucagon	Increases glycogenolysis; increases glucogeneogenesis, increases lipolysis
Epinephrine	
Thyroxine	Increases glycogenolysis; increases gluconeogenesis; increases protein synthesis
Growth hormone	Increases amino-acid uptake into cells; increases protein synthesis; increases glycogenolysis; increases lipolysis
Cortisol	Increases gluconeogenesis; increases lipolysis; increases the breakdown of proteins
Testosterone	Increases protein synthesis

19.14 Briefly discuss the effects of diabetes mellitus (Problems 12.24–12.26) on (a) carbohydrate, (b) protein, and (c) lipid, metabolism.

(a) In diabetes, there is a decreased entry of glucose into many tissues and an increased release of glucose from the liver into the circulation. This causes an extracellular glucose excess and an intracellular deficiency. With a lack of intracellular glucose, energy requirements are met by greatly increasing the catabolism of fat and protein.

(b) There is a decrease in the uptake of amino acids into tissues, decreased protein synthesis, and acceleration of protein catabolism to CO_2 and H_2O and to glucose. The net effect is a negative nitrogen balance, protein depletion, and wasting.

(c) There is decreased conversion of glucose to fatty acids (because of the low intracellular glucose level), and an acceleration of lipid catabolism. Because of the increased lipid catabolism, the plasma level of free fatty acids may more than double, with a corresponding jump in the formation of acetyl CoA (see Fig. 19-2) and consequent production of ketone bodies (Problem 19.10).

19.15 Explain how diabetic ketosis can lead to acidosis, and give the clinical consequences.

Like any acids, ketone bodies liberate hydrogen ions. Thus, in untreated diabetes, the plasma pH drops, and hydrogen ions are secreted into the kidney tubules and excreted in the urine. When the kidney's ability to remove H^+ from the plasma is exceeded, it removes Na^+ and K^+ instead. The resultant electrolyte and water losses may lead to dehydration, *hypovolemia* (decreased volume of blood), depressed consciousness, and, finally, coma.

OBJECTIVE F   *To learn the methods used to measure the energy content of foods and the terms used to express body energy expenditure.*

Survey   **Calorie.** The basic unit of energy is the *joule* (J); but, for energy as heat, one frequently uses the *calorie* (cal), where

$$1 \text{ cal} = 4.184 \text{ J}$$

The "calorie" of dieticians' charts is equal to 1000 cal, or 1 kcal.

Bomb calorimeter. The energy obtainable from a sample of food can be determined by placing it in a sealed chamber surrounded by a jacket filled with water of known volume and temperature, the whole being thermally insulated from the environment. As the food is completely oxidized (burned), heat is liberated into the water. The temperature rise of the water gives the caloric value of the food sample. Some typical values obtained by this method are:

Food	Energy Content
Carbohydrate (1 g)	4.1 kcal
Lipid (1 g)	9.5 kcal
Protein (1 g)	5.3 kcal

Direct calorimetry. The energy liberated by oxidation of foods in the body is measured by placing a person in a chamber that is sensitive to the heat loss from the body. (Same principle as bomb calorimeter.)

Indirect calorimetry. The oxygen taken in by way of the respiratory system is consumed in cellular oxidations. Therefore, standard tables may be used to translate the minute respiratory volume (Problem 17.21) into the rate of heat production.

Respiratory quotient (R.Q.). The ratio of volume of carbon dioxide produced (in a given period) to volume of oxygen consumed (in the same period).

Metabolic rate. The energy expenditure of the body per unit time. The BMR (Chapter 12) is the metabolic rate as measured when the person is resting but awake. The test is usually taken in the morning before rising, the person having fasted for at least 12 hours and having slept for 8 hours.

19.16 If a person had an R.Q. of 0.72, what type of food was being utilized for energy?

Fat. The R.Q. for carbohydrates (strictly: for glucose) is $6/6 = 1$, because in the metabolic equation,

$$C_6H_{12}O_6 + 6O_2 \longrightarrow 6CO_2 + 6H_2O$$

the coefficients may be interpreted as numbers of moles or numbers of liters. The R.Q. for protein is 0.80, since protein contains less oxygen than carbohydrates and therefore requires more oxygen in burning. The R.Q. for fat is 0.70, since fat contains even less oxygen.

19.17 What factors affect the metabolic rate?

Metabolic rate increases with increasing *body size*, *body temperature*, *activity*, *levels of thyroid hormones*, or *sympathetic stimulation*. It is 10% higher in men than in women, and it decreases with increasing age.

OBJECTIVE G   *To identify the essential vitamins and minerals, their sources in the diet, uses in the body, and the syndromes of their deficiency.*

Survey   In Tables 19-2 and 19-3, the second column gives the Recommended Dietary Allowance, in milligrams per day.

19.18 Classify vitamins according to solubility.

Fat-soluble vitamins are A, D, E, and K. They are absorbed from the gastrointestinal tract along with lipids and can be stored in the body. *Water-soluble vitamins* are the B-vitamins and C. They are absorbed from the intestines along with water and are generally not stored in the body.

19.19 Can vitamins be synthesized in the body?

No: vitamins are obtained only from external sources, such as ingested foods. (Vitamin K is synthesized by bacteria in the gastrointestinal tract.)

OBJECTIVE H   *To describe the body's thermal equilibrium.*

Survey   Heat is continually being produced as a by-product of metabolism and is continually being lost to the surroundings: the body normally keeps these in balance.

Heat Gain	Heat Loss
1. Metabolism (see Problem 19.17)	1. Radiation
2. Muscular activity (shivering)	2. Evaporation
	3. Conduction

Table 19-2

Vitamin	RDA	Dietary Sources	Major Body Functions	Deficiency Syndromes
Vitamin A (Retinol)	1	Green vegetables, dairy products	Synthesis of rhodopsin	Night blindness; skin disorders
Vitamin B_1 (Thiamine)	1.5	Organ meats, whole grains, legumes	Coenzyme in cellular respiration	Heart disease; peripheral nerve changes; beriberi
Vitamin B_2 (Riboflavin)	1.8	Meats, leafy vegetables, milk	Component of FAD	Lesions of lips, mouth, and tongue
Vitamin B_6 (Pyridoxine)	2	Meats, vegetables, whole-grain cereals	Coenzyme in amino-acid metabolism	Irritability; muscle twitching; seizures
Vitamin B_{12} (Cobalamine)	0.003	Meats, eggs, dairy products	Coenzyme in nucleic-acid metabolism	Pernicious anemia; nervous disorders
Niacin	20	Meats, grains, legumes	Component of NAD	Pellagra (lesions in skin and GI tract); nervous disorders
Pantothenic acid	5–10	Liver, eggs, yeast	Component of coenzyme A	Fatigue; skin inflammation; nausea
Biotin	0.3	Meats, vegetables, legumes	Component of coenzyme	Fatigue; muscle pains; depression; nausea
Folic acid	0.4	Liver, green vegetables, eggs, whole-grain cereals	Coenzyme in nucleic-acid metabolism	Anemia; diarrhea
Vitamin C (Ascorbic acid)	45–50	Citrus fruits, tomatoes, salad greens	Formation of intercellular material in connective tissues	Scurvy
Vitamin D (Cholecalciferol)	0.001	Eggs, dairy products, cod-liver oil	Growth of bones; absorption of calcium	Rickets in children; osteomalacia in adults
Vitamin E (Tocopherol)	15	Meats, dairy products, leafy vegetables	Antioxidant to prevent damage to cell membranes	Anemia (possibly)
Vitamin K (Phylloquinone)	0.03	Meats, fruits, leafy vegetables	Synthesis of clotting factors	Hemorrhage

Table 19-3

Mineral	RDA	Dietary Sources	Major Body Functions	Deficiency Syndromes
Calcium	800	Eggs, milk, cheese, vegetables	Formation of bones and teeth; clotting; nerve and muscle activity; many cellular functions	Rickets; tetany; osteoporosis
Chlorine	2.5	Table salt, most foods	Water-electrolyte balance; acid-base balance; formation of HCl in stomach	Fluid imbalance
Cobalt	1	Most foods	Component of vitamin B_{12}	Anemia
Copper	2	Most foods	Synthesis of hemoglobin; component of enzyme involved in melanin formation	Anemia
Fluorine	1–2	Seafood, drinking water	Component of bones, teeth, and other tissues	Dental caries
Iodine	0.1	Seafood, table salt	Component of thyroid hormones	Hypothyroidism
Iron	10–15	Meat, egg yolk, legumes, nuts, cereals	Component of hemoglobin, myoglobin, and cytochromes	Anemia
Magnesium	300–350	Many foods	Bone formation; nerve and muscle function	Tetany
Manganese	300–350	Meats	Activates several enzymes; reproduction; lactation	Infertility
Phosphorus	800	Meat, dairy products, fish, poultry	Formation of bones and teeth; component of buffer system and nucleic acid	Weakness
Potassium	2	Meat, banana, seafood, milk	Nerve conduction; electrolyte balance	Skeletal and cardiac muscle weakness
Sodium	2000–5000	Most foods, table salt	Nerve conduction; electrolyte balance	Cramps; weakness; dehydration
Zinc	15	Most foods	Component of several enzymes	Reduced growth; hair loss, vomiting

19.20 What is the normal range of resting body temperature?

    For oral temperatures, see Fig. 19-4; rectal temperatures run 1 °F (0.55 °C) higher.

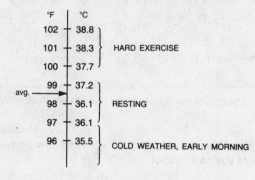

Fig. 19-4

19.21 Briefly describe the mechanisms of body heat loss.

    Radiation: transfer of heat, in the form of electromagnetic waves, from the surface of the body to the surrounding environment. *Conduction*: molecule-to-molecule transfer of heat from the surface of the body to objects (clothing) in contact with it. *Evaporation*: heat is lost as water evaporates from the body surfaces (580 cal per mL of water). Water loss from skin and lungs is normally about 600 mL per day.

Review Questions

Multiple Choice

1. Which of the following chemical reactions can occur anaerobically? (*a*) glycolysis, (*b*) Krebs cycle, (*c*) conversion of lactic acid to pyruvic acid, (*d*) conversion of pyruvic acid to acetyl CoA.

2. The synthesis of glycogen molecules for cellular storage is referred to as: (*a*) glycogenolysis, (*b*) beta oxidation, (*c*) glyconeogenesis, (*d*) glycogenesis.

3. Between meals, the blood glucose level is maintained by: (*a*) insulin, (*b*) glycogenolysis, (*c*) lipogenesis, (*d*) glycogenesis.

4. Within the cell, Krebs cycle reactions occur in the: (*a*) Golgi complex, (*b*) ribosomes, (*c*) nucleolus, (*d*) mitochondria.

5. If oxygen is lacking, how many molecules net of ATP are produced by the catabolism of one molecule of glucose? (*a*) 1, (*b*) 2, (*c*) 8, (*d*) 38.

6. The least preferred energy source for cells is: (*a*) glucose, (*b*) protein, (*c*) fats, (*d*) vitamins.

7. During which phase of metabolism is the greatest amount of energy generated? (*a*) glycolysis, (*b*) conversion of pyruvic acid to acetyl CoA, (*c*) glycogenesis, (*d*) electron transport.

8. Anabolic metabolism includes: (*a*) processes by which substances are synthesized, (*b*) changes of larger molecules into smaller ones, (*c*) glycolysis, (*d*) all processes needed to maintain life.

9. Aerobic respiration acts to increase body: (a) CO_2, (b) water, (c) ATP, (d) all the above.

10. Vitamins are essential in metabolism because they: (a) serve as structural components, (b) serve as sources of energy, (c) act as coenzymes, (d) cannot be stored in the body.

11. Beriberi is caused by a deficiency of vitamin: (a) A, (b) B_1, (c) B_{12}, (d) B_6.

12. How many ATP molecules does the cell gain from aerobic metabolism of one glucose molecule? (a) 2, (b) 4, (c) 36, (d) 56.

13. What fraction of the energy released during catabolism of glucose is captured by ATP? (a) 25% (b) 50%, (c) 75%, (d) 100%.

14. The synthesis of glucose from protein or lipid is referred to as: (a) glycogenesis, (b) glucose oxidation, (c) glucosynthesis, (d) gluconeogenesis.

15. Which of the following hormones does not stimulate gluconeogenesis? (a) cortisol, (b) epinephrine, (c) estrogen, (d) thyroxine.

16. Which hormone increases amino acid uptake by cells, protein synthesis, and glycogenolysis? (a) cortisol, (b) growth hormone, (c) glucagon, (d) epinephrine.

17. The ratio of volume of carbon dioxide produced to volume of oxygen consumed is called: (a) bomb calorimeter, (b) metabolic rate, (c) direct quotient, (d) respiratory quotient.

18. A respiratory quotient of 0.70 would indicate that the main source of food was: (a) carbohydrate, (b) fat, (c) protein, (d) carbohydrate and protein mixture.

19. Reduced growth, hair loss, and vomiting may result from a deficiency of: (a) iron, (b) copper, (c) potassium, (d) zinc.

20. The respiratory quotient: (a) is the ratio of the volume of carbon monoxide produced to the volume of oxygen consumed, (b) is 1.0 when glucose is the substrate being metabolized, (c) is not dependent upon the type of food being metabolized, (d) will rise during prolonged periods of starvation.

21. Important mechanisms for heat transfer include: (a) evaporation, (b) conduction, (c) radiation, (d) all the above.

22. Per unit weight, which contains the most energy? (a) carbohydrates, (b) proteins, (c) fats, (d) vitamins.

23. When the ambient temperature is very high, say 105 °F, the body will lose heat by: (a) radiation, (b) conduction, (c) evaporation, (d) increased metabolism.

Matching I

1. Vitamin A   (a) Deficiency most commonly produces pernicious anemia
2. Vitamin B_1 (b) Deficiency may result in scurvy
3. Vitamin B_6 (c) Involved in synthesis of rhodopsin
4. Vitamin B_{12} (d) Necessary for synthesis of clotting factors
5. Vitamin C   (e) Also referred to as pyridoxine
6. Vitamin D   (f) Promotes absorption of calcium
7. Vitamin K   (g) Deficiency may result in beriberi

Matching II

1. Calcium
2. Cobalt
3. Copper
4. Iodine
5. Iron

(a) Component of vitamin B_{12}
(b) Component of thyroid hormones
(c) Required for clotting, bone formation, and muscle contraction
(d) Component of hemoglobin and myoglobin
(e) Involved in melanin formation

Matching III

1. Ascorbic acid
2. Vitamin A
3. Vitamin D
4. Folic acid
5. Vitamin K
6. Niacin

(a) Cheese
(b) Green, leafy vegetables
(c) Cod-liver oil
(d) Citrus fruits
(e) Legumes
(f) Whole-wheat products

Chapter 20

Urinary System

OBJECTIVE A   *To give the components of the urinary system and their functions.*

Survey   See Fig. 20-1. The urinary system consists of the *kidneys*, which form urine; the *ureters*, which transport urine to the *urinary bladder*, where it is stored; and the *urethra*, which carries urine to the outside of the body.

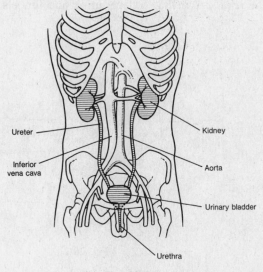

Ureter

Inferior
vena cava

Kidney

Aorta

Urinary bladder

Urethra

Fig. 20-1

20.1   What are the specific functions of the urinary system?

   Like most body systems, the urinary system is involved in maintaining homeostasis. (1) The urinary system plays a critical role in regulating the composition of the body fluids (water balance, electrolyte balance, and acid-base balance). (2) The urinary system rids the body of the wastes of metabolism and phagocytosis (cf. Problem 13.14), and also removes foreign chemicals, drugs, and food additives. (3) The kidneys serve as minor endocrine organs, as described in Problem 13.11.

20.2   Is the name "excretory system" appropriate for the urinary system?

   Although the urinary system has a major excretory function, other body systems are also excretory. Carbon dioxide is eliminated through the respiratory system; excessive water, salts, nitrogenous wastes, and even excessive metabolic heat are removed through the integumentary system; and various bile digestive wastes are eliminated through the digestive system. It is therefore best to call the urinary system by that name.

20.3   Does *micturition* mean the same as *urination*?

   No: *urination* is the voiding of urine from the urinary bladder, whereas *micturition* is the physiological process of urination, including nerve impulses and muscular responses.

OBJECTIVE B   *To give an anatomical description of the* kidneys.

Survey   Refer to Fig. 20-2. The kidneys are located on either side of the vertebral column in the abdominal cavity, at the level of the twelfth thoracic to the third lumbar vertebrae. Each kidney is approximately 11.25 cm (4 in) long, 5.5–7.5 cm (2–3 in) wide, and 2.5 cm (1 in) thick. The concave medial surface is called the *hilum*, and through it the renal artery enters and the renal vein and ureter exit. The kidney is embedded retroperitoneally in an adipose and fibrous capsule. The outer portion of the kidney, or the *cortex*, contains capillary tufts and convoluted tubules; the inner portion, or the *medulla*, is composed of a series of triangular masses, called *renal pyramids*, separated by *renal columns*. The pyramids are bundles of *papillae*, with each papilla projecting into a small depression, called a *minor calyx* (pl., *calyces*). Several minor calyces unite to form a *major calyx*. In turn, the major calyces join to form the *renal pelvis* that collects urine and funnels it to the ureter.

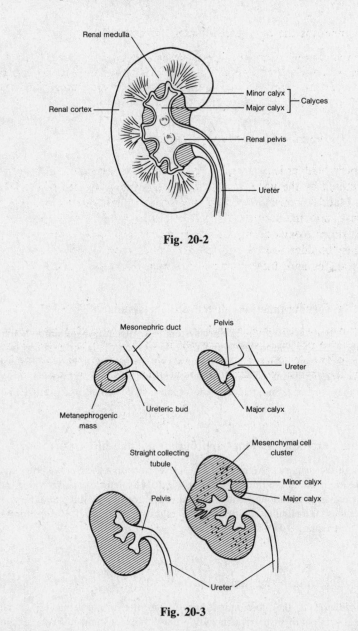

Fig. 20-2

Fig. 20-3

20.4   Describe the innervation of the kidneys.

The kidneys have an autonomic nerve supply derived from the tenth, eleventh, and twelfth thoracic nerves. Sympathetic stimulation of the renal plexus produces a vasomotor response in the kidneys that affects the circulation of the blood by regulating the diameters of arterioles.

OBJECTIVE C   *To trace the embryonic development of the kidneys.*

Survey   Refer to Fig. 20-3. Either permanent kidney is derived from a *metanephros*, which begins to form during the fifth week of development. The metanephros has two mesodermal sources: (1) a *ureteric bud*, which forms the drainage system for urine, and the stalk of which forms the ureters; and (2) a *metanephrogenic mass*, which develops from the urogenital ridge and forms the meaty portion of the kidney.

20.5   Is the metanephros the only precursor of the kidney?

No: the most primitive kidney, the *pronephros*, develops during the fourth week and persists only through the sixth week. It never functions as a kidney, but its ductal system gives rise to portions of the metanephros and the *mesonephros*. The latter develops from the urogenital ridge toward the end of the fourth week and lasts through the eighth week.

OBJECTIVE D   *To describe the anatomy of the* urinary bladder.

Survey   The urinary bladder is posterior to the symphysis pubis and anterior to the rectum. Its shape is determined by the volume of urine it contains. Like that of the GI tract, the wall of the urinary bladder consists of four layers: the (innermost) *mucosa*, the *submucosa*, the *muscularis*, and the *serosa*. When the urinary bladder is empty, the mucosa forms many folds, or *rugae*, which disappear when the urinary bladder becomes distended. The floor of the urinary bladder is a triangular area, called the *trigone*, which has an opening at each of its three angles and, lacking rugae, is smooth in appearance.

20.6   What are the functions of the tunicae of the urinary bladder?

The mucosa, of transitional epithelium, permits distension of the organ; the submucosa, of vascular tissue, provides a rich blood supply; the muscularis, whose three interlaced smooth-muscle laminas are collectively called the *detrusor muscle*, aids in micturition; and the serosa, of simple squamous epithelium, is a physical and functional continuation of the peritoneum.

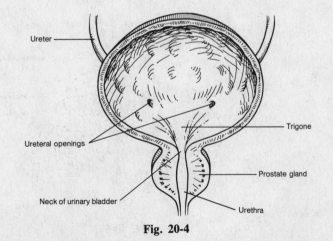

Fig. 20-4

20.7   Does the urinary bladder have sympathetic or parasympathetic innervation?

Both—see Table 10-3. *Sympathetic fibers*, from T12, L1, and L2 (Fig. 10-6), innervate the trigone, ureteral openings (Fig. 20-4), and blood vessels of the urinary bladder; *parasympathetic fibers*, from S2–S4, innervate the detrusor muscle. In addition, specialized stretch receptors respond to distension and relay sensory impulses (sympathetic) to the brain via the pelvic splanchnic nerve.

OBJECTIVE E   *To compare the* ureters *and the* urethra *as to structure and function.*

Survey   The *ureters* transfer urine from the renal pelvises of the kidneys to the urinary bladder. The ureters are retroperitoneal, and each consists of three layers: mucosa, muscularis, and fibrous layer. Movement of urine through the ureters is by peristaltic waves.

The *urethra* conveys urine from the urinary bladder to the outside of the body. The *internal urethral sphincter*, of smooth muscle, and the *external urethral sphincter*, of skeletal muscle, constrict the lumen of the urethra, causing the urinary bladder to fill. The urethra of a female is about 4 cm (1.5 in) long, and that of a male is about 20 cm (8 in) long.

20.8   Does the urethra belong solely to the urinary system?

Not in the male, where it conveys semen from the reproductive organs during ejaculation.

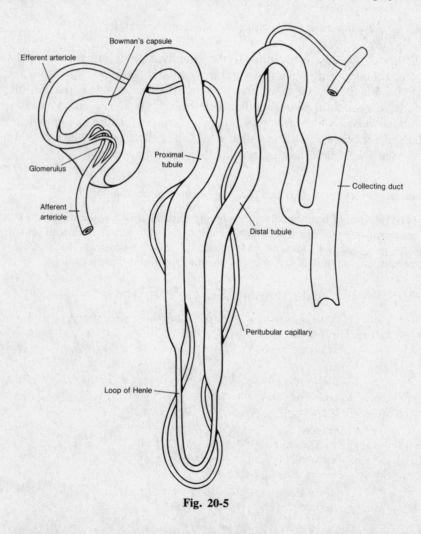

Fig. 20-5

OBJECTIVE F   *To learn the parts of a nephron.*

Survey   Figure 20-5 depicts the structural and functional (urine-forming) unit of the kidney, the *nephron*. There are over one million nephrons per kidney. The components of a nephron are: (1) *glomerulus*, (2) *glomerular capsule* (*Bowman's capsule*), (3) *proximal convoluted tubule*, (4) *loop of Henle*, (5) *distal convoluted tubule*, and (6) *collecting duct*.

20.9   Describe the glomerulus and surrounding capsule.

As shown in Fig. 20-6(*a*), the glomerulus is a network of about 50 capillaries. The capillary endothelium lining the glomerulus has many circular fenestrations, or pores, of between 50 and 100 nm in diameter. This makes the glomerulus 100 to 1000 times more permeable than ordinary capillaries.

The glomerular (Bowman's) capsule is a double-walled cuplike structure composed of squamous epithelium. The outer, parietal layer is continuous with the epithelium of the proximal tubule, whereas the inner, visceral layer is composed of modified cells called *podocytes* [Fig. 20-6(*b*)] which are closely associated with the glomerular capillaries.

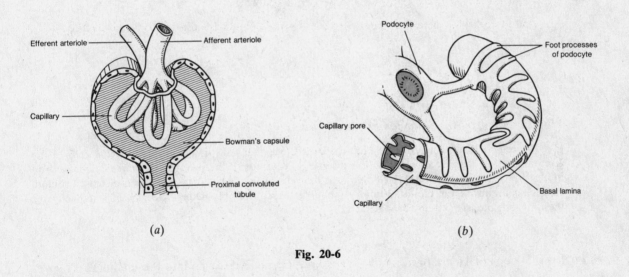

(*a*)   (*b*)

Fig. 20-6

20.10 Describe the structural components of the tubular segments of the nephron.

The *proximal convoluted tubule* is continuous with the parietal epithelium of the glomerular capsule; it consists of a single layer of cuboidal cells with many microvilli (as a *brush border*) which increase the surface area. It terminates in the thin segment of the descending portion of the loop of Henle.

The *loop of Henle* has descending and ascending thin limbs and an ascending thick portion (Fig. 20-5). The thin segments are lined with flat squamous cells which lack microvilli, as do the cuboidal cells that compose the thick segment, which runs between the afferent and efferent arterioles.

The *distal convoluted tubule* begins in the *macula densa*, a mass of specialized epithelial cells of the tubule wall, which is adjacent to the afferent arteriole (Fig. 20-7). It is shorter than the proximal tubule; has some microvilli (which do not, however, compose a brush border); and terminates as it empties into a collecting duct.

A *collecting duct* is formed by the confluence of several distal tubules; collecting ducts, in a renal pyramid, drain urine into the renal pelvis.

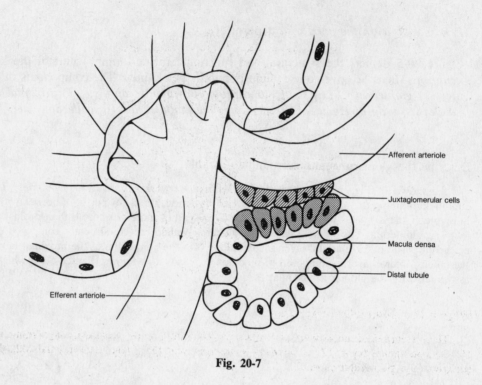

Fig. 20-7

20.11 What is meant by the *juxtaglomerular apparatus*?

The cells of the macula densa (Problem 20.10), together with special *juxtaglomerular cells* of the afferent arteriole (see Fig. 20-7), compose a sensory apparatus. If the juxtaglomerular cells sense a drop in blood pressure in the afferent arteriole or if the cells of the macula densa sense an increased sodium chloride concentration in the distal tubule, renin is released from the juxtaglomerular cells and activates the renin-angiotensin system (Problem 12.22).

20.12 Do the loops of Henle of all nephrons extend the same distance into the medulla?

No: *cortical nephrons*, which are close to the outer surface of the kidney, have very short, thin loops; whereas *juxtamedullary nephrons*, located deep in the cortex adjacent to the medulla, have long loops of Henle that extend deep into the medulla.

OBJECTIVE G   *To identify the three basic components of kidney (i.e., nephron) function.*

Survey   Refer to Fig. 20-8. (1) **Glomerular filtration**: fluid and solutes in the blood plasma of the glomerulus pass into the glomerular capsule. The portion of the plasma that enters the capsule is referred to as the *glomerular filtrate*; it amounts to some 180 L per day (multiple filtration). (2) **Tubular reabsorption**: approximately 99% of the filtrate is transported (actively or passively) out of the tubular lumen, into the interstitial fluid, and then into the peritubular capillaries. (3) **Tubular secretion**: noxious substances (Problem 20.19) are actively transferred from the peritubular capillaries, into the interstitial fluid, and then into the tubular lumen.

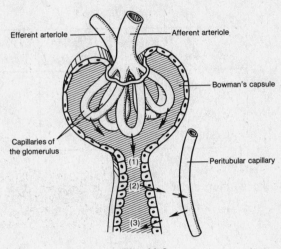

Efferent arteriole
Afferent arteriole
Bowman's capsule
Capillaries of
the glomerulus
Peritubular capillary
(1)
(2)
(3)

Fig. 20-8

20.13 What is the *glomerular membrane*?

The membrane of the glomerular capillaries is referred to as the glomerular membrane; it consists of (1) the endothelial layer, (2) a *basement membrane*, and (3) a layer of epithelial cells that line the surface of the glomerular capsule.

20.14 Why may a larger volume of fluid move across the glomerular membrane than across the membrane of other capillaries?

See Problem 20.9. There is also the fact that hydrostatic pressure within the glomerular capillaries (50 to 60 mmHg) is greater than in other capillaries (10 to 30 mmHg).

20.15 What is the composition of the glomerular filtrate?

Red or white blood cells are generally not filtered, nor are plasma proteins; therefore, the glomerular filtrate is the same as plasma, except that it has no significant amount of protein. The presence of red blood cells or protein in the urine indicates that the hydrostatic pressure in the glomerular capillaries is excessively high or that there is a defect in the glomerular membrane.

20.16 Define the *glomerular filtration rate*.

The GFR is the volume of filtrate formed by all the nephrons of both kidneys each minute. In the adult female, the GFR is about 110 mL/min; in the male, about 125 mL/min. Thus, about 7.5 L/h, or 180 L/day, is formed.

20.17 What fraction of the glomerular filtrate is reabsorbed?

Approximately 99% of the filtrate is reabsorbed out of the tubules and returned to the vascular system, while about 1% is excreted as urine (see the average values given in Table 20-1). The urine volume is regulated according to the needs of the body. Most of the solutes are reabsorbed completely or almost completely, again depending upon the body's need for that particular substance.

Table 20-1

Substance	Kg/day Filtered	Kg/day Excreted	Percent Reabsorbed
Water	180	1.8	99
Glucose	0.180	0.180	100
Sodium	0.630	0.0032	99.5
Urea	0.056	0.028	50

20.18 What percent of the glomerular filtrate water is reabsorbed in each segment of the tubule?

From Fig. 20-9, it is seen that, e.g., about 80% of the reabsorption takes place in the proximal tubule.

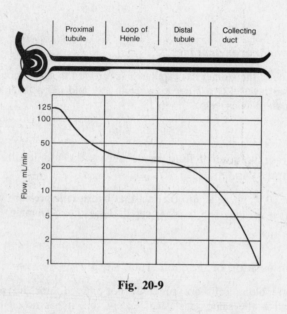

Fig. 20-9

20.19 What substances are actively transported from the peritubular capillaries to the tubule lumen?

Hydrogen, potassium, penicillin, poisons, drugs, metabolic toxins, and chemicals which are not normally present in the body.

OBJECTIVE H   *To explain how the kidneys regulate the concentration of urine.*

Survey   The kidneys produce either a concentrated or dilute urine depending on the operation of a *countercurrent exchange mechanism*, and the amount of circulating antidiuretic hormone (Problem 12.14).

20.20 The urine-concentrating mechanism is a combination of osmotic and diffusive transfers among the *medullary interstitium* (the fluid-filled space into which the tubules project), the tubules, and the *vasa recta* (medullary capillaries). Outline these transfers.

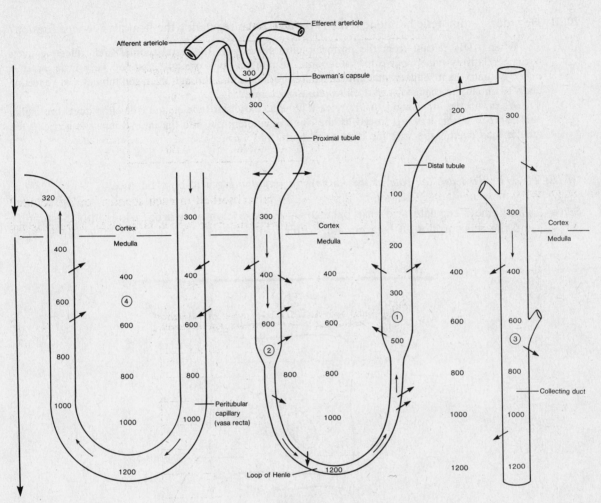

Fig. 20-10

With reference to Fig. 20-10:

1. The thick portion of the ascending limb of the loop of Henle actively transports negatively charged chloride ions out of the tubular fluid and into the medullary interstitium, establishing a difference in electric potential across the tubular wall. This potential causes positively charged sodium ions to pass out into the interstitium. The ascending limb is impermeable to water, and as sodium and chloride ions move out, the fluid in the ascending limb becomes more dilute as it passes toward the cortex.

2. Sodium and chloride ions diffuse into the descending limb, causing the fluids in the descending limb to become more concentrated. The descending limb is permeable to water, and as water diffuses out into the interstitium as a result of the osmotic gradient, the tubular fluid in the descending limb becomes more concentrated as it approaches the bend in the loop.

3. Ions are actively transported into the interstitium from the collecting duct; urea passively diffuses out of the collecting duct into the interstitium.

4. The vasa recta have hairpin loops in parallel with the loops of Henle. Sodium, chloride, and water diffuse into the descending vasa recta, and sodium and chloride diffuse out of the ascending vasa recta; thus, these vessels function as countercurrent exchangers. In addition, only a small quantity (1 to 2%) of the total renal blood flow passes through the vasa recta. As a result, vasa recta circulation carries only a minute amount of the medullary interstitial solutes away from the medulla.

20.21 How does antidiuretic hormone (ADH) participate in regulating the final urine concentration?

When ADH release from the posterior pituitary is low, the distal tubules and collecting ducts become relatively impermeable to water, and, despite the high osmotic gradient, very little water is pulled out into the medullary interstitium. Therefore, fluids pass through the distal tubules and collecting ducts essentially unchanged, and a dilute urine is excreted.

When, on the other hand, ADH release is high, the distal tubules and collecting ducts are highly permeable to water, which is forced by the osmotic gradient out into the interstitium. As a result, the tubular fluid equilibrates with the interstitial fluids, and a concentrated urine is excreted.

OBJECTIVE I   *To learn the role of the kidneys in maintaining* acid-base balance.

Survey   The kidneys regulate acid-base balance by the secretion of hydrogen ions into the tubules and the reabsorption of bicarbonate. Refer to Fig. 20-11.

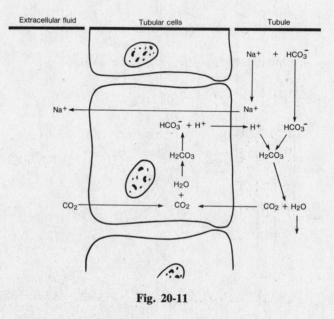

Fig. 20-11

In *acidosis* (Problem 17.28), the ratio of CO_2 to HCO_3^- in the extracellular fluid is increased because of increased production of CO_2 or increased H^+ formation from metabolites. The renal response is:

1. Increased amounts of CO_2 enter the tubular cells from the extracellular fluid.
2. Increased amounts of H^+ are secreted into the tubular lumen. Some of the H^+ combines with HCO_3^- in the tubular fluid, while the remainder combines with buffers in the tubular lumen. One Na^+ ion is reabsorbed for each H^+ ion secreted.
3. Bicarbonate ions (HCO_3^-) in the tubular lumen are reabsorbed into the extracellular fluid via the above reactions.

The net result is that hydrogen ions are excreted from the body, and sodium and bicarbonate ions are conserved by the body.

In *alkalosis* (Problem 17.28), the ratio of HCO_3^- to CO_2 increases and the pH rises. The renal response is:

1. Decreased amounts of CO_2 enter the tubular cells from the extracellular fluid.

2. Decreased amounts of H^+ are secreted into the tubular lumen; and with less H^+ to combine with HCO_3^-, less HCO_3^- is absorbed.

The net result is that hydrogen ions are retained and bicarbonate ions are excreted.

20.22 Identify the buffer systems in the tubular fluid that carry excess H^+ ions into the urine and prevent the pH from dropping too low.

   Phosphate buffer system ($HPO_4^{2-} + H^+ \rightarrow H_2PO_4^-$). The quantity of HPO_4^{2-} in the tubular fluid that has been filtered and not reabsorbed is about four times that of $H_2PO_4^-$. Excess H^+ ions in the filtrate combine with HPO_4^{2-} to form $H_2PO_4^-$.

   Ammonia buffer system ($NH_3 + H^+ \rightarrow NH_4^+$). Ammonia ($NH_3$) is formed by the tubular cells and diffuses into the tubular lumen, where it reacts with H^+ to form ammonium ions.

20.23 Over what range does the urinary pH change under severe acidosis or alkalosis?

   4.5–8.0.

OBJECTIVE J *To give the elementary events of* micturition.

Survey Refer to Fig. 20-12.

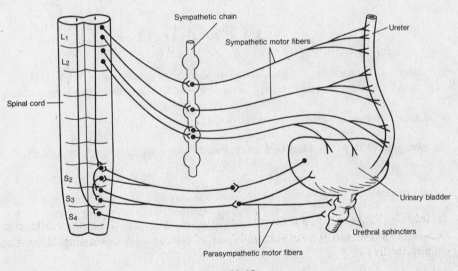

Fig. 20-12

1. Urinary bladder fills with urine and becomes distended.
2. Stretch receptors in the bladder wall discharge impulses via sensory neurons to the sacral spinal cord.
3. Sensory neurons send impulses up the spinal cord to the higher brain centers.
4. Parasympathetic nerve impulses stimulate the detrusor muscle and the internal urethral sphincter.
5. The detrusor muscle contracts rhythmically, and the internal urethral sphincter relaxes.
6. The need to urinate is increased.

7. Urination can be prevented by voluntary contraction of the external urethral sphincter and by inhibition of the micturition reflex by impulses from brain centers.

8. If the decision is to urinate, the external urethral sphincter is relaxed and the micturition reflex is facilitated by impulses from brain centers.

9. The detrusor muscle contracts and urine is released through the urethra.

10. Neurons of micturition reflex centers are inactivated, detrusor muscle relaxes, and the urinary bladder begins to fill again with urine.

20.24 Why are infants unable to contain their urine?

Voluntary inhibition of micturition requires the maturation of portions of the brain and spinal cord, which is accomplished only after several years of life.

20.25 Define (adult) *incontinence*.

Incontinence is the inability to retain urine in the urinary bladder, resulting in its continuous emptying. Incontinence may be caused by central or peripheral nerve damage, various urinary diseases, or tissue damage within the urinary bladder or urethra.

Key Clinical Terms

Acute renal failure Sudden loss of kidney function, usually associated with shock or intense renal vasoconstriction, that lasts from a few days to as long as three weeks. In most cases the kidney damage is repairable.

Azotemia Excessive blood nitrogen compounds.

Blood urea nitrogen (BUN) An index of the accumulation of urea and other nitrogenous wastes in the urine, and thus of renal dysfunction. Normal range: 8–25 g/L.

Chronic renal failure Progressive destruction and shrinking of the kidneys; they become incapable of producing urine. May be caused by chronic *glomerulonephritis* (q.v.) or *pyelonephritis* (q.v.). Early symptoms are *polyuria* (q.v.) and *nocturia* (q.v.); later the patient develops weakness, insomnia, loss of appetite, nausea, acidosis, and azotemia. Because of the permanent damage, the options for sustaining life are *hemodialysis* (q.v.) or kidney transplantation.

Cystitis Inflammation of the urinary bladder.

Cystoscopy Inspection of the urinary system by use of the *cystoscope*. Tissue and urine samples are obtained for diagnosis and for detection of obstructions.

Dysuria Painful urination.

Glomerulonephritis Inflammation of the glomeruli; generally caused by bacterial (streptococcal) infection elsewhere in the body. As toxins are given off by the streptococci, the antigen-antibody complexes accumulate in the glomeruli, producing the inflammation. If the infection is untreated, the glomeruli are replaced by fibrous tissue, and chronic renal disease may develop.

Hematuria Blood in the urine.

Hemodialysis   A technique for purifying the blood outside the body.

Nephrolithiasis   Kidney stones (tiny particles to large calculi) which form as a result of infections, metabolic disorders, or dehydration. They may cause obstruction and intense pain as they pass through the urinary system.

Nocturia   Night urination.

Oliguria   Diminished quantity of urine.

Polyuria   Excessive urine output.

Pyelography   Intravenous injection of a radiopaque dye which permits X-ray examination of the kidney, ureters, and bladder, as the dye passes through the urinary system.

Pyelonephritis   Bacterial infection and inflammation in the renal pelvis, which, if untreated, spreads progressively into the calyces and tubules of the nephrons.

Renal clearance   The volume of plasma per minute that is cleared of a given substance. Let

   $U \equiv$ concentration of substance in urine, mg/mL

   $P \equiv$ concentration of substance in plasma, mg/mL

   $F \equiv$ urine flow, mL/min

Then   Renal clearance $= UF/P$

Uremia   Retention of urinary constituents in the blood, owing to kidney dysfunction.

Urethritis   Inflammation of the urethra.

Urinalysis   Measurement of urine volume (750–2000 mL/day), pH, specific gravity, protein, mucin, ketone bodies, bilirubin, glucose, blood cells, epithelial cells, and casts.

Review Questions

Multiple Choice

1. If blood volume rises: (a) the signals to the kidneys from the parasympathetic division are stimulated, (b) the signals to the kidneys from the sympathetic division are stimulated, (c) the signals to the kidneys from the sympathetic division are inhibited, (d) the kidneys secrete more erythropoetin.

2. Insufficiency of ADH is the condition called: (a) Addison's disease, (b) diabetes insipidus, (c) kidney disease, (d) none of these.

3. Aldosterone causes excessive salt reabsorption in the: (a) proximal tubules, (b) distal tubules, (c) loops of Henle, (d) efferent arterioles.

4. Antidiuretic hormone is secreted by the: (a) kidney, (b) adrenal glands, (c) thyroid, (d) hypothalmus.

5. Increased permeability of the collecting ducts is due to: (a) ADH, (b) renin, (c) aldosterone, (d) angiotension I.

6. The kidneys are involved in the: (*a*) bicarbonate buffer system, (*b*) phosphate buffer system, (*c*) ammonia buffer system, (*d*) all the above.

7. The epithelial cells of the distal tubules, proximal tubules, and collecting ducts all secrete:

      (*a*) aldosterone   (*c*) bicarbonate ions

      (*b*) ADH   (*d*) hydrogen ions

into the tubular fluid.

8. The basic functional unit of the kidney is the: (*a*) glomerulus, (*b*) cortex, (*c*) nephron, (*d*) medulla.

9. Renal failure will *not* result in: (*a*) generalized edema, (*b*) high concentration of nonprotein nitrogens, (*c*) low concentration of necessary electrolytes, (*d*) acidosis.

10. The single diuretic that is most important in inhibiting the secretion of ADH is: (*a*) water, (*b*) alcohol, (*c*) vitamin B_{12}, (*d*) Na^+.

11. Which of the following cells secretes renin? (*a*) macula densa, (*b*) glomerular epithelical, (*c*) juxtaglomerular, (*d*) basement membrane.

12. Which of the following structures functions as a countercurrent exchanger? (*a*) afferent arterioles, (*b*) urethra, (*c*) trigon, (*d*) vasa recta.

13. The distal portions of the tubules are relatively impermeable to (*a*) aldosterone, (*b*) antidiuretic hormone, (*c*) renin, (*d*) macula densa.

14. Sodium balance is controlled by: (*a*) GFR and sodium reabsorption, (*b*) GFR and renin production, (*c*) sodium reabsorption and potassium secretion, (*d*) none of these.

15. Which of the following does not tend to increase osmolality in the medullary interstitial fluid? (*a*) active transport of ions into the interstitium, (*b*) passive diffusion of large amounts of urea from collecting ducts into the interstitium, (*c*) transport of additional sodium and chloride into the interstitium, (*d*) movement of ions into the interstitium by osmosis.

16. Glucose reabsorption: (*a*) is dependent on insulin bonds, (*b*) occurs completely within the proximal tubules, (*c*) varies with metabolism, (*d*) is regulated by the liver.

17. Capillaries in the glomerulus, at the glomerular capsule: (*a*) have larger pores, (*b*) are shorter to accommodate more in the space, (*c*) are innervated, (*d*) can absorb excess sodium ions.

18. What fraction of the total cardiac output goes to the kidneys? (*a*) 1/5, (*b*) 1/8, (*c*) 1/10, (*d*) 1/3.

19. In what cells would be found the greatest concentration of mitochondria? (*a*) skeletal muscle, (*b*) liver, (*c*) proximal convoluted tubule, (*d*) cardiac muscle.

20. Renal diabetes: (*a*) makes glucose appear in the urine while at low concentration in the blood, (*b*) occurs after excessive hemorrhage, (*c*) results from a lack of ADH, (*d*) is more common in youth.

21. For each H^+ ion secreted into the tubules: (*a*) the pH is doubled; (*b*) an Na^+ ion is reabsorbed; (*c*) there is a decrease in the release of aldosterone; (*d*) the pH decreases, causing acidosis.

22. To compensate for alkalosis, the kidney tubules: (a) reabsorb more Na^+; (b) having less CO_2 from the blood for reaction with HCO_3^-, allow HCO_3^- to pass out in the urine; (c) secrete more angiotensin I, which stimulates the hypothalamus to secrete more ADH to cause increased filtration in the glomeruli; (d) absorb more K^+.

23. The mechanism by which the kidneys produce concentrated urine is called: (a) bicarbonate buffer system, (b) renin-angiotensin system, (c) countercurrent exchange, (d) urinary densification.

24. With a lack of ADH, the: (a) distal convoluted tubules and collecting ducts become more impermeable to water, (b) kidney becomes more sensitive to H^+, (c) proximal tubules reabsorb K^+ instead of Na^+, (d) hypothalamus cannot regulate O_2 levels.

25. Nitrogenous excretion products include: (a) amino acid, (b) ammonia, (c) HNO_3^-, (d) thionitrate.

26. The nitrogen that must be excreted comes from: (a) the metabolic breakdown of carbohydrates, (b) the nitric buffer system, (c) the deamination of proteins, (d) the atmosphere.

27. Glomerular filtrate is identical to plasma, except in respect to: (a) pH, (b) glucose content, (c) amounts of albumin and globulins, (d) formed-element and protein content.

28. Cl^- ions are reabsorbed because of: (a) the permeability of the tubules to this ion caused by aldosterone, (b) the electrostatic attraction of Na^+ ions, (c) their involvement in calcium retention, (d) their high osmotic gradient.

29. Increased glomerular filtration results from: (a) increased cardiac output, (b) decrease in blood pressure, (c) decreased fluid intake, (d) environmental temperature increase.

30. Pick out the false statement: (a) the first step in urine formation is filtration of blood plasma from the glomerulus into Bowman's capsule; (b) the second step in urine formation is the reabsorption of water and some of the solutes by the tubular cells; (c) the tubular cells also help to form urine by secreting variable substances such as H^+ and NH_3^+; (d) slightly less than half of the water filtered from the blood in urine formation is reabsorbed into the blood; (e) normally, blood proteins are not filtered from the blood into the capsule.

31. Kidney innervation regulates: (a) composition of the glomerular filtrate, (b) diameter of the arterioles, (c) water reabsorption, (d) activity of the brush border cells.

32. The collecting ducts in the kidney: (a) can secrete water molecules actively into the urine, (b) are responsible for most of the reabsorption of water that occurs in the kidney, (c) determine to a large extent the final osmolality of urine, (d) are rendered impermeable to water by ADH.

33. Aldosterone: (a) is produced mainly in the juxtaglomerular apparatus, (b) increases sodium reabsorption by the nephron, (c) increases potassium reabsorption by the nephron, (d) tends to increase the hydrogen-ion concentration in the blood.

34. Renal clearance: (a) is calculated as UP/F, where U is urinary concentration of the material, F is urine

volume/min, and P is plasma concentration of the material; (b) of inulin provides an estimate of glomerular filtration rate; (c) of creatinine provides an estimate of renal plasma flow; (d) of phosphate is decreased by parathormone (parathyroid hormone).

35. When a patient is treated with an aldosterone antagonist, there is likely to be a fall in: (a) urine volume, (b) plasma potassium concentration, (c) blood viscosity, (d) blood volume.

36. Capillary pressure in the renal glomeruli: (a) is lower than pressure in the efferent arterioles, (b) rises when the afferent arterioles constrict, (c) is higher than in most other capillaries in the body, (d) falls by about 10% when arterial pressure falls 10% below the normal level.

37. The macula densa is part of the: (a) proximal renal tubule, (b) afferent renal arteriole, (c) distal renal tubule, (d) efferent renal arteriole, (e) none of the above.

38. A patient has a plasma urea concentration of 26 mg per 100 mL and a urinary urea concentration of 18.2 mg/mL. Urine flow is 2.0 mL/min. The clearance rate for urea is: (a) 35 mL/min, (b) 70 mL/min, (c) 105 mL/min, (d) 140 mL/min, (e) 175 mL/min.

True/False

1. The kidney is important in activating vitamin D, which is used in the synthesis of blood.

2. Bilirubin, a by-product of the destruction of RBCs, can be found in the urine.

3. The afferent arterioles bring the blood supply to the glomeruli.

4. Most reabsorption occurs in the distal convoluted tubules.

5. The kidneys synthesize and secrete glucose during prolonged fasting.

6. The kidneys help to counter alkalosis by reabsorbing excess HCO_3^-.

7. The countercurrent mechanism helps the kidney dilute urine to counter acidosis.

8. Aldosterone increases the permeability of the distal convoluted tubules to water only.

9. Cells of the proximal convoluted tubule have even more mitochondria than do muscle cells.

10. The kidneys regulate glucose levels by secreting any excess in the urine.

11. The ascending loop of Henle actively transports Cl^- out of the tubule.

12. One symptom of diabetes is polyurea.

13. A difference in hydrostatic pressure is one of the mechanisms for pushing blood fluid through the glomerulus to form the filtrate.

14. ADH is necessary for the reabsorption of Na^+.

Matching

1. Aldosterone
2. Antidiuretic hormone
3. Juxtaglomerular cells
4. Loop of Henle
5. Vasa recta
6. pH of the blood
7. pH of the urine
8. Nephron
9. Ureter
10. Angiotensin II
11. Renin
12. Micturition

(a) Adjacent to the macula densa of the capillaries
(b) Network of capillaries that descend around the loop of Henle
(c) Tube running from kidney to bladder
(d) Composed of Bowman's capsule, proximal and distal tubules, loop of Henle, glomerulus and collecting duct
(e) Act of passing urine
(f) Tubule descending from the cortex to the medulla in the kidney
(g) 7.0–7.4
(h) Stimulates the zona glomerulosa of the adrenal glands
(i) 4.5–6.0
(j) Secreted by zona glomerulosa
(k) 5.5–7.8
(l) Tube running from urinary bladder out of the body
(m) Secreted by juxtaglomerular cells
(n) Synthesized in the hypothalamus
(o) Necessary for deamination of proteins

Chapter 21

Water and Electrolyte Balances

OBJECTIVE A *To give the distribution of water among the body compartments.*

Survey In Fig. 21-1, B.W. ≡ total body weight.

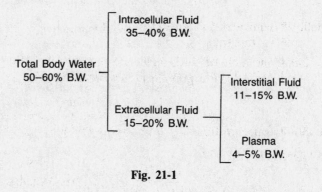

Fig. 21-1

21.1 List the major functions of water in the body.

    Universal solvent and suspending medium; helps regulate body temperature; participates in hydrolysis reactions; lubricant.

21.2 How does total body water vary with age, sex, and body weight?

    Age. Total body water constitutes up to 75–80% of the B.W. in infants and young children. This percentage decreases with age, and in older people total body water may be as low as 40–50% of the B.W.

    Sex. Women usually have less total body water than men, because of the increased proportion of fatty tissues which contain lesser amounts of water.

    Body weight. Obese people have less water per B.W., again because of the relative abundance of fatty tissues.

OBJECTIVE B *To give the terms ordinarily used to state solute concentrations (in body fluids).*

Survey ***Percent solutions***
$$\% \text{ solution} \equiv (\text{grams solute})/(100 \text{ mL solution})$$
$$\equiv (\text{grams solute})/(\text{dL solution})$$

    Example: If 200 mL of solution contains 5 g of dissolved NaCl, the solution is termed <u>2.5% NaCl.</u>
$$\text{mg} \% \equiv (\text{milligrams solute})/(100 \text{ mL solution})$$
$$\equiv 1000 \times (\% \text{ solution})$$

Molarity (molar concentation)
Let MW denote the molecular weight of the solute (e.g., for NaCl, MW = 23 + 35.5 = 58.5). Then *one mole* (1 mol) of solute weighs MW grams, whence:

Moles solute $\equiv$ (grams solute)/MW

and

Molarity $\equiv$ (moles solute)/(L solution)

Example: The previously considered solution is (5/58.5) moles of NaCl in 0.200 liters, giving a molarity of 0.427. The solution would be termed 0.427 *M* NaCl.

Normality

For a solute that dissociates into ions of valence $\pm v$:

Equivalents solute $\equiv v \times$ (moles solute)

and

Normality $\equiv$ (equivalents solute)/(L solution)
$\equiv v \times$ molarity

Example: Since one mole of NaCl dissociates into one mole of Na^+ (valence +1) and one mole of Cl^- (valence -1), the previously considered solution is 0.427 *N* NaCl.

21.3 What is the molar concentration of a 0.9% NaCl solution?

The NaCl content is

$$\frac{0.9 \text{ g}}{1 \text{ dL}} = \frac{9 \text{ g}}{1 \text{ L}} = \frac{9 \text{ g}}{1 \text{ L}} \times \frac{1 \text{ mol}}{58.5 \text{ g}} = 0.154 \text{ mol/L}$$

or 154 mmol/L.

21.4 A sample of intracellular fluid was found to be 0.06-normal in Mg^{2+} (cf. Table 21-1). What mg % is this equivalent to?

The ionic valence is +2, and so

$$\text{molarity} = \frac{\text{normality}}{2} = 0.03 \text{ mol/L} = 0.03 \text{ mmol/mL}$$

Then, since MW = 24 for Mg,

$$\frac{0.03 \text{ mmol}}{1 \text{ mL}} \times \frac{24 \text{ mg}}{1 \text{ mmol}} = \frac{0.72 \text{ mg}}{1 \text{ mL}} = \frac{72 \text{ mg}}{100 \text{ mL}}$$

The sample is thus 72 mg % Mg^{2+}.

21.5 List mean concentrations of the more important chemicals in the extracellular and intracellular fluids.

All concentrations in Table 21-1, except those of glucose, are normalities, in milliequivalents per liter.

Table 21-1

	Na^+	K^+	Ca^{2+}	Mg^{2+}	Cl^-	Amino Acids	Glucose, mg %
Extracellular	142	4	5	3	103	5	90
Intracellular	10	140	1	58	4	40	0–20

21.6 How are the volumes of the fluid compartments measured?

Using an *indirect dilution technique*, one introduces a specific quantity of an exogenous substance (dye, radioisotope, etc.) which because of its chemical properties becomes evenly distributed within a certain fluid compartment. One then removes an aliquot of the fluid from that compartment and determines the concentration of the substance.

$$\text{Compartment volume (mL)} = \frac{\text{Quantity of substance introduced (mg)}}{\text{Substance concentration in compartment (mg/mL)}}$$

Compartment	Substances Used
Total	^3H_2O (radioactive water), antipyrine
Extracellular	Thiosulfate
Plasma	Evans blue

The intracellular and interstitial volumes are determined indirectly, as differences.

OBJECTIVE C *To define the* fluid balance *of the body.*

Survey Under normal conditions, fluid intake equals output, so that the body maintains a constant volume. A typical water budget is given in Fig. 21-2.

Water Intake		Water Output	
Ingested liquids	1400 mL	Urine	1500 mL
Solid and semisolid foods	800	Skin	500
Oxidation of food	300	Lungs	350
	2500	Feces	150
			2500

Fig. 21-2

21.7 What are the causes and symptoms of *dehydration*, or *hypovolemia*, and what is the body's response?

Causes. Decreased intake (lack of water, psychotic refusal to drink) and/or increased output (vomiting, diarrhea, loss of blood, drainage from burns, lack of ADH owing to diabetes insipidus).

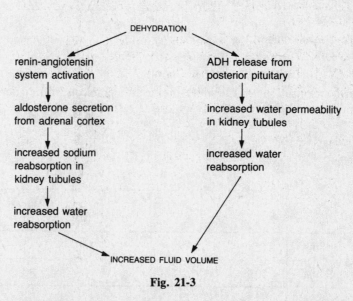

Fig. 21-3

Symptoms. Loss of weight, rise in body temperature, increase in heart rate and cardiac output, decrease in blood pressure, sunken eyeballs, decreased urinary output (lack of ADH).

Response. There is a decrease in salivary secretion and consequent drying of the mouth and pharynx. This dryness is interpreted by the brain as thirst, and drinking is stimulated. There is also an increased release of ADH and aldosterone, which leads to a conservation of body fluids via the mechanism indicated in Fig. 21-3.

21.8 What are the causes and symptoms of *water intoxication*, or *hypervolemia*, and what is the body's response?

Causes. Excessive IV administration of fluids, psychotic drinking episodes, decreased urinary output because of renal failure.

Symptoms. Decrease in body temperature, increased blood pressure, edema, weight gain, vomiting, convulsions, coma.

Response. The decrease in osmolarity of fluids in the hypothalamus causes an inhibition of thirst and a decreased release of ADH and aldosterone. With decreased ADH and aldosterone secretion, there is an increased urinary output (inverse to the process of Fig. 21-3).

OBJECTIVE D   *To distinguish between* electrolytes *and* nonelectrolytes.

Survey   *Electrolytes* are chemicals formed by ionic bonding, which dissociate into electrically charged ions (cations and anions) when they dissolve in the body fluids. Examples of electrolytes are acids, bases, and salts. *Nonelectrolytes* are formed by covalent bonding; they do not ionize when dissolved in the body fluids. Most organic compounds are nonelectrolytes.

21.9 List the general functions of body electrolytes.

(1) Control of the osmolarity in the fluid compartments; (2) maintenance of the acid-base balance in body fluids; (3) metabolization as essential minerals; (4) participation in all cellular activities.

21.10 Identify some of the common electrolyte-imbalance disorders.

Hyponatremia (*low blood sodium level*). *Causes*: sweating, diarrhea, certain diuretics, Addison's disease, excessive water intake. *Symptoms*: muscular weakness, headache, hypotension, circulatory shock, mental confusion.

Hypernatremia (*high blood sodium level*). *Causes*: diabetes insipidus, inadequate water intake, Cushing's syndrome. *Symptoms*: nervous system disorders, mental confusion, coma.

Hypokalemia (*low blood potassium level*). *Causes*: vomiting, diarrhea, Cushing's syndrome, kidney failure. *Symptoms*: muscle weakness, paralysis, shallow breathing, cardiac arrythmias.

Hyperkalemia (*high blood potassium level*). *Causes*: kidney disease, Addison's disease. *Symptoms*: muscle weakness, paralysis, cardiac arrythmias, cardiac arrest.

Hypochloremia (*low blood chloride level*). *Causes*: vomiting, diarrhea, dehydration. *Symptoms*: muscle spasms, alkalosis, depressed breathing.

Review Questions

Multiple Choice

1. Total body water would be relatively greatest in a: (a) newborn male, (b) teenage female, (c) middle-aged male, (d) 100-year-old female.

2. To make 300 mL of a 10% solution, how much solute should be added? (a) 10 g, (b) 10 mg, (c) 3 mg, (d) 30 g.

3. Loss of weight, rise in body temperature, decrease in blood plasma, sunken eyeballs, and decreased urinary output are all symptoms of: (a) hypervolemia, (b) positive water balance, (c) hypovolemia, (d) Cushing's syndrome.

4. The normality, in milliequivalents per liter (meq/L), of a 60 mg % solution of $CaCl_2$ (MW = 110, valence = 2) is: (a) 5.4, (b) 10.9, (c) 15.3, (d) 111.1.

5. The feeling of thirst arises from: (a) increased osmolarity of the body fluids, (b) increased plasma volume, (c) hypervolemia, (d) hypersecretion of ADH.

6. The interstitial fluid is: (a) part of the extracellular fluid and makes up 5% of the body weight, (b) part of the intracellular fluid and makes up 20% of the body weight, (c) part of the extracellular fluid and makes up 15% of the body weight, (d) that fluid in the cell nucleus.

7. Extracellular fluid differs from intracellular fluid in that it: (a) forms the major proportion of total body water, (b) is composed mainly of transcellular fluids, (c) has a higher sodium/potassium ratio, (d) contains less glucose.

8. Hyperkalemia differs from hypokalemia in that it: (a) causes muscle weakness, (b) causes cardiac arrythmias, (c) causes paralysis, (d) may be caused by Addison's disease.

9. If a laboratory animal excretes 12.6 grams of NaCl per day, what would be the rate of Na^+ excretion, expressed in meq/day? (a) 125, (b) 215, (c) 375, (d) 400.

10. How many milliequivalents are there in 230 mL of a 0.1 M KCl solution? (a) 13, (b) 18, (c) 23, (d) 133.

11. Which of the following has the smallest volume? (a) extracellular fluid, (b) plasma, (c) interstitial fluid, (d) intracellular fluid.

12. The most abundant cation in the extracellular fluid is: (a) Na^+, (b) K^+, (c) Ca^{2+}, (d) Mg^{2+}.

13. There is only a slight difference in composition between: (a) plasma and intracellular fluid, (b) plasma and interstitial fluid, (c) intracellular and extracellular fluids, (d) plasma and water.

14. The average adult female is:

> (a) 40–50%   (c) 30–40%
> (b) 50–60%   (d) 60–70%

fluid, by weight.

15. In a 200-pound male, the intracellular fluid would weigh about: (a) 100 pounds, (b) 60 pounds, (c) 10 pounds, (d) 80 pounds.

16. Approximately how many milliliters of fluid are lost per day via the lungs? (*a*) 800, (*b*) 500, (*c*) 350, (*d*) 150.

17. Which of the following individuals would be most seriously affected by severe diarrhea? (*a*) 15-year-old male, (*b*) 30-year-old male, (*c*) 35-year-old female, (*d*) 80-year-old female.

18. The barrier that separates the interstitial fluid from the intracellular fluid is the: (*a*) cell membrane, (*b*) capillary wall, (*c*) diaphragm, (*d*) mesentery.

19. Potassium is the primary cation of the: (*a*) plasma, (*b*) interstitial fluid, (*c*) transcellular fluids, (*d*) intracellular fluid.

20. Diabetes insipidus may induce: (*a*) hyperkalemia, (*b*) hypernatremia, (*c*) isotonic dehydration, (*d*) hypervolemia.

Chapter 22

Reproductive Systems

OBJECTIVE A *To appreciate the biological value of* sexual reproduction.

Survey The male and female reproductive systems are specialized to produce offspring that have *genetic diversity*, inherited through a *gamete* from each parent.

22.1 What are *gametes*?

Gametes, also called *germ cells* or *sex cells*, are the functional reproductive cells (ova or spermatozoa). They are *haploid* cells: each contains a half-complement—or 23 single chromosomes—of the genetic material. Fertilization of an ovum by a spermatozoon produces a normal *diploid* cell, the *zygote*, in which the chromosomes of the ovum have been paired with those of the spermatozoon. Thus is genetic diversity realized.

22.2 How is the sex of a child determined?

One out of the 23 pairs of human chromosomes determines sex. Sex chromosomes are of two types, X and Y. A female pair of sex chromosomes consists of two X-chromosomes; consequently, all female gametes, or ova, contain a single X-chromosome. The male pair of sex chromosomes is an X- and a Y-chromosome; thus, equal numbers of X and Y male gametes, or spermatozoa, are produced. It follows that an offspring will be female or male according to whether the fertilizing spermatozoon is of X-type or of Y-type. The two possibilities will be equally likely.

OBJECTIVE B *To understand the processes of* spermatogenesis *and* oogenesis.

Survey Spermatogenesis is the process by which sperm cells (which become free-swimming spermatozoa) are produced in the testes of a male, and oogenesis is the process by which ova are produced in the ovaries of a female. Both processes involve a special kind of cell division, called *meiosis*.

22.3 Describe meiosis and how it differs in oogenesis and spermatogenesis.

Refer to Fig. 22-1. In meiosis, each chromosome duplicates itself as in mitosis (Fig. 3-9). However, unlike mitosis, the homologous chromosomes are attached to each other and come to lie alongside one another in pairs, producing a *tetrad* of four *chromatids*. Two *maturation divisions* are required to effect the separation of the tetrad into four daughter cells, each with one-half the original number of chromosomes. In these divisions, maternal and paternal chromosomes become freely assorted, yielding a great number of possible combinations in the haploid gametes.

The nuclear aspects of meiosis are similar in males and females. There is, however, a marked difference in the cytoplasmic aspects. The two meiotic divisions of the primary oocyte do not result in four equally mature gametes, as in the male, but in only one mature ovum.

OBJECTIVE C *To differentiate between* primary *and* secondary *sexual organs*.

Survey The primary sexual organs, or *gonads*, are the testes (in the male) and the ovaries (in the female). They are mixed glands (Problem 12.2) in that they produce both sex hormones and gametes. Secondary, or accessory, sexual organs are those structures (in the male or the female) that mature at puberty and are essential in caring for and transporting gametes.

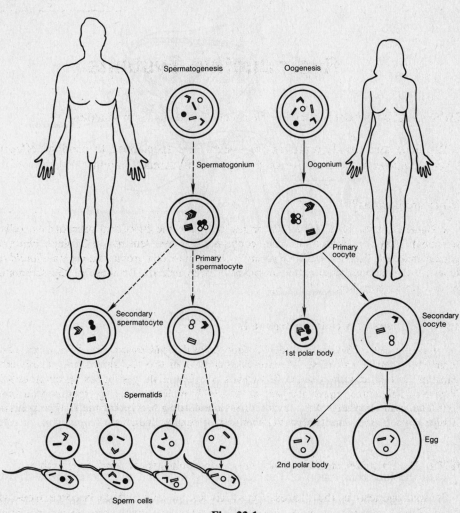

Fig. 22-1

22.4 What is responsible for the development of the secondary sexual organs?

Increased release of sex hormones—testosterone in the male, estradiol and other estrogens in the female—at puberty. (Review Chapter 12.)

22.5 Which of the following are secondary sexual organs, and which are *secondary sexual characteristics*? (*a*) uterine tubes, (*b*) pubic hair, (*c*) epididymides, (*d*) breasts.

The *uterine tubes* of the female and the *epididymides* (two masses of tubules attached to the testes) of the male are *secondary sexual organs* because they mature during puberty and are essential for successful sexual reproduction. Enlarged *breasts* in the female are considered a sexual attractant and therefore a *secondary sexual characteristic*; the same is true of *pubic hair*, in either sex.

OBJECTIVE D   *To identify the organs of the male reproductive system.*

System   The male reproductive organs are: the *testes*, within the *scrotum*; the *epididymides*, attached to the testes; spermatozoa-transporting ducts called *ductus* (*vasa*) *deferentia*; accessory glands (*seminal vesicles*, *prostate gland*, and *bulbourethral gland*); *penis*; *ejaculatory ducts*; and the *urethra*. See Fig. 22-2.

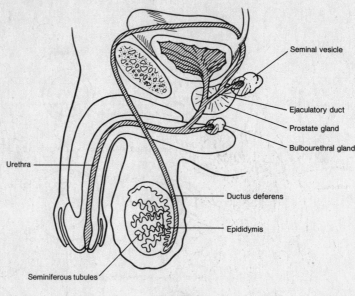

Fig. 22-2

22.6   Which organs constitute the male external genitalia?

The *penis* and the scrotum-enclosed *testes* and associated *epididymides*. The penis contains erectile tissue and serves as an intromittent organ during sexual intercourse (*copulation*, or *coitus*).

22.7   Detail the interconnection of the accessory glands and the spermatozoa-transporting ducts.

The *ejaculatory duct* opens into the prostatic portion of the *urethra*. This duct is formed by the union of the *ductus deferens* and the duct of the *seminal vesicle*. The *prostate gland*, which surrounds the junction of the ejaculatory duct and the urethra, drains directly into both of these ducts. The *bulbourethral gland* empties into the urethra at the base of the penis.

22.8   Which are unpaired organs? (*a*) penis, (*b*) ductus deferens, (*c*) ejaculatory duct, (*d*) prostate gland, (*e*) seminal vesicle.

(*a*) and (*d*). The paired ductus deferentia carry sperm from the epididymides of the testes to the paired ejaculatory ducts. Each ejaculatory duct receives secretions from a seminal vesicle and the prostate gland. The ejaculatory ducts join the urethra at the level of the prostate gland, near the urinary bladder.

OBJECTIVE E   *To examine the* testes, *with particular regard to their descent into the* scrotum.

Survey   The testes, formed in the pelvic cavity, come down to lodge within the protective scrotum, where optimal conditions exist for their dual functions of spermatogenesis and hormone (testosterone) secretion. Figure 22-3(*a*) and (*b*) shows a testis in coronal and transverse section, respectively.

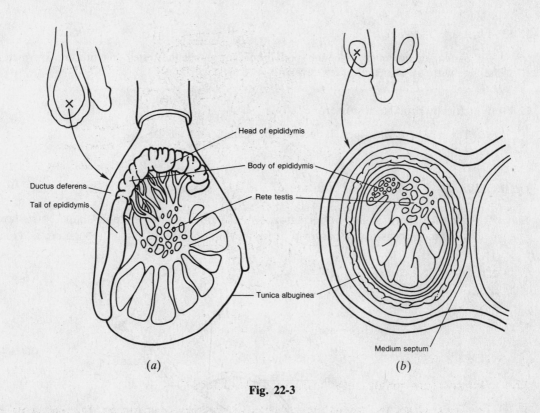

Fig. 22-3

22.9 Describe the gross structure of a testis.

The testis is an ovoid, whitish organ, measuring about 4 cm by 2.5 cm, which consists of 250 to 300 wedge-shaped lobules. It is covered by the fibrous *tunica albuginea* and encapsuled by the *tunica vaginalis*. A *medium* (*scrotal*) *septum* separates the two testes.

22.10 When do the testes descend into the scrotum, and what may happen if they do not?

The descent starts during the twenty-eighth week of prenatal development and usually is completed in the twenty-ninth week.

If one or both testes are not in the scrotal sac at birth, it may be possible to induce descent by administering certain hormones. If this procedure does not work, surgery is necessary and is generally performed before age five. Failure to correct the situation may result in sterility and/or a tumorous testicle.

22.11 Describe the thermostatic action of the scrotum.

A temperature of 96 °F (35 °C) is optimal for production and storage of sperm. At suboptimal temperatures, scrotal muscles contract involuntarily, to bring the testes closer to the heat of the body; at superoptimal temperatures, they relax.

22.12 Where are the sperm cells produced, and where are they stored, within the testis?

Spermatogenesis occurs in the seminiferous tubules. Once the sperm are produced, they move through the seminiferous tubules and enter the *rete testis* (Fig. 22-3) for further maturation. The sperm are then transported out of the testis through a series of efferent ductules into the epididymis, where further maturation and storage occurs. Mature sperm are also stored in the lower portion of the ductus deferens. The entire process from production to maturation requires approximately two months.

22.13 What are *Sertoli cells*?

Sertoli, or *interstitial*, *cells* are specialized cells within the testes that produce spermatozoa and secrete nutrients to them as they develop.

22.14 Which cells produce testosterone?

The *cells of Leydig*, located between the seminiferous tubules.

OBJECTIVE F   *To describe the structure of a mature* sperm cell.

Survey   A mature sperm cell, or spermatozoon, is a microscopic, tadpole-shaped structure about 60 μm long, consisting of an oval head, a cylindrical body, and an elongated tail (Fig. 22-4).

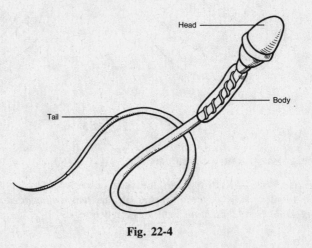

Fig. 22-4

22.15 What are the functions of the separate parts of a sperm cell?

The *head* contains a nucleus with a chromatin of 23 chromosomes (haploid cell). The *acrosome* of the head contains an enzyme that assists penetration of the spermatozoon into the ovum. The *body* contains mitochondria which provide the energy for locomotion via the *tail*, a flagellum that propels the sperm in a lashing movement, at about 3 mm per hour.

OBJECTIVE G   *To give the structure and functions of the* penis.

Survey   The penis is composed mainly of erectile tissue and, when distended, serves as the intromittent, or copulatory, organ of the male. The urethra of the flaccid penis serves the urinary system as a conduit of urine from the urinary bladder.

22.16 How is the penis attached to the pelvic floor (*perineum*)?

See Fig. 22-5(*a*). The root of the penis is expanded posteriorly to form the *bulb* and the *crus* of the penis. The bulb is located in the urogenital triangle of the perineum, where it is attached to the under surface of the perineal membrane and enveloped by the *bulbospongiosus muscle*. The crus, in turn, attaches the root of the penis to the pubic arch and to the perineal membrane. The crus is positioned superior to the bulb and is enveloped by the *ischiocavernosus muscle*.

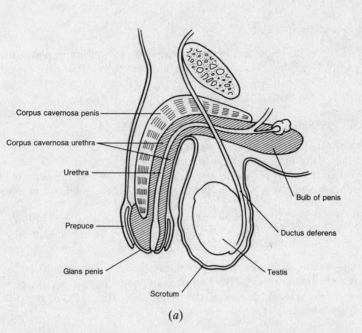

Corpus cavernosa penis

Corpus cavernosa urethra

Urethra

Prepuce

Glans penis

Scrotum

Bulb of penis

Ductus deferens

Testis

(a)

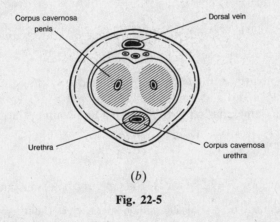

Corpus cavernosa penis

Dorsal vein

Urethra

Corpus cavernosa urethra

(b)

Fig. 22-5

22.17 Which parts of the penis are erectile, and which are not?

Erectile: three longitudinal columns of tissue (Fig. 22-5); namely, two *corpora cavernosa penis*, composing the dorsal portion of the penis, and a *corpus cavernosa urethra* (or *corpus spongiosum*), which is ventral and surrounds the penile urethra. *Nonerectile*: the *glans penis*, or enlarged terminal portion of the corpus cavernosa urethra, and its covering (in an uncircumcised male), the retractible *prepuce*, or *foreskin*.

22.18 How is the penis erected?

Erection of the penis depends upon a surplus of blood entering the arteries of the penis as compared to the volume exiting through venous drainage. Normally, a constant sympathetic stimulation of the arterioles of the penis maintains a partial constriction of smooth muscles within the arteriolar walls, so that an even flow of blood occurs throughout the penis. During sexual excitement, however, parasympathetic impulses cause marked vasodilation of the arterioles, and, with more blood entering than leaving, the penis grows turgid. A simultaneous, slight vasoconstriction of the dorsal veins of the corpora cavernosa and the corpus spongiosum furthers the effect.

22.19 What is meant by *ejaculation*?

Ejaculation is the expulsion of semen through the urethra of the penis. In contrast to erection, ejaculation involves sympathetic innervation of accessory reproductive organs.

22.20 Define *impotence*.

Impotence is the inability of a sexually mature male to sustain an erection until ejaculation. The causes of impotence may be physical (e.g., a structural abnormality of the penis, vascular irregularities, neurological disorders, or certain diseases). Generally, however, the cause of impotence is psychological, and skilled counseling by a sex therapist is necessary if the condition is to be corrected.

22.21 Are there erectile tissues in the female genitalia?

Yes: the erectile structures of the female genitalia are homologous to those of the male, and include the *clitoris* and *vestibular bulbs*. The clitoris is the homolog of the glans penis; the vestibular bulbs are homologous to the erectile tissues of the penile shaft. In addition, there is erectile tissue within the *areolae* of the breasts.

OBJECTIVE H *To learn the composition and volume of* semen, *and the factors involved in male fertility.*

Survey Semen is the mixture of fluids that is ejaculated from the erect penis. It consists of sperm from the epididymides and ductus deferentia, and additives from the seminal vesicles and prostate gland. Semen has a pH of about 6.5 and contains large quantities of prostaglandins (Problem 12.4).

22.22 Estimate the concentration of spermatozoa (*sperm count*) in semen.

There will normally be 200 to 500 million sperm ejected during an ejaculation. This amounts to roughly 100 million per mL of ejaculate. If the sperm concentration is less than 10 to 20 million per mL, the male is likely to be infertile. A male who has had a *vasectomy* (removal of a portion of each vas deferens) can still ejaculate seminal fluid, but it contains no spermatozoa.

22.23 Describe the condition called *varicocele*.

Varicocele exists when one or both of the testicular veins draining from the testes are swollen (varicosed), resulting in poor vascular circulation in the testes and hence interference with spermatogenesis. Male infertility is primarily due to varicocele.

OBJECTIVE I *To classify the components of the female reproductive system.*

Survey **Primary organs** (Fig. 22-6) are the *ovaries*, which produce the ova (egg cells). **Internal accessory organs** (Fig. 22-6) include the *uterine tubes* (or *oviducts* or *fallopian tubes*), the *uterus*, and the *vagina*. **External structures** (Fig. 22-7) include the *labia majora*, *labia minora*, *clitoris*, *vestibule*, and *vestibular glands*.

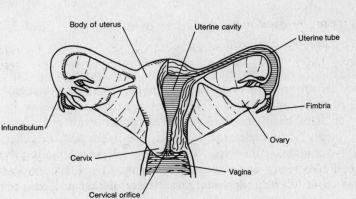

Fig. 22-6

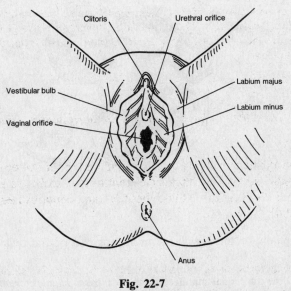

Fig. 22-7

22.24 Describe the major functions of the female internal accessory organs.

The *vagina* acts as an excretory duct for uterine secretions and the menstrual flow, receives seminal fluid from the male, and serves as the lower part of the birth canal. The *uterus* provides protection and nourishment to the developing embryo and fetus. The mucosal lining of the uterus (*endometrium*) undergoes changes associated with the menstrual cycle (see Problem 22.35). The muscular layer of the uterus (*myometrium*) produces powerful rhythmic contractions that result in parturition, or birth of the baby. The uterine tubes transport the ova from the ovaries to the uterus.

22.25 What is the function of the vestibular glands?

They secrete a mucous fluid during periods of sexual excitement, which moistens and lubricates the vestibule and lower regions of the vagina.

22.26 List the anatomical subdivisions of the uterus.

The dome-shaped portion above the entrance of the uterine tubes is the *fundus*; the major, tapered, central region is the *body*; and the lower, narrow portion that protrudes into the vagina is the *cervix*. (The interior of the body is the *uterine cavity*, and the interior of the cervix is the *cervical canal*.)

OBJECTIVE J *To describe the structure of an* ovary *and the cyclical development of the* follicle, ovum, *and* corpus luteum.

Survey Refer to Fig. 22-8. The ovaries are located in the upper pelvic cavity, one on each side of the uterus, held in position by several ligaments. In the outer region of each ovary are tiny masses of cells, called *ovarian follicles*; each follicle contains an immature egg. As many as twenty follicles begin to develop at the beginning of a 28-day *ovarian cycle*; however, normally only one follicle reaches full development and the others undergo degeneration. About the middle of the cycle, the mature follicle containing the ovum bulges from the surface of the ovary and releases the ovum, in the process known as *ovulation*. After ovulation, the follicle cells undergo a structural change (*luteinization*) to form the corpus luteum.

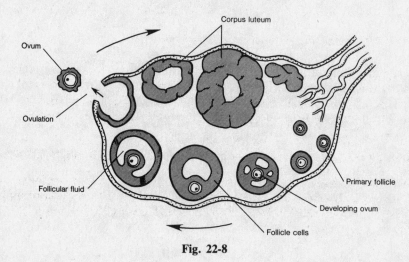

Fig. 22-8

22.27 One usually speaks of the *menstrual cycle*, instead of the *ovarian cycle*. Are these distinct cycles?

No; rather, they are dual manifestations of a basic *female hormonal cycle* (Objective L). When "ovarian cycle" is used, the focus is on the changes in the ovaries effected by the hormonal changes; when "menstrual cycle" is used, the focus is on the flow or nonflow of blood from the vagina, resulting from the very same hormonal changes.

22.28 What causes a primary follicle to mature?

FSH (Tables 12-1 and 12-2), along with subsequent small quantities of LH. After about the fifth day of the menstrual cycle, the two ovaries between them contain just one maturing follicle.

22.29 What is responsible for ovulation and luteinization?

A sharp rise in LH secretion causes (i) the mature follicle to rupture, releasing its egg; (ii) the ruptured follicle to fill with blood; (iii) the clotted blood to be replaced with lipid-rich luteal cells, forming the corpus luteum.

22.30 What are some of the indicators of ovulation?

Body temperature. There is usually an increase in the basal body temperature upon ovulation, which is maintained during the rest of the menstrual cycle. To detect this increase, the body temperature is taken with a *basal thermometer* each morning.

Cervical mucus. Cervical mucus, when smeared on a glass slide, dried, and examined under a microscope, shows different patterns at different times during the cycle. In the middle of the cycle (when ovulation occurs), the mucus exhibits a fernlike pattern.

Abdominal pain (*mittelschmerz*). The rupture of the follicle at the time of ovulation may cause some hemorrhage and local inflammation, which in turn can induce low abdominal pain.

22.31 What is the length of the fertile period in the monthly ovarian cycle?

The sperm can live about 48 to 72 hours in the female genital tract, whereas the egg can survive only about 24 hours after ovulation. Therefore, coitus must occur no earlier than 48–72 hours before ovulation and no later than 24 hours after ovulation, i.e., the fertile period ranges from $48 + 24 = 72$ to $72 + 24 = 96$ hours in length.

OBJECTIVE K   *To detail the events of the menstrual cycle.*

Survey   Menstruation starts at puberty and continues until the *menopause*, approximately 36 years later. The day in which discharge of blood from the vagina begins is taken as the first day of the cycle; the cycle ends on the last day prior to the next menstrual flow. Normally the cycle runs for about one lunar month, or 28 days, but it can vary from 22 to 35 days.

22.32 List the four phases of a menstrual cycle and indicate their lengths in a 28-day cycle.

(1) *Menstrual phase* or *menses*: days 1 to 5 ± 2. (2) *Proliferative* or *preovulatory* or *follicular phase*: day 5 to day of ovulation (14 ± 2 days before the onset of the next menstruation). (3) *Secretory* or *progesterone* or *luteal phase*: day of ovulation to day 28. (4) *Ischemic phase* (cf. Problem 14.15): days 27 and 28.

22.33 Define *menarche*.

The epoch of the first menstrual flow, or *period*, is called *menarche*; it marks the beginning of a female's reproductive life and is one of the most obvious signals of the onset of puberty. Menarche usually occurs between the ages of 9 and 17 (the average age is 12.5).

22.34 Which has the more constant length, the preovulatory phase or the secretory phase?

The preovulatory phase is quite variable, lasting longer in long (29- to 35-day) cycles and ending sooner in short (22- to 27-day) cycles. The length of the secretory phase is usually constant at about 14 days.

OBJECTIVE L   *To delineate the female hormonal cycle.*

Survey   Under the control of the hypothalamus, the anterior pituitary and the ovaries secrete, in a cyclical time pattern, steroid hormones which regulate all female reproductive activities.

22.35 Analyze the female hormonal cycle as a progression of events. Start with the CNS and ignore feedback mechanisms.

1. Hypothalamus begins to release *luteinizing-releasing hormone* (LRH); target organ is anterior pituitary.
2. LRH stimulates secretion of FSH and LH.

3. FSH and LH affect the ovaries as detailed in Table 12-2. (This constitutes the ovarian cycle, which terminates in the reabsorption of the corpus luteum, leaving a white scar, or *corpus albicans*.)

4. The mature ovarian follicle secretes estrogens, which provoke a thickening of the endometrium (proliferative phase of the menstrual cycle).

5. The corpus luteum secretes progesterone and estrogens, which act to level off the endometrium and ready it for *implantation* (Problem 22.39; secretory phase of the menstrual cycle).

6. With formation of the corpus albicans, the levels of progesterone and estrogens drop, the endometrium decomposes, and a new menstrual cycle begins (ischemic phase of the current menstrual cycle; see Problem 22.36).

22.36 What initiates menstrual flow?

If there is no pregnancy—and therefore no placenta—the corpus luteum begins to degenerate on about day 22 to 24 of the cycle, because of the lack of HCG (Table 12-4). As the corpus luteum ceases to function, the levels of estrogens and progesterone drop rapidly; in response, the uterine blood vessels constrict and the endometrium becomes ischemic. The endometrium degenerates and blood that escapes from damaged vessels, carrying tissue fragments, passes out through the vagina as the menstrual flow.

22.37 How do oral contraceptives work?

By inhibiting the release of LH and FSH from the anterior pituitary, they prevent maturation of the developing follicle and thereby block ovulation.

22.38 How does the fertility drug *clomiphene* work?

It is *believed* to increase the release of LRH, which in turn stimulates the release of LH and FSH (which further ovulation).

OBJECTIVE M   *To trace the paths of an ovum and spermatozoa through the female reproductive tract and to describe the early embryonic development of the fertilized egg.*

Survey   After being deposited in the vagina, sperm cells must move up through the cervical canal and uterine cavity into a uterine tube (the correct one), in order to encounter the ovum on its way down from the ovary. Fertilization normally occurs in the distal one-third of the tube. Following fertilization, the zygote undergoes mitosis during its approximately 3-day journey down the uterine tube to the uterine cavity (Fig. 22-9). There the developing *blastocyst* remains free for about another 3 days, before it begins to implant into the endometrium.

22.39 Define the terms *cleavage, morula, blastocyst,* and *implantation*.

Cleavage—early successive divisions of the zygote; *morula*—solid globular mass of cells in embryonic development; *blastocyst*—an early embryonic stage having the form of a hollow ball of cells; *implantation*—the embedding of the blastocyst into the endometrium.

22.40 How many sperm reach the uterus?   the uterine tube?   the egg?

Of the 200 to 500 million sperm ejected into the vagina (Problem 22.22), only about 1 million reach the uterus, and just a few thousand reach the mouth of the uterine tube. Only about 10 to 100 sperm reach the upper part of the tube, where fertilization takes place.

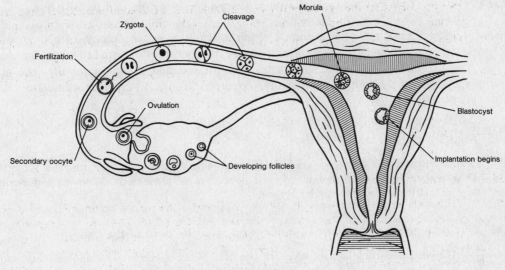

Fig. 22-9

22.41 What is an *ectopic pregnancy*, and which are the common sites of occurrence?

A pregnancy is *ectopic* when the fertilized egg becomes implanted at a site other than in the uterine cavity. (Compare *ectopic pacemaker*, Chapter 14.) The most frequent ectopic site is in a uterine tube; one then speaks of a *tubular pregnancy*. Other sites are the cervix (*cervical pregnancy*) and on the linings of abdominal viscera (*abdominal pregnancies*).

OBJECTIVE N   *To summarize the hormonal and other changes that occur during pregnancy.*

Survey   Under the influence of HCG (Problem 12.28), the corpus luteum is maintained and continues to secrete progesterone and estrogens. The placenta itself secretes large quantities of these same hormones, which: (i) sustain the endometrium; (ii) inhibit release of FSH and LH (thereby halting the menstrual cycle); (iii) stimulate the development of mammary glands; (iv) inhibit (progesterone) or stimulate (estrogens) uterine contractions; (v) increase (estrogens) uteroplacental blood flow and also enlarge the uterus, breasts, vagina, and vaginal orifice.

The placenta secretes *placental lactogen* (PL), which has milk-fostering and other activities.

Maternal cardiac output, blood volume, and caloric requirements all increase.

22.42 Is it only the sex hormones whose levels rise during pregnancy?

No: there are (lesser) rises in glucocorticoids, thyroxine, and parathyroid hormone (Table 12-1).

22.43 What forms the basis of most methods for detecting pregnancy?

Pregnancy is indicated by the presence of HCG, which shows up in the urine.

22.44 Give an accounting of the weight gained during pregnancy?

The pregnant mother gains on the average 20 to 25 lb. Typically, the fetus accounts for 7 lb; uterus, 2 lb; placenta and membranes, 2.5 lb; breasts, 2 lb; and the balance in fat, extracellular fluid, and blood.

OBJECTIVE O   *To become aware of what is known about the mechanism of labor.*

Survey   *Labor* comprehends the mother's physical activities in the process of giving birth (*parturition*). It is thought that a *positive feedback* mechanism is essential to labor: the more dilated the cervix and the more contractile the uterine and abdominal walls, the greater the secretion of hormones—principally, oxytocin (Table 12-2)—that stimulate dilation and contractility.

22.45   What are the two stages of labor?

First stage: the cervix undergoes gradual and progressive dilation until the birth canal is large enough to accommodate the head of the fetus. *Second stage*: successive waves of abdominal and uterine contractions propel the fetus down the birth canal.

22.46   Why do uterine contractions continue even after the birth of the baby?

The postpartum contractions sever the connections between the uterus and placenta, and cause the uterine vessels to constrict. This greatly reduces hemorrhage and brings about uterine *involution* (shrinkage to the former size).

22.47   Is the baby always expelled headfirst?

The head is expelled first in about 95% of normal births. In the remaining 5%, the buttocks are expelled first (*breech presentation*).

OBJECTIVE P   *To examine the hormonal control of mammary gland development and lactation.*

Survey   The mammary glands (breasts) are accessory reproductive organs that are specialized to produce milk after pregnancy. At the onset of puberty, the ovarian hormones stimulate *alveolar glands* and *alveolar ducts* (Fig. 22-10) of the breast to develop. During pregnancy, further glandular and ductile development takes place under the influence of progesterone and estrogens, respectively. Several other hormones (see Problem 22.42) are believed to be necessary in readying the mammary glands for milk production. Prolactin is inhibited during pregnancy by the high levels of progesterone and estrogens.

After childbirth estrogens and progesterone dwindle, and secretion of prolactin is no longer inhibited. This hormone, as its name implies, stimulates milk production. Nursing stimulates the nipple and areola, sending sensory input via the spinal cord to the hypothalamus, which releases oxytocin. Oxytocin stimulates contraction of the myoepithelial cells, which causes the ejection, or letdown, of milk.

22.48   What is *colostrum*?

Milk is not produced during the first day or two after parturition; however, a few milliliters of a clear, fat-free, yellowish-white fluid, known as *colostrum*, is secreted at this time. It serves as the baby's first food.

22.49   How much milk is produced during lactation?

Up to 1.5 L of milk is secreted daily at the peak of nursing activity. This volume requires the daily metabolization of about 100 g of lactose, 50 g of fat, and 2–3 g of calcium phosphate.

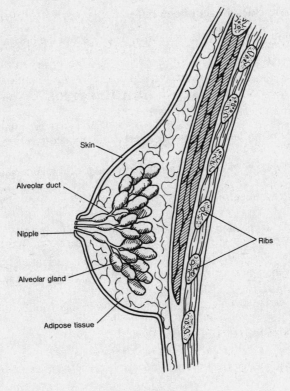

Fig. 22-10

22.50 Can a nursing mother become pregnant?

> The release of prolactin during nursing usually inhibits the menstrual cycle and therefore ovulation; however, a nursing mother can become pregnant, so that nursing does not constitute a reliable contraceptive method.

OBJECTIVE Q   *To list the common methods of* contraception.

Survey   Anything that interferes with fertilization of the egg or implantation of the embryo can be termed a means of contraception. The standard methods follow. *Rhythm method*: abstinence from sexual intercourse for a few days before and a few days after ovulation. *Coitus interruptus*: withdrawal of the penis prior to ejaculation. *Condom*: sheathing the penis to prevent sperm from being deposited in the vagina. *Diaphragm, douche, spermicidal solutions*: techniques that prevent the sperm from entering the cervix. *Oral contraceptives* ("the pill"): drugs that inhibit the release of gonadotropins (LH and FSH) and, therefore, prevent ovulation. *Intrauterine devices* (*IUDs*): contrivances that prevent implantation of the fertilized egg. *Vasectomy*: cutting and tying off the vasa deferentia, thus preventing release of sperm. *Tubal ligation*: cutting and tying off the uterine tubes, thus preventing ova from reaching the uterus.

22.51 What substances are contained in oral contraceptives?

> Synthetic estrogens and progesterone, which are not degraded in the gastrointestinal tract and act like the natural hormones. (High levels of these hormones will preserve the endometrium, causing ovulation to cease.)

22.52 What are possible side effects of the pill?

Water retention, acne, headaches, nausea, tenderness of the breasts, and formation of intravascular clots.

Key Clinical Terms

Amenorrhea A suppression or absence of menstruation. Irregular or interrupted menstruation may be a natural response to stress of various kinds and not necessarily related to pregnancy.

Caesarean section Extraction of a fetus through an incision of the abdominal wall and the uterus.

Cryptorchidism Arrested descent of a testis.

Dysmenorrhea Difficult or painful menstruation, generally due to obstruction, inflammation, or some disease.

Episiotomy Incision of the perineum to facilitate delivery and prevent excessive laceration.

Gestation The normal period of pregnancy (period of intrauterine fetal development).

Hysterectomy Removal of the uterus.

Laparoscopy Visual examination of abdominal organs, especially the uterus, uterine tubes, and ovaries.

Oophorectomy Removal of an ovary.

Orchitis Inflammation of the testis; may be the result of certain diseases such as mumps.

Papanicolaou smear A *Pap smear* is a diagnostic test for cervical or endometrial cancer. Shed epithelial cells of vaginal and uterine regions are screened for malignance. After age 20, a woman should have a Pap smear at least yearly.

Prolapsus uteri Inferior displacement of the uterus, which may distend from the vagina.

Prostatitis Inflammation of the prostate gland.

Pruritus vulvae Severe itching of the vulva; frequently accompanies *vaginitis* (q.v.).

Vaginitis Inflammation of the vagina; may be due to *venereal disease* (q.v.) or to certain bacterial or fungus infections.

Venereal disease (VD) There are five principal types, all contracted by sexual contact with an infected person of either sex:

   (1) *Gonorrhea* initially infects the urethral mucous membrane. Advanced stages involve the genitalia and the conjunctiva of the eyes and joints. The etiological agent is a gonococcus, *Neisseria gonorrhoea*.
   (2) *Syphilis* is less common than gonorrhea, but it is the more serious of the two diseases. It is caused by a spirochete, *Treponema pallidum*, and may be chronically degenerative if not treated.
   (3) *Chancroid* is an infectious venereal ulcer caused by a bacillus, *Haemophilus ducreyi*.
   (4) *Lymphogranuloma venereum* is a venereal infection caused by a virus and characterized by a genital ulcer.
   (5) *Genital herpes* is an incurable venereal disease caused by type II herpes simplex virus. During the recurring infectious stage, the afflicted person will have numerous clusters of painful genital blisters.

Review Questions

Multiple Choice

1. The portion of the female reproductive system that is homologous to the glans penis of the male is the: (*a*) labia majora, (*b*) clitoris, (*c*) ovary, (*d*) vagina.

2. The soft mucosal lining of the uterus is the: (*a*) peritoneum, (*b*) endometrium, (*c*) mediastinum, (*d*) synovial membrane, (*e*) mesentery.

3. Weakening of the suspensory ligaments would directly affect the position of the: (*a*) mammary glands, (*b*) uterus, (*c*) uterine tubes, (*d*) ovaries.

4. Fertilization normally occurs in the: (*a*) uterine tube, (*b*) vagina, (*c*) uterus, (*d*) ovary.

5. Which of the following does not belong to the external genitalia? (*a*) clitoris, (*b*) fornix, (*c*) labia minora, (*d*) vaginal orifice, (*e*) hymen.

6. Testosterone is responsible for maintenance of: (*a*) a functional male reproductive system, (*b*) regular ovulation, (*c*) mature endometrium, (*d*) cells of the interstitial spaces, (*e*) all the above.

7. The function of the male secondary sex organs is to: (*a*) transfer sperm to the female, (*b*) regulate sperm production, (*c*) produce sperm, (*d*) all the above.

8. The filling of venous sinuses under the influence of sexual stimulation is most closely associated with: (*a*) spermatogenesis, (*b*) menstruation, (*c*) ovulation, (*d*) erection.

9. The duct that leads the sperm up over the pubic arch and to the side of the urinary bladder is the: (*a*) epididymis, (*b*) urethra, (*c*) convoluted tubule, (*d*) ductus deferens, (*e*) spermatic cord.

10. The tightly convoluted tubule that lies along the posterior surface of the testis is the: (*a*) seminiferous tubule, (*b*) rete testis, (*c*) epididymis, (*d*) ductus deferens, (*e*) seminal vesicle.

11. Spermatozoa are discharged through the male ducts in the order: (*a*) epididymis, ductus deferens, ejaculatory duct, urethra; (*b*) ejaculatory duct, epididymis, ductus deferens, urethra; (*c*) epididymis, ductus deferens, urethra, ejaculatory duct; (*d*) ductus deferens, epididymis, ejaculatory duct, urethra.

12. The development of the ovarian follicle is influenced by: (*a*) prolactin, (*b*) FSH, (*c*) testosterone, (*d*) insulin, (*e*) none of the above.

13. Which of the following secretes progesterone? (*a*) anterior pituitary gland, (*b*) corpus luteum, (*c*) corpus luteum and ovarian follicle, (*d*) hypothalamus, (*e*) posterior pituitary gland.

14. After the menopause, blood concentration of: (*a*) estrogens remains low, (*b*) FSH remains high, (*c*) progesterone remains low, (*d*) all the preceding, (*e*) both (*a*) and (*c*).

15. Formation and maintenance of the corpus luteum is mainly effected by: (*a*) FSH, (*b*) LH, (*c*) progesterone, (*d*) estrogens.

16. A woman's body temperature: (*a*) rises at the onset of menstruation, (*b*) drops two days before ovulation, (*c*) rises at the time of ovulation, (*d*) drops abruptly two days after ovulation.

17. Which of the following secretes estrogens: (*a*) anterior pituitary, (*b*) corpus luteum and ovarian follicle, (*c*) ovarian follicle only, (*d*) hypothalamus, (*e*) posterior pituitary.

18. Menstruation is initiated by: (*a*) sudden release of FSH from the pituitary, (*b*) lack of estrogens and progesterone due to degeneration of the corpus luteum, (*c*) increased release of estrogens and progesterone from the corpus luteum, (*d*) sudden drop in LH.

19. Which hormone stimulates testosterone secretion? (*a*) LH, (*b*) progesterone, (*c*) FSH, (*d*) ACTH.

20. In a normal, healthy, 25-year-old woman with a menstrual cycle of 28 days: (*a*) the proliferative phase of the uterus is caused by estrogens produced by ovarian follicles; (*b*) menstruation is caused by progesterone from the corpus luteum; (*c*) injections of estrogens and/or progesterone will cause an enlargement of the ovaries and an increase in the production of mature follicles; (*d*) the concentration of estradiol in the plasma begins to fall prior to ovulation and continues to decrease until menstruation; (*e*) all the above.

21. Sperm cells are stored prior to ejaculation in the: (*a*) penile urethra, (*b*) prostate gland, (*c*) epididymides, (*d*) seminal vesicles, (*e*) ejaculatory ducts.

22. Oral contraceptive treatment with mixtures of estrogens and progesterone: (*a*) if given daily throughout the year, would tend to prevent menstruation from occurring; (*b*) is thought to act mainly by preventing implantation of the fertilized ovum; (*c*) is thought to depress anterior pituitary secretion of gonadotropic hormones; (*d*) may cause a decrease in body weight.

23. An embryo with XX chromosomes develops female secondary sex organs because of: (*a*) estrogens, (*b*) androgens, (*c*) absence of androgens, (*d*) absence of estrogens.

24. The milk ejection reflex is stimulated by: (*a*) oxytocin, (*b*) estrogen, (*c*) prolactin, (*d*) progesterone.

25. The principal cause of male infertility is: (*a*) dilute seminal fluid, (*b*) a varicocele, (*c*) acidic seminal fluid, (*d*) low testosterone levels, (*e*) blocked tubules in the epididymides.

True/False

1. The ovaries and uterus are the female primary sex organs.

2. Seminal vesicles, bulbourethral glands, and the prostrate gland are all accessory glands of the male reproductive system.

3. The labia majora in the female are homologous to the scrotum in the male.

4. Meiosis is peculiar to the gonads.

5. Feminization of the female genitalia is caused by the production of estrogens by the gonads during early fetal development.

6. *Fallopian tube*, *uterine tube*, and *oviduct* refer to the same organ.

7. The ostium is the opening into the fimbria.

8. Sympathetic stimulation of the arteries within the penis causes engorgement of the erectile tissue as arterial flow increases and venous drainage decreases.

9. The ejaculatory ducts store spermatozoa and additives to produce semen prior to ejaculation.

10. Mammary glands are modified sebaceous glands.

11. Prolactin causes the breasts to enlarge and the mammary glands to mature during puberty.

12. The first menstrual discharge is referred to as menarche.

13. The female vestibular glands keep the pH of the vagina constant.

14. Cells of Leydig produce spermatozoa and secrete nutrients to developing spermatozoa within the testes.

15. The secretory phase of menstruation is characterized by discharge of the menses.

Matching

1.	Vestibular glands	(a)	Protective sheath
2.	Prepuce	(b)	Produce testosterone
3.	Ovarian follicle	(c)	Storage of sperm
4.	Scrotum	(d)	Meiosis
5.	Cells of Leydig	(e)	Secretes estrogens
6.	Uterine tube	(f)	Secretes lubricant
7.	Clitoris	(g)	Encloses testes
8.	Penile urethra	(h)	Transports ova
9.	Interstitial cells	(i)	Erectile tissue
10.	Epididymis	(j)	Transports seminal fluid

Answers to Review Questions

CHAPTER 1

Multiple Choice

1. (*b*)   2. (*c*)   3. (*a*), (*b*)   4. (*d*)   5. (*a*)   6. (*a*)   7. (*b*)   8. (*d*)   9. (*b*)   10. (*a*)
11. (*c*)   12. (*a*)   13. (*e*)   14. (*b*)   15. (*a*)   16. (*c*)   17. (*e*)   18. (*b*)   19. (*a*)   20. (*c*)
21. (*d*)   22. (*b*)   23. (*c*)   24. (*c*)   25. (*b*)

True/False

1. T   2. T   3. T   4. T   5. F   6. T   7. F   8. F; but sweating following exercise is a feedback phenomenon   9. F   10. T

Matching

1. (*j*)   2. (*e*)   3. (*l*)   4. (*g*)   5. (*b*)   6. (*m*)   7. (*a*)   8. (*f*)   9. (*k*)   10. (*h*)
11. (*d*)   12. (*i*)   13. (*c*)

CHAPTER 2

Multiple Choice

1. (*a*)   2. (*b*)   3. (*a*)   4. (*d*)   5. (*a*)   6. (*d*)   7. (*c*)   8. (*b*)   9. (*c*)   10. (*d*)
11. (*b*)   12. (*c*)   13. (*b*)   14. (*a*)   15. (*b*)   16. (*c*)   17. (*d*)   18. (*a*)   19. (*c*)
20. (*b*)   21. (*c*)   22. (*d*)   23. (*d*)   24. (*c*)   25. (*b*)

True/False

1. F   2. F   3. T   4. T   5. F   6. T   7. T   8. F   9. T   10. F   11. F   12. F

Matching

1. (*c*)   2. (*e*)   3. (*g*)   4. (*n*)   5. (*h*)   6. (*m*)   7. (*l*)   8. (*o*)   9. (*k*)   10. (*b*)
11. (*j*)   12. (*i*)   13. (*d*)   14. (*f*)   15. (*a*)

CHAPTER 3

Multiple Choice

1. (*d*)   2. (*b*)   3. (*a*)   4. (*b*)   5. (*d*)   6. (*b*)   7. (*d*)   8. (*a*)   9. (*c*)   10. (*b*)
11. (*e*)   12. (*a*)   13. (*a*)   14. (*c*)   15. (*c*)   16. (*b*)   17. (*b*)   18. (*c*)   19. (*c*)   20. (*a*)
21. (*e*)   22. (*e*)   23. (*b*)   24. (*a*)   25. (*a*)   26. (*d*)

Matching

1. (*a*)   2. (*d*)   3. (*g*)   4. (*c*)   5. (*f*)   6. (*d*)   7. (*e*)

Labeling

A. nucleus   B. ribosomes   C. nucleolus   D. smooth ER   E. rough ER   F. mitochondria
G. Golgi apparatus   H. secretory granules   I. plasma membrane   J. microtubules   K. centriole
L. lysosome

CHAPTER 4

Multiple Choice

1. (c)   2. (d)   3. (a)   4. (a)   5. (b)   6. (c)   7. (d)   8. (a)   9. (b)   10. (b)
11. (c)   12. (d)   13. (a)   14. (c)   15. (b)   16. (d)   17. (c)   18. (d)   19. (a)
20. (a)

True/False

1. T   2. F; pseudostratified...   3. T   4. F   5. F   6. F; hyaline cartilage   7. T   8. T
9. F   10. T   11. T   12. F; only that of the epidermis   13. F; only the lipid content is lost and gained
14. T   15. F

Completion

1. Histology   2. simple squamous epithelial   3. stratified   4. endothelium   5. smooth
6. Keratin   7. merocrine   8. cancellous   9. Plasma   10. reticular   11. edema
12. mesoderm   13. Skeletal   14. dendrites   15. myelin

Matching I

1. (b)   2. (d)   3. (e)   4. (f)   5. (c)   6. (g)   7. (a)

Matching II

1. (c)   2. (e)   3. (a)   4. (b)   5. (d)   6. (g)   7. (f)

Matching III

1. (e)   2. (d)   3. (b)   4. (a)   5. (f)   6. (c)

CHAPTER 5

Multiple Choice

1. (c)   2. (b)   3. (e)   4. (b)   5. (a)   6. (d)   7. (c)   8. (b)   9. (b)   10. (a)
11. (b)   12. (e)   13. (c)   14. (b)   15. (e)   16. (a)   17. (c)   18. (c)   19. (e)
20. (e)

True/False

1. T   2. F; the skin is an organ   3. T   4. F; third-degree   5. T   6. T   7. F   8. F
9. F; sudoriferous   10. T   11. F; the skin is virtually waterproof   12. F; apocrine sweat glands do not
mature until puberty   13. T   14. F; alopecia is not a disease   15. F; acne is an inflammatory
condition of sebaceous glands

CHAPTER 6

Multiple Choice

1. (*c*)   **2.** (*c*)   **3.** (*a*)   **4.** (*b*)   **5.** (*d*)   **6.** (*b*)   **7.** (*a*)   **8.** (*c*)   **9.** (*b*)   **10.** (*d*)
11. (*d*)   **12.** (*a*)   **13.** (*a*)   **14.** (*c*)   **15.** (*c*)   **16.** (*a*)   **17.** (*d*)   **18.** (*b*)   **19.** (*c*)
20. (*a*)   **21.** (*e*)   **22.** (*d*)   **23.** (*b*)   **24.** (*c*)   **25.** (*a*)   **26.** (*a*)   **27.** (*b*)   **28.** (*a*)
29. (*a*)   **30.** (*d*)

True/False

1. F; only the tibia   **2.** F   **3.** T   **4.** F; rotational   **5.** F; red bone marrow   **6.** F; calcium and phosphorus   **7.** T   **8.** T   **9.** F; transverse foramina   **10.** T   **11.** F; fibula   **12.** F; endochondral
13. T   **14.** T   **15.** F   **16.** F; "lessening the angle at a hinge joint"   **17.** F; osteoclasts
18. T   **19.** T   **20.** F; generally, but not always

CHAPTER 7

Multiple Choice

1. (*b*)   **2.** (*a*)   **3.** (*d*)   **4.** (*c*)   **5.** (*a*)   **6.** (*d*)   **7.** (*c*)   **8.** (*d*)   **9.** (*c*)   **10.** (*b*)
11. (*a*)   **12.** (*d*)   **13.** (*b*)   **14.** (*b*)   **15.** (*a*)   **16.** (*a*)   **17.** (*c*)   **18.** (*b*)   **19.** (*c*)
20. (*d*)   **21.** (*d*)   **22.** (*c*)   **23.** (*d*)   **24.** (*d*)   **25.** (*a*)   **26.** (*c*)   **27.** (*b*)   **28.** (*b*)
29. (*a*)

True/False

1. T   **2.** F   **3.** F   **4.** T   **5.** F   **6.** T   **7.** T   **8.** F   **9.** F   **10.** T   **11.** F   **12.** T
13. T   **14.** F   **15.** T   **16.** F   **17.** F   **18.** T   **19.** F   **20.** T   **21.** F   **22.** T   **23.** F
24. F   **25.** T

Matching

1. (*b*)   2. (*c*)   3. (*g*)   4. (*d*)   5. (*e*)   6. (*f*)   7. (*h*)

Labeling

A. trapezius   B. pectoralis major   C. deltoid   D. biceps brachii   E. serratus anterior
F. rectus abdominis   G. external oblique   H. brachioradialis   I. palmaris longus   J. sartorius
K. gracilis   L. rectus femoris   M. vastus lateralis   N. vastus medialis   O. tibialis anterior
P. occipitalis   Q. trapezius   R. deltoid   S. infraspinatus   T. triceps brachii   U. latissimus
dorsi   V. external oblique   W. flexor carpi ulnaris   X. gluteus medius   Y. gluteus maximus
Z. semimembranosus   AA. semitendinosus   BB. biceps femoris   CC. gastrocnemius
DD. soleus

CHAPTER 8

Multiple Choice

1. (*d*)   **2.** (*c*)   **3.** (*d*)   **4.** (*c*)   **5.** (*d*)   **6.** (*b*)   **7.** (*b*)   **8.** (*b*)   **9.** (*a*)   **10.** (*c*)
11. (*b*)   **12.** (*a*)   **13.** (*b*)   **14.** (*c*)   **15.** (*d*)   **16.** (*a*)   **17.** (*d*)   **18.** (*a*)   **19.** (*a*)
20. (*c*)   **21.** (*d*)   **22.** (*d*)   **23.** (*c*)   **24.** (*d*)   **25.** (*d*)

Matching

1. (*e*)   2. (*d*)   3. (*a*)   4. (*c*)   5. (*b*)

True/False

1. T   2. F; away from   3. T   4. T   5. F   6. F   7. F   8. F   9. T   10. F
11. T   12. T   13. F   14. F   15. F   16. F   17. F   18. F   19. T   20. F   21. F
22. T

Completion

1. dendrites, cell body   2. neurons, glial   3. multipolar   4. diameter, myelinated
5. saltatory conduction   6. Schwann cells   7. synapse   8. neurovesicles   9. synaptic cleft
10. spatial summation

CHAPTER 9

Multiple Choice

1. (*c*)   2. (*b*)   3. (*d*)   4. (*a*)   5. (*c*)   6. (*b*)   7. (*e*)   8. (*a*)   9. (*d*)   10. (*a*)
11. (*c*)   12. (*e*)   13. (*b*)   14. (*d*)   15. (*a*)   16. (*b*)   17. (*e*)   18. (*b*)   19. (*a*)   20. (*b*)
21. (*a*)   22. (*e*)   23. (*a*)   24. (*d*)   25. (*d*)   26. (*d*)   27. (*c*)   28. (*b*)   29. (*a*)   30. (*b*)
31. (*b*)   32. (*b*)   33. (*c*)   34. (*c*)   35. (*b*)   36. (*a*)   37. (*e*)   38. (*a*)   39. (*c*)   40. (*d*)

True/False

1. T   2. T   3. F   4. F   5. F   6. T   7. F   8. T   9. T   10. T   11. T   12. T
13. F

Matching

1. (*f*)   2. (*c*)   3. (*a*)   4. (*b*)   5. (*i*)   6. (*e*)   7. (*d*)   8. (*g*)   9. (*j*)   10. (*h*)

CHAPTER 10

Multiple Choice

1. (*a*)   2. (*c*)   3. (*c*)   4. (*c*)   5. (*c*)   6. (*c*)   7. (*b*)   8. (*a*)   9. (*d*)   10. (*d*)
11. (*c*)   12. (*a*)   13. (*c*)   14. (*c*)   15. (*b*)   16. (*c*)   17. (*c*)   18. (*b*)   19. (*c*)   20. (*d*)
21. (*e*)   22. (*d*)   23. (*a*)   24. (*b*)   25. (*b*)   26. (*c*)   27. (*b*)   28. (*c*)   29. (*b*)   30. (*c*)
31. (*d*)   32. (*a*)   33. (*b*)   34. (*c*)   35. (*e*)   36. (*a*) (1), (*b*) (4), (*c*) (3), (*d*) (4), (*e*) (1),
(*f*) (2), (*g*) (1), (*h*) (2)

True/False

1. F   2. T   3. T   4. F   5. F; accessory   6. T   7. T   8. T   9. F   10. T

CHAPTER 11

Multiple Choice

1. (*c*)   **2.** (*c*)   **3.** (*a*)   **4.** (*c*)   **5.** (*b*)   **6.** (*d*)   **7.** (*a*)   **8.** (*d*)   **9.** (*c*)   **10.** (*a*)
11. (*b*)   **12.** (*a*)   **13.** (*d*)   **14.** (*a*)   **15.** (*d*)   **16.** (*b*)   **17.** (*b*)

True/False

1. T   **2.** F   **3.** T   **4.** T   **5.** T   **6.** T   **7.** F   **8.** T   **9.** F   **10.** T   **11.** F; aqueous
12. F   **13.** F   **14.** T   **15.** T   **16.** F   **17.** F

CHAPTER 12

Multiple Choice

1. (*d*)   **2.** (*c*)   **3.** (*a*)   **4.** (*e*)   **5.** (*c*)   **6.** (*b*)   **7.** (*c*)   **8.** (*a*)   **9.** (*b*)   **10.** (*c*)
11. (*a*)   **12.** (*b*)   **13.** (*d*)   **14.** (*d*)   **15.** (*c*)   **16.** (*a*)   **17.** (*c*)   **18.** (*a*)   **19.** (*d*)   **20.** (*a*)
21. (*a*)   **22.** (*c*)   **23.** (*d*)   **24.** (*d*)   **25.** (*a*)   **26.** (*b*)   **27.** (*b*)   **28.** (*d*)   **29.** (*c*)   **30.** (*c*)
31. (*e*)   **32.** (*a*)   **33.** (*b*)   **34.** (*a*)

True/False

1. T   **2.** T   **3.** F   **4.** F   **5.** T   **6.** T   **7.** F   **8.** T   **9.** T

Matching

1. (*c*)   2. (*b*)   3. (*h*)   4. (*a*)   5. (*e*)   6. (*g*)   7. (*f*)   8. (*d*)

CHAPTER 13

Multiple Choice

1. (*b*)   **2.** (*c*)   **3.** (*d*)   **4.** (*c*)   **5.** (*c*)   **6.** (*c*)   **7.** (*d*)   **8.** (*a*)   **9.** (*d*)   **10.** (*d*)
11. (*a*)   **12.** (*b*)   **13.** (*d*)   **14.** (*d*)   **15.** (*d*)   **16.** (*d*)   **17.** (*d*)   **18.** (*c*)   **19.** (*b*)   **20.** (*b*)
21. (*c*)   **22.** (*a*)   **23.** (*a*)   **24.** (*c*)   **25.** (*b*)   **26.** (*d*)   **27.** (*b*)   **28.** (*d*)   **29.** (*a*)   **30.** (*b*)
31. (*c*)   **32.** (*b*)   **33.** (*c*)   **34.** (*a*)   **35.** (*d*)

Matching

1. (*f*)   2. (*a*)   3. (*d*)   4. (*e*)   5. (*i*)   6. (*c*)   7. (*g*)   8. (*h*)   9. (*b*)

CHAPTER 14

Multiple Choice I

1. (*a*)   **2.** (*d*)   **3.** (*d*)   **4.** (*b*)   **5.** (*a*)   **6.** (*c*)   **7.** (*d*)   **8.** (*b*)   **9.** (*a*)   **10.** (*c*)
11. (*a*)   **12.** (*c*)   **13.** (*a*)   **14.** (*a*)   **15.** (*d*)   **16.** (*a*)   **17.** (*c*)   **18.** (*b*)   **19.** (*c*)   **20.** (*b*)
21. (*a*) 4, (*b*) 6, (*c*) 1, (*d*) 3, (*e*) 7, (*f*) (5), (*g*) 2

True/False

1. F; at 25 days   **2.** T   **3.** T   **4.** T   **5.** T   **6.** F; increases both   **7.** F   **8.** T   **9.** T
10. F   **11.** F   **12.** F

Matching

1. (*a*)   2. (*e*)   3. (*d*)   4. (*f*)   5. (*b*)   6. (*c*)

Labeling

A. aorta   B. sinoatrial node   C. right atrium   D. atrioventricular node   E. right ventricle
F. pulmonary artery   G. pulmonary veins   H. left ventricle   I. interventricular septum

Multiple Choice II

1. B   2. G   3. D   4. H   5. E   6. C   7. C

CHAPTER 15

Multiple Choice

1. (*c*)   **2.** (*d*)   **3.** (*c*)   **4.** (*d*)   **5.** (*c*)   **6.** (*a*)   **7.** (*c*)   **8.** (*b*)   **9.** (*c*)   **10.** (*b*)
11. (*a*)   **12.** (*d*)   **13.** (*d*)   **14.** (*c*)   **15.** (*a*)   **16.** (*b*)   **17.** (*b*)   **18.** (*b*)   **19.** (*a*)
20. (*c*)   **21.** (*b*)   **22.** (*d*)   **23.** (*c*)   **24.** (*b*)   **25.** (*c*), (*e*)   **26.** (*e*)   **27.** (*c*)   **28.** (*b*), (*d*)
29. (*e*), (*f*)   **30.** (*a*), (*c*)

CHAPTER 16

Multiple Choice

1. (*d*)   **2.** (*b*)   **3.** (*d*)   **4.** (*a*)   **5.** (*c*)   **6.** (*b*)   **7.** (*c*)   **8.** (*c*)   **9.** (*b*)   **10.** (*d*)
11. (*c*)   **12.** (*b*)   **13.** (*a*)   **14.** (*c*)   **15.** (*a*)   **16.** (*a*)   **17.** (*d*)   **18.** (*b*)   **19.** (*d*)

True/False

1. T   **2.** F   **3.** F   **4.** T   **5.** T   **6.** F   **7.** F; IgM, not IgL   **8.** T   **9.** F   **10.** F
11. T   **12.** T

Completion

−	+	+	+
−	−	+	+
−	+	−	+
−	−	−	−

CHAPTER 17

Multiple Choice

1. (*d*)　　**2.** (*b*)　　**3.** (*a*)　　**4.** (*d*)　　**5.** (*d*)　　**6.** (*a*)　　**7.** (*b*)　　**8.** (*d*)　　**9.** (*b*)　　**10.** (*c*)
11. (*d*)　　**12.** (*b*)　　**13.** (*b*)　　**14.** (*d*)　　**15.** (*c*)　　**16.** (*d*)　　**17.** (*a*)　　**18.** (*b*)　　**19.** (*d*)

True/False

1. F; nasal fossae　　**2.** T　　**3.** F; less than　　**4.** T　　**5.** F; diffusion mechanisms　　**6.** T　　**7.** T
8. F; vomer and ethmoid　　**9.** F; alkaline　　**10.** T　　**11.** T　　**12.** F　　**13.** T　　**14.** F　　**15.** F
16. F; A decreased　　**17.** F; detach from　　**18.** T

CHAPTER 18

Multiple Choice

1. (*d*)　　**2.** (*c*)　　**3.** (*b*)　　**4.** (*d*)　　**5.** (*a*)　　**6.** (*d*)　　**7.** (*b*)　　**8.** (*a*)　　**9.** (*d*)　　**10.** (*a*)
11. (*c*)　　**12.** (*a*)　　**13.** (*c*)　　**14.** (*c*)　　**15.** (*b*)　　**16.** (*d*)　　**17.** (*a*)　　**18.** (*a*)　　**19.** (*c*)
20. (*b*) and (*c*)　　**21.** (*b*)　　**22.** (*d*)　　**23.** (*a*)

True/False

1. T　　**2.** F; increase　　**3.** T　　**4.** F; skeletal muscles　　**5.** F; symptomatic of a disease　　**6.** T
7. T　　**8.** F; vitamin B_{12}　　**9.** T　　**10.** F; emulsification　　**11.** T　　**12.** F; plicae circulares　　**13.** F
14. F; mucosa　　**15.** T

CHAPTER 19

Multiple Choice

1. (*a*)　　**2.** (*d*)　　**3.** (*b*)　　**4.** (*d*)　　**5.** (*b*)　　**6.** (*b*)　　**7.** (*d*)　　**8.** (*a*)　　**9.** (*d*)　　**10.** (*c*)
11. (*b*)　　**12.** (*c*)　　**13.** (*b*)　　**14.** (*d*)　　**15.** (*c*)　　**16.** (*b*)　　**17.** (*d*)　　**18.** (*b*)　　**19.** (*d*)
20. (*b*)　　**21.** (*d*)　　**22.** (*c*)　　**23.** (*c*)

Matching I

1. (*c*)　　2. (*g*)　　3. (*e*)　　4. (*a*)　　5. (*b*)　　6. (*f*)　　7. (*d*)

Matching II

1. (*c*)　　2. (*a*)　　3. (*e*)　　4. (*b*)　　5. (*d*)

Matching III

1. (*d*)　　2. (*a*)　　3. (*c*)　　4. (*f*)　　5. (*b*)　　6. (*e*)

CHAPTER 20

Multiple Choice

1. (*c*) 2. (*b*) 3. (*b*) 4. (*d*) 5. (*a*) 6. (*d*) 7. (*d*) 8. (*c*) 9. (*c*) 10. (*a*)
11. (*c*) 12. (*d*) 13. (*b*) 14. (*a*) 15. (*d*) 16. (*b*) 17. (*a*) 18. (*a*) 19. (*c*)
20. (*a*) 21. (*b*) 22. (*b*) 23. (*c*) 24. (*a*) 25. (*b*) 26. (*c*) 27. (*d*) 28. (*b*)
29. (*a*) 30. (*d*) 31. (*b*) 32. (*c*) 33. (*b*) 34. (*b*) 35. (*d*) 36. (*c*) 37. (*c*)
38. (*d*)

True/False

1. F 2. T 3. T 4. F 5. T 6. F; acidosis 7. F; hypervolemia 8. F 9. T
10. F 11. T 12. T 13. T 14. F

Matching

1. (*j*) 2. (*n*) 3. (*a*) 4. (*f*) 5. (*b*) 6. (*g*) 7. (*i*) 8. (*d*) 9. (*c*) 10. (*h*)
11. (*m*) 12. (*e*)

CHAPTER 21

Multiple Choice

1. (*a*) 2. (*d*) 3. (*c*) 4. (*b*) 5. (*a*) 6. (*c*) 7. (*c*) 8. (*d*) 9. (*b*) 10. (*c*)
11. (*b*) 12. (*a*) 13. (*b*) 14. (*b*) 15. (*d*) 16. (*c*) 17. (*d*) 18. (*a*) 19. (*d*)
20. (*b*)

CHAPTER 22

Multiple Choice

1. (*b*) 2. (*b*) 3. (*a*) 4. (*a*) 5. (*b*) 6. (*a*) 7. (*a*) 8. (*d*) 9. (*d*) 10. (*c*)
11. (*a*) 12. (*b*) 13. (*b*) 14. (*a*) 15. (*a*) 16. (*c*) 17. (*b*) 18. (*b*) 19. (*a*)
20. (*a*) 21. (*c*) 22. (*c*) 23. (*c*) 24. (*a*) 25. (*b*)

True/False

1. F 2. T 3. T 4. T 5. F; by the absence of androgens 6. T 7. F; into the uterine tube
8. F; parasympathetic 9. F; ducts receive 10. F; sweat 11. F; estrogens 12. T
13. F 14. F; of Sertoli 15. F; menstrual phase

Matching

1. (*f*) 2. (*a*) 3. (*e*) 4. (*g*) 5. (*b*) 6. (*h*) 7. (*i*) 8. (*j*) 9. (*d*) 10. (*c*)

Index